The Political Turn in Analytic Philosophy

Eide

—

Foundations of Ontology

Edited by
Otávio Bueno, Javier Cumpa Arteseros, John Heil,
Peter Simons, Erwin Tegtmeier, and Amie L. Thomasson

Volume 11

The Political Turn in Analytic Philosophy

Reflections on Social Injustice and Oppression

Edited by
David Bordonaba Plou, Víctor Fernández Castro,
José Ramón Torices

DE GRUYTER

ISBN 978-3-11-135277-0
e-ISBN (PDF) 978-3-11-061231-8
e-ISBN (EPUB) 978-3-11-061299-8
ISSN 2198-1841

Library of Congress Control Number: 2021952041

Bibliographic information published by the Deutsche Nationalbibliothek
The Deutsche Nationalbibliothek lists this publication in the Deutsche Nationalbibliografie;
detailed bibliographic data are available on the Internet at http://dnb.dnb.de.

© 2023 Walter de Gruyter GmbH, Berlin/Boston
This volume is text- and page-identical with the hardback published in 2022.
Printing and binding: CPI books GmbH, Leck

www.degruyter.com

Acknowledgements

This volume is the result of the individual and joint reflection of the three editors but, first and foremost, it is the result of many years of discussion and collaboration with some of the members of the research group Filosofía y Análisis (HUM-975) and the Unidad de Excelencia FiloLab from the Universidad de Granada, Spain. We are deeply grateful to all of them for playing their part in making this volume possible. Additionally, we do not want to leave unmentioned all those authors who, in recent years, have placed political interests at the center of the discussion of analytic philosophy and many of whom have contributed to this volume.

We also want to thank Audrey Yap, José Medina, Alessandra Tanesini, Manuel de Pinedo, Neftalí Villanueva, Cristina Borgoni, Emily McWilliams, Deborah Mühlebach, Bianca Cepollaro, Saray Ayala-López, Esa Díaz-León, Manuel Almagro-Holgado, Alba Moreno-Zurita, William J. Berger, Daniel J. Singer, Aaron Bramson, Patrick Grim, Jiin Jung, and Bennett Holman for their collaboration, without which this volume would not have been possible. We would also like to acknowledge the invaluable feedback from all those who helped us to review the contributions included in this book: Michel Croce, Matthew Cull, María José Frapolli, Ann Garry, Javier González de Prado, Jesús Navarro, Eduardo Pérez-Navarro, Uwe Peters, Manuel de Pinedo, Jennifer Saul, Josefa Toribio, Jesús Vega, and Neftalí Villanueva. Their insightful comments have greatly improved the quality of the volume.

The EIDE Series of De Gruyter has produced this volume. We hope that our enthusiasm for the topic and, above all, the high quality of the contributions contained in it will do due honor. We are particularly grateful to one of the editors of the series, Javier Cumpa, for encouraging us to submit our proposal. We would also like to thank all those at De Gruyter for their patience, support, and editorial guidance. In particular, we would like to thank Christoph Schirmer, Aaron Sanborn-Overby, and Anne Hiller.

This volume is an essential step in consolidating an existing concern among analytic philosophers for political issues. We hope that more and more of them will apply the tools of analytic philosophy to make the world we live in a more just world.

https://doi.org/10.1515/9783110612318-001

Table of Contents

Part III: **Meaning, Politics, and Identity**

Part IV: **Epistemology and Polarization**

David Bordonaba-Plou, Víctor Fernández-Castro,
and José R. Torices
Editor's Introduction

The analytic tradition in philosophy is often characterized by the goal of clarity
and the insistence on explicit argumentation (ESAP 2021), but also by the recognition of names like Frege, Wittgenstein, Russell, and Carnap among the figures
that established the disciplines and that spread their ideas, methodologies, and
positions in the English-speaking world during the twentieth century (Beaney
2011). This tradition has framed its activity within disciplines such as the philosophy of language, science, metaphysics and epistemology and has oriented its
research to central concepts like truth, objectivity, knowledge, reality, and meaning. In recent decades, we have witnessed a growing interest in the political
meaning and consequences of some of these key concepts. In this sense, questions such as "What is knowledge?" and "How do our speech acts mean?"
have given rise to questions like "Who has the voice and power to transmit
knowledge?," "Who is being unfairly disbelieved in our epistemic practices?,"
and "Can speech acts be distorted due to the speaker's membership in a disadvantaged group?"

This change of perspective aims at elucidating how fundamental concepts of
philosophy of language, science, metaphysics and epistemology can shed light
on different sources of oppression and injustice is what we call the political
turn in analytical philosophy. This volume aims to be a collection of original essays providing extensive contributions to this political turn by using relevant
tools in contemporary analytic philosophy for social and political change.
Since the first chapter of the volume is devoted to elucidating the characteristics
and historical frame of the political turn, we will devote this introduction to survey the content of the volume.

Part I: "Analytic Philosophy and Social Involvement" contains two chapters dedicated to foundational and historical issues regarding the political articulation of analytic philosophy. In the first chapter, "Analytic Philosophy as Philosophical Activism," David Bordonaba-Plou, Víctor Fernández-Castro, and José
R. Torices aim at characterizing the political turn and how it can be distinguished from its historical precursor and other intersections between politics and analytic philosophy. They argue that we can distinguish the contemporary political
interest of analytic philosophers as a form of political activism, that is, as being
motivated by the ambition of resisting oppression and injustices and being committed to orient their theories or analysis to such a purpose.

https://doi.org/10.1515/9783110612318-002

In chapter 2, "Conceptual Engineering and Neurath's Boat," Audrey Yap returns to one of the most prominent figures of the Vienna Circle to secure helpful tools for the political turn in contemporary analytic philosophy. Specifically, Yap advocates two ideas. First, she defends the view that Carnapian conceptual engineering can alleviate some problems of ameliorative projects. Second, she contends that a Neurathian approach can be better for feminist empiricism than a naturalized epistemology, because it can help us decide between alternative argumentative strategies and competing scientific theories.

Part II: "Mind, Knowledge, and the Social World" offers different insights from epistemology and philosophy of mind to oppression, activism, and injustice. In particular, these chapters explore how the understanding of mind and epistemic agency can illuminate different phenomena involved in oppression and injustice like implicit bias, narcissism, servility, and value monism. Furthermore, this part proposes different strategies for epistemic resistance like epistemic de-platforming, self-empowerment, structural transformation, and the use of the ameliorative project as a way to confront conservative epistemologies.

In chapter 3, "Political Epistemology," José Medina reflects on the recent shift in the literature on epistemic injustice from an ethics of knowing to a politics of knowing. Miranda Fricker's (2007) seminal work on epistemic injustice is focused on the epistemic life of individuals. In contrast, Medina's work normatively assesses the supra-individual level of epistemic injustice, that is, the epistemic dynamics of institutions and groups. Concerning the former, Medina distinguishes three criteria to assess the epistemic behavior of an institution: action protocols, staff training, and accountability procedures. Then, he defends two kinds of epistemic activism to fight institutional epistemic injustices: epistemic self-empowerment and epistemic structural transformation.

In "Intellectual Vices in Conditions of Oppression" (chapter 4), Alessandra Tanesini explores the idea of epistemic oppression and the psychological effects that this kind of oppression has both in the privileged, depicted as vices of superiority such as arrogance and narcissism, and the oppressed, depicted as vices of inferiority such as timidity and servility. Epistemic oppression, a broader concept than epistemic injustice, involves "unfair distributions of epistemic goods (such as information) and resources (like education) and/or failures to recognise fully the epistemic abilities of epistemic subjects belonging to some underprivileged social groups." She applies Iris Young's (1990) taxonomy to distinguish five different types of epistemic oppression: epistemic exploitation, epistemic marginalization, epistemic powerlessness, epistemic subordination (or cultural imperialism), and epistemic violence.

In chapter 5, "Epistemic De-Platforming," Neftalí Villanueva and Manuel de Pinedo begin by showing how paradoxical it is to consider, as recommended by epistemic contextualism, that we should reject certain epistemic possibilities when the context becomes more or less demanding and that it is considered bad epistemology to ignore certain kinds of arguments when they come from a group that we have good reason to ignore, for example, groups with fascist worldviews. In contrast to the outlined view, the authors argue that we should consider the epistemic politics of ignoring and opposing a theory if it is associated with people or other theories that you have every reason to distrust not only a good epistemic practice but also a good political practice.

Cristina Borgoni argues that implicit attitude does not need to be biased, and thus, a complete account of the nature of implicit bias needs to incorporate questions about what is biased and what is political in them. In other words, a constitutive account of the nature of implicit biases requires understating the political role of implicit bias in perpetuating injustice. As such, "Philosophy of Mind after Implicit Biases" (chapter 6) attempts to provide some methodological and programmatic insights from the philosophy of mind regarding the study of the implicit bias as biased.

Finally, Emily McWilliams addresses the prevalent belief that, while feminist work is important, it is not a part of epistemology proper, but responds instead to a largely orthogonal set of social and political concerns. According to her, the prevalence of this belief has partly to do with the methodologies used in epistemic theorizing. In "Ameliorative Inquiry in Epistemology" (chapter 7), McWilliams argues that a particular type of methodology—ameliorative inquiry—opens up possibilities for seeing feminist work as not orthogonal, but central to theories of epistemic kinds and epistemic value. Furthermore, ameliorative methodology opens the way toward endorsing pluralism about epistemic value and motivates the claim that the concerns of feminist and liberatory epistemology are a core part of epistemology proper.

Part III: "Meaning, Politics, and Identity" addresses different issues related to the social and political dimension of the meaning of our verbal expressions, concepts, and social practices. More specifically, this part focuses on the oppressive and unjust dimension of the use of derogatory terms and the practices that are embedded in them, but also, about the necessity of providing substantive theories of meaning (e. g. of the term 'woman') and conceptual landscaping in order to redress hermeneutical injustices and other forms of oppression.

Deborah Mühlebach argues that the recent interest in analyzing the pernicious political and moral aspects of problematic terms like slurs or derogatory terms is not sufficient to illuminate the sources of injustice or oppression in our social practices. Elaborating on Robert Brandom's inferentialism, "Tackling Verbal Derogation" (chapter 8) analyzes how the level of social practices and meanings can feed into pernicious linguistic exchanges, and thus, how constructive contestation at these levels could be more efficacious than focusing on verbal actions as the current literature in philosophy of language and semantics do.

"The Power to Shape Context" (chapter 9) considers how presuppositions can mobilize some prejudice and negative attitude, and thus, promote injustice and discrimination via a back door type of testimony even using harmless language. This model presented by Rae Langton, however, cannot accommodate value talk and informative evaluative presuppositions. As such, Bianca Cepollaro presents a model that can account for taste, moral and aesthetic value judgments and dispenses with the notion of testimony.

In "Hermeneutical Injustice and Conceptual Landscaping" (chapter 10), Saray Ayala-López proposes that conceptual landscaping, the practice of crafting our available conceptual resources or generating new ones when a particular worldly phenomena resists conceptualization, needs to go beyond failure-fixing. In other words, we should not limit conceptual engineering to cases where we detect a failure of the existing dominant collective available resources. Two arguments motivate this claim. First, hermeneutical injustice does not always reveal gaps in the collective conceptual sources but also distortions of information. Second, there are contexts that need creativity in order to generate new concepts that help to make sense of new realities such as the contexts of gender open babies.

Finally, E. Díaz-León goes against the idea that feminism does not need to provide a substantive account of the meaning of 'woman' in order to describe and fight the oppression of women. As such, "The Meaning of 'Woman' and the Political Turn in Philosophy of Language" (chapter 11) aims at providing a new version of contextualism that is able to address some of the challenges in the debate between trans-inclusive and trans-exclusive accounts of the meaning of woman.

Part IV "Epistemology and Polarization" deals with a notion that has received a more than a significant amount of attention from both the research community and the public: polarization. The concept of polarization refers to the abnormal distribution of opinions on a particular issue and seems to be a fundamental phenomenon to understand some political problems such as the rise of populism or the inability of some democratic governments to make their decision-making mechanisms work. This last part of the volume is devoted to exploring the philosophical and epistemological consequences of this concept as well as its role in the generation of certain injustices.

"Affective Polarization and Testimonial and Discursive Injustice" (chapter 12) explores the connection between polarization and testimonial and discursive injustice. Manuel Almagro-Holgado and Alba Moreno-Zurita defend the claim that political polarization can be understood in two ways, as ideological polarization, that is, as extremism, or as affective polarization, that is, as radicalism. Next, they argue that, only by understanding polarization in this second sense, we can make sense of the relation between polarization and testimonial and discursive injustices.

In chapter 13, "Philosophical Considerations of Political Polarization," William J. Berger, Daniel J. Singer, Aaron Bramson, Patrick Grim, Jiin Jung, and Bennett Holman illustrate how philosophy and political science can inform one another by providing an overview of the philosophical contributions they have made elsewhere on the topic of political polarization, particularly of the epistemic kind. First, the authors consider ways to provide a more explicit terminology to understand how to measure polarization; second, they discuss precise mechanistic accounts of polarization; and third, they examine a novel normative view about a possible source of polarization that casts polarization as a possible outcome of rational, but limited, agents interacting. Contrary to recent work on this topic, this last contribution illustrates how dynamics akin to epistemic bubbles and echo chambers can develop without associated epistemic vices."

Part I: **Analytic Philosophy and Social Involvement**

David Bordonaba-Plou, Víctor Fernández-Castro,
and José R. Torices

Analytic Philosophy as Philosophical Activism

Abstract: This chapter characterizes the idea of the political turn in analytic philosophy. It first examines possible connections between analytic philosophy and politics, situating analytic feminism as the precursor of the turn. We then consider some of the attempts that have been made in analytic philosophy to make explicit the political turn. Specifically, we explore whether the ideal/non-ideal theory distinction serves to elucidate what is distinctive about the political turn, concluding that this distinction is of little help in undertaking such a task. Finally, we propose to characterize the political turn as philosophical activism. That is, the political turn is committed to understanding and identifying particular forms of social injustice and intervening to eradicate them.

1 Introduction

Over the last two decades, we have witnessed how many influential authors within the analytic tradition working in areas such as epistemology, philosophy of language, and metaphysics have used different conceptual apparatuses and theories to shed light on some politically significant phenomena. These phenomena encompass different aspects of politics, such as the various sorts of injustices to which some groups are exposed. To give an example, Miranda Fricker has developed different concepts and arguments within epistemology to illuminate the experience of certain disadvantaged groups whose testimonies and voices are systematically stripped of credibility precisely by virtue of the identity of the group to which they belong. Such phenomena generate a particular type of injustice that Fricker calls testimonial injustice, which she considers a subtype of a broader phenomenon called epistemic injustice.[1] Fricker's work is far from being the only example of this concern with the political sphere in philosophical

---

[1] While Fricker's work on epistemic injustice is a groundbreaking contribution to analytic epistemology, it is necessary to remark, following Dotson (2011 and 2012) and Anderson (2017), that we can already find invaluable contributions on the nefarious role of prejudices and stereotypes in epistemic practices in the works of black feminist, non-analytic thinkers such as Audre Lorde, Hazel Carby, Patricia Hills Collins, and Anna Julia Cooper, to name just a few.

https://doi.org/10.1515/9783110612318-003

discussions of language, knowledge, or the mind, where we can find approaches to the silencing and harmful function of pornography and slurs[2], the use of propaganda by neo-fascism[3], analyses of how misogynist or racist structures affect people's ability to exercise their autonomous agency[4], studies analyzing the different senses in which the term 'polarization' can be understood[5], the relationship between disagreement and polarization[6], and investigations of the consequences of echo chambers to epistemology[7], and aesthetics[8]. As a result, concepts such as epistemic injustice and oppression, discursive injustice, implicit bias, polarization, gender, and race, among others, have become central in contemporary analytic philosophy. However, like any novel philosophical movement, this political turn in analytic philosophy raises different questions about its nature, origin, and scope. Can we capture this diversity of issues and problems under the same movement? How does this turn differ from applying the analytic methods to political philosophy in the tradition of John Rawls or Robert Nozick? Does it make sense to differentiate the political turn from analytic feminism?

This chapter aims to characterize the political turn in analytic philosophy. In our view, what distinguishes the political turn from other intersections between analytic philosophy and politics is the interest in understanding the different forms of injustice and unfair power relations experienced by disadvantaged groups and the use of particular theories and concepts within different areas of analytic philosophy like the philosophy of language, mind, epistemology, and metaphysics. Moreover, we will argue, this particular way of politically orienting the conceptual instruments of the philosophy of language, epistemology, and metaphysics first appeared, at least systematically, in analytic feminism, and therefore, we understand the contemporary political turn as an extension of it. In the first section, we sketch what we mean by analytic philosophy, and introduce some historical examples of intersections between analytic methods and political interests that differ from what we mean by the political turn. Section 2 will discuss the relation between analytic philosophy and political philosophy by introducing several examples that, although they are cases of politically

---

2 See, e.g., Langton and West 1999; Saul 2006; Mikkola 2011; Anderson and Lepore 2013; Maitra 2013; and Kukla 2018.

3 See, e.g., Stanley 2018.

4 See, e.g., Haslanger 2012; Mackenzie and Stoljar 2000; and Manne 2017.

5 See Bramson et al. 2017 and Bordonaba 2019.

6 See Kelly 2008; Kenyon 2014; and Pennebaker 2015.

7 See Baumgaertner 2014; Nguyen 2020a, 2020b; and Elzinga 2020.

8 Robson 2014.

concerned analytic philosophy, cannot be characterized as belonging to the political turn. Section 3 introduces our definition the political turn and how it arises within so-called feminist analytic philosophy or analytic feminism. In section 4, we present a possible objection to the attempt to characterize the political turn or to distinguish it from standard practices within contemporary analytic philosophy. Finally, in section 5, we respond to this objection while proposing a distinction that will help us clarify how the political turn is a genuinely original movement within the analytic tradition. We conclude that the political turn in analytic philosophy must be understood as an explicit commitment of certain philosophers to gear their theories or analysis towards social and political activism.

2 Analytic Philosophy and Politics

The characterization of analytic philosophy as a historical school of thought, a methodological approach or style of philosophizing, has been proven to be highly problematic (Beaney 2013; and Glock 2008). A common way of characterizing it often appeals to its methodological clarity; for instance, according to the European Society of Analytic Philosophy, "Analytic philosophy is characterized above all by the goal of clarity, the insistence on explicit argumentation in philosophy, and the demand that any view expressed be exposed to the rigors of critical evaluations and discussions by peers" (quoted in Beaney 2013). Other definitions, however, invoke a historical perspective that appeals to the linguistic turn (Gerrard 1997) or the contrast that arises in Anglophone philosophy departments between analytic and continental approaches during the mid-twentieth century, where analytic philosophers often recognize themselves as successors of figures like Frege, Moore, Russell, Wittgenstein, and the logical positivists (see Garavaso 2018 and Garry 2018a, 2018b). The points of disagreement focus not only on these divergent views but also on understanding each one. For instance, some scholars insist that some oft-mentioned authors like Frege should be excluded as founders of analytic philosophy (see Hacker 2007 and Soames 2003a,2003b; for a reply, see Beaney 2006) or argue that only Wittgenstein in the *Tractatus* can be considered as embracing a linguistic turn (Hacker 2013).

Be that as it may and recognizing these divergent views, we follow Ann Garry (2018b, pp. 17–18) on characterizing analytic philosophy not as a doctrine or specific method but as a tradition that, more or less flexibly, is recognized as (a) having philosophers like Frege, Moore, Russell, Wittgenstein, and the logical positivists among their canonical figures; (b) desiring clarity and precision in the use of concepts like truth, objectivity, moral agency, etc.; and (c) contrasting with other traditions like absolute idealism, phenomenology, and poststructuralism.

This definition is very rough and has, as Garry herself argues, intentionally loose boundaries. This leaves the door open to debates about the nature of the tradition or the prominent figures of the canon but at the same time allows us to trace analytic philosophy in a great diversity of areas such as the philosophy of language, epistemology, metaphysics, philosophy of science, and philosophy of mind.

However, in what sense can we speak of an intersection between analytic philosophy and politics? A first point of contact can be found precisely in their canonical figures. Bertrand Russell is more than well-known both for his activism and for his social and political publications. For instance, he is famous for his strong pacifism and activism against the participation of the United Kingdom in the First World War (Vellacott 1980), which led to his imprisonment, or his struggle against the manufacture of nuclear weapons, which included the writing of the *Russell-Einstein Manifesto* (Irvine 2021). Moreover, he wrote several influential books on political and social issues like the *Principles of Social Reconstruction* (1916), *On Education* (1926), *The Scientific Outlook* (1931), and *Power: A New Social Analysis* (1938), where he analyzes different issues like the role of power or education in a society and the importance of scientific understanding for fighting against the negative consequences of superstitions and religion. Likewise, Michael Dummett is well known for his anti-racist activism and his famous book *On Immigration and Refugees*, where he analyzes the inability of British governments to meet basic moral standards in relation to immigrants and refugees.

Another example of how the intersection between politics and analytic philosophy is manifested through its canonical figures is the Vienna Circle. The Vienna Circle is known primarily for propelling logical empiricism. Logical empiricism, logical positivism, or neo-positivism[9] is defined as "a scientific world conception" (Carnap, Hahn, and Neurath 1973), free of metaphysical and theological thinking. Some of the most basic ideas that define logical empiricism are: (a) a consideration of experience as the sole source of knowledge; (b) a conception of philosophical work as the clarification of statements through the use of the method of logical analysis; (c) the application of an empiricist criterion of meaning; (d) an explicit rejection of metaphysics; (e) the rejection of the existence and possibility of a priori synthetic judgments; and (f) the defense of unified science and collective work.

---

9 For a detailed explanation of the subtle differences between each of these terms, see Uebel (2019).

As we can see, none of these ideas are aimed directly at the political sphere. However, many of the Vienna Circle members had deep political concerns throughout their lives, in some cases even holding political office. For example, Otto Neurath was "president of the Central Economic Office in the short-lived socialist government of Bavaria" (Reisch 2005, p. 32). However, some authors have argued that the private political convictions of the members of the Vienna Circle were not reflected in their philosophical work. In this line, Richardson (2009a and 2009b) argues that, although there were political concerns among some of the Circle members, it is not accurate to say that their philosophical work was politically engaged. For instance, although Carnap was a convinced leftist militant all his life, he "was a firm political neutralist, however, in his philosophical work" (Richardson 2009a, p. 17).

For these logical empiricists, such expressions were not an empirically adequate way of categorizing people, and, because of this lack of empirical adequacy, society should avoid using such categorizations.

Other authors[10], however, have defended the existence of a left wing within the Vienna Circle, consisting of Otto Neurath, Rudolf Carnap, Hans Hahn, and Phillip Frank, that would have developed a "form of logical empiricism that was critical and politically engaged" (Uebel 2005, p. 755). For instance, Stadler (1991, p. 165) argues that Neurath's pluralistic conception of philosophy of science was embodied not only in his encyclopedism and the idea of unified science but also in his proposal of using these ideas to democratize all levels of society. In a similar vein, Romizi (2012) contends that the scientific world conception created a "corporate identity under which the Vienna Circle acted as a unitary social actor, not only within the scientific and philosophical communities but also within the sociopolitical context of that time" (Romizi 2012, p. 210). Bright (2017) argues that some logical empiricists (Carnap, Neurath, Schlick, Reichenbach, Waismann, Hahn, Frank, and Lewin) express both in the ethical and the scientific sphere their attitude against using "racial categorizations" such as "Rudson is black, and all black people like the music of Bruce Springsteen" (Bright 2017, p.9)

In summary, several authors have argued that the separation between politics and philosophy in the Vienna Circle was not as sharp as tradition suggests.[11] In fact, this could be the case for Neurath. In addition to the marked political

---

10 Stadler 1991; Reisch 2005 and 2009; Uebel 2005; Howard 2009; Romizi 2012; Bright 2017; and Yap 2022.
11 Besides, it should be noted that the philosophy of science developed by some members was purged of its political and social commitments in the United States during the Cold War (see Reisch 2005).

and democratizing character of the unified science project, his philosophical ideas often had a marked emancipatory character. An unmistakable sign of this is, for example, his development of the ISOTYPE system (see Resich 2005, pp. 32–33), an instance of democratization of knowledge that can still be seen today in streets, museums, and airports all over the world. However, it seems possible that Neurath is an isolated case, and other figures of the Vienna Circle, such as, for example, Carnap, do not exhibit similar political concerns. It is no coincidence that most authors who defend a politically engaged reading of the philosophy of the Vienna Circle focus on the figure of Neurath. Be that as it may, this is something under debate that will have to be elucidated from a historical perspective.

The second group of intersections between analytic philosophy and politics is manifested in what is known as analytic political philosophy (Wolff 2006). As "a philosophical reflection on how best to arrange our collective life" (Miller 1998, para. 1), political philosophy focuses on the application of ethical concepts to the social sphere (see Moseley 2002). It deals with various theories regarding forms of government, economic systems and institutions, and social existence that people could live in, but also with concepts like justice, equality, freedom, democracy, and utopia. Although many of the canonical figures of analytic philosophy kept their philosophical work relatively separate from their politics, we can find examples of analytic philosophers dealing with political concepts and problems almost from the beginning (Wolff 2006). Two paradigmatic examples are the philosophers Herbert L. A. Hart and Margaret Macdonald. They applied conceptual elements of so-called ordinary language philosophy to legal and political concepts as early as 1951 in the volume *Essays in Logic and Language* edited by Anthony Flew (Wolff 2006). These works had continuity in the 1970s with the celebrated works of philosophers such as Rawls (see Rawls 1999) and Nozick (see Nozick 1974). They addressed political issues such as the configuration of the state, the notion of utopia, and the idea of justice from a clear analytical position since argumentation, clear distinctions, and the attempt to reach rigor and clarity characterize their work. Similarly, in Gerald Cohen (1978) or Jon Elster (1985), the authors call themselves analytic Marxists to make explicit both their attraction to Marx's thought and their dissatisfaction with the standards of rigor and clarity of other Marxist theorists like Louis Althusser (see Wolff 2006). Finally, we can also find many works from more contemporary authors, for example, Nussbaum (2006), Sen (2010), or Sandel (1982), which are recognized by their style and methodology as belonging to the analytic tradition. In brief, all those authors can be regarded as being part of the analytic tradition in the sense that they are identified with clarity, argumentative rigor, and the use of analytical tools such as mental experiments or conceptual analysis, while its main research focus is on problems that can be called purely political.

Despite this, however, there are two significant differences between these authors, with perhaps[12] the exception of Neurath, and those we consider to be members of the contemporary political turn in analytic philosophy. First, while these authors are concerned with general political issues such as democracy and justice, the political turn deals with particular instances of injustices and forms of oppression. Secondly, while both are committed to analytical rigor and clarity and their tradition when dealing with the political, the political turn has as a more particular feature: the use of specific theories and concepts from areas of philosophy that are not traditionally considered political, such as the philosophy of language and mind, epistemology and metaphysics, to name a few.

3 From Analytic Feminism to the Political Turn

In the previous section, we advanced the idea that the political turn is distinguished from other intersections between analytic philosophy and politics or other political works by analytic philosophers through two fundamental dimensions. First, the political turn is not only characterizable as analytic by its aspiration to a certain conceptual rigor and clarity but also by the explicit use of theoretical tools from disciplines traditionally associated with analytic philosophy such as the philosophy of language, metaphysics, and epistemology. Second, the political interest and objectives of the turn do not manifest themselves in general interests in how we can shape our collective lives, institutions, and economic systems to have a more just and democratic society, but in particular political interests associated with identifying and addressing particular injustices

---

12 As will become evident throughout the chapter, as we conceive the political turn in analytic philosophy, cases such as the Vienna Circle, Michael Dummett, and Bertrand Russell are hardly conceivable as being part of that turn or even as predecessors. This does not mean, of course, that the importance of their activism should not be recognized or that there are not particular cases—such as Neurath or ordinary language philosophers like Margaret Macdonald, who addressed political issues with their philosophical tools—that can be considered as examples or intermediate cases of the requirements we will discuss below. We admit and are open to historical considerations of this kind. However, and for the purpose of this chapter, we consider the political turn in contemporary analytic philosophy as being different and contrastable with the various intersections between politics and analytic philosophy considered in this section.

of disenfranchised groups.[13] Third, as we shall see below, this way of doing philosophy involves a particular mode of philosophical activism, in that it attempts to change the world through giving us a better understanding of certain unjust phenomena and power relations and equips us with certain tools for resistance and awareness.

In this section, we discuss and exemplify these fundamental aspects in order to try to come up with a more or less tractable definition of the contemporary political turn. To do so, we will begin by discussing analytic feminism. As we have seen, we can track several examples of activism potentially identifiable with the turn in the history of analytic philosophy. However, we believe that it is not until the advent of so-called analytic feminism that these central features of doing politically oriented philosophy became systematic. In this sense, we submit the idea that the contemporary political turn in analytic philosophy should be understood as a generalization of analytic feminism. However, what is analytic feminism in the first place?

Keeping in mind what we said in section 2, analytic feminism can be understood as a tradition within feminism that identifies itself with the methods, core concepts, and canonical philosophers of the analytic tradition (Garavaso 2018 and Garry 2018a and 2018b). Thus, analytic feminism shares with other types of feminisms the realization that women, as a group, are in a position of subordination in relation to men, the aim of accounting for the sources and causes of such subordination and the commitment to activism—understood as the objective of ending the subordination and offering different strategies for such a purpose (Callahan 1996, p. 186). However, contrary to other types of feminisms, analytic feminism embraces the goals and values of analytic philosophy that attempts to be governed by high standards of clarity and rigorousness and to focus on notions such as truth, knowledge, rationality, and objectivity (Cudd 1995, 2005; Garavaso 2018; and Jeske 2002). This definition is sufficiently broad to encompass different ways of understanding the connection between feminism and analytic philosophy. For instance, the Society for Analytical Feminism characterizes its fundamental mission as the promotion of "the study of issues in Feminism by methods broadly construed as analytic, to examine the use of analytic methods as applied to feminist issues and to provide a means by which those interested in Analytic Feminism may meet and exchange ideas" (Cudd and Norlock 2018, p. 39). A broader definition is presented by

---

13 Of course, the ultimate goal of this change of emphasis is to live in more just and democratic societies. But the starting point of the turn, as we shall see, is injustice understood as a phenomenon that deserves attention in its own right, not as an accidental phenomenon arising from a mere lack of justice.

Cudd (1995 and 2005), who emphasizes that analytic feminism not only takes advantage of the analytic methods and theories but also applies their insights and values to issues that have traditionally been of interest to analytic philosophers. She presents this point as follows:

> analytic feminists insist on seeing how sexism, androcentrism, and the domination of the profession of philosophy by men distorts philosophers' pursuit of truth and objectivity. Analytic feminism often attempts to reclaim these notions from androcentric biases: to find what is epistemically compelling in these concepts and what is morally good in their application and to separate that from the sexist baggage that has traditionally accompanied them. Some analytic feminists (and this has been the focus of much of my own work) argue that, properly analyzed, these concepts can be used to undermine androcentrism or unjust gendered social institutions. (Cudd 1995, p. 3)

This emphasis on reclaiming the notions of analytic philosophy from androcentric biases has motivated the aspects of feminist work that we consider fundamental for understanding the political turn. As we understand it, the political turn must be regarded as a generalization of analytic feminism. That is, the political turn in analytic philosophy is based on a shift of political interest and application of tools that appears systematized in feminism, where the center of research is the experience of women, but that later has been extended to other groups that are oppressed for reasons of race, sexual orientation, or social status. To analyze these fundamental aspects, let us start with a paradigmatic example: the feminist theories of agency.

The current feminist theories of agency and autonomy, also known as relational theories of agency (see Mackenzie and Stoljar 2000), opposed standard views in autonomy and agency, according to which the key notion for understanding agency is intentional agency, characterized as a course of action motivated by reasons. Different standard theories provide different characterizations of this connection between reasons and intentional actions (Schlosser 2019), for instance, causal theories regard the connection in causal terms. Such a view of agency is also closely tied to the standard notion of autonomy, which could be traced back to Kant (1785), and often emphasizes individual characteristics of autonomy like preferences, self-government, one's reasons for action, self-control, and independence. Feminist criticism of the notion of autonomy and agency lies precisely on casting into question the individualistic nature of the standard view of agency and autonomy. Agents are not isolated from their social environment— e.g., social relations, institutions—which strongly influences and shapes their values, preferences, and other features involved in determining agency and autonomy. For instance, being an autonomous agent requires behaving in a way that promotes one's preferences and values. However, those preferences and values are instituted by the social environment of the individual. This key point leads us to perceive how the social environment may influence the way we perceive certain cases as examples of autonomy or not. Can we, for example, consider cases of autonomies where women behave to comply with choices and preferences that are unconsciously accommodated to oppressive social conditions? What about cases of self-abnegation, where women may exhibit excessive deference to others' wishes because of their acquired social role? These cases serve as obvious counterexamples to the standard view and highlight the importance of refining our theories of agency and autonomy so that they can account for the experience of the oppressed groups. At the same time, these counterexamples have motivated the appearance of relational theories that attempt to account for these aspects and place emphasis on the non-individual features of autonomy that allow us to better understand some sources of gender-based abuses, such as how social expectations can undermine an agent's autonomy despite their own individual reasons for action.

We are now in a good position to look in more detail at the two aspects that we consider fundamental to analytic feminism and, by extension, to the political turn in analytic philosophy. First, as the example manifests, the turn exhibits an interest in identifying and understanding power relations and oppressions toward a disenfranchised group—women in this case. This move entails an orientation to political issues, so injustices, the experience of the oppressed and their problems became research objectives and yardsticks for assessing particular theories in the area. For instance, before the contributions of the feminist epistemologists, little had been written in analytic epistemology about epistemic practices and how power relations interfere with them. As Fricker notes, "epistemology as it has traditionally been pursued has been impoverished by the lack of any theoretical framework conducive to revealing the *ethical and political aspects of our epistemic conduct.*" (Fricker 2007, p. 2, emphasis added).

In the same vein, Sally Haslanger's (2000, 2012) work in metaphysics has focused on the analysis of categories that are socially and politically significant (such as gender, race), thus moving away from more traditional metaphysical approaches in which the social and political were seen as belonging to other spheres of knowledge (see, e. g. Haslanger 2000, p. 107). The same, of course, goes for the philosophy of language, where traditionally, the type of expressions studied by philosophers has been expressions of everyday use without any political significance, and the social and political have not occupied much space in philosophy of language discussions until recently (see, e. g., Saul 2018, p. 360). A crucial feature of this political orientation is that they exhibit a particular and concrete character in at least two senses. On the one hand, the object of research is particular in contrast to general political features—e. g. democracy, institutions, and so on. The turn is concerned with more concrete targets, disadvantaged groups as victims of unfair power relations. On the other hand, this is translated into an approach to the object of research as concrete epistemic, linguistic, and intentional rather than ideal agents. Furthermore, such political orientation entails a pragmatic move, evaluating philosophical tools in terms of their capacities for explaining and revealing different forms of injustices.

The second fundamental aspect is precisely the use of particular theoretical tools from traditional analytic areas like the philosophy of language, mind, action, epistemology, and metaphysics to account for the injustices and unfair power relations that are relevant from the perspective of women or another disenfranchised group. In our example, the theoretical tools are relational theories in the philosophy of action. However, we can find examples in areas like the philosophy of language, epistemology, and metaphysics. For instance, we can see how speech act theory helps to illuminate the relationship between pornography and silencing of women[14], how certain views on testimony explain the barriers that someone encounters when giving or receiving knowledge due to their social identity[15] or how different accounts on language use may facilitate the understanding of mechanisms behind political manipulation, the appearance of linguistic injustices or the dangers of harmful speech[16], to name a few but paradigmatic examples.

In a nutshell, we understand the political turn as a movement within the analytic tradition that uses theoretical tools from classical disciplines such as the philosophy of language, mind, epistemology, and metaphysics to detect and understand different forms of injustice and unfair power relations faced by disadvantaged groups in their experiences as epistemic, social, and linguistic agents. In this sense, the political turn can be understood as a form of activism in that it uses these theoretical tools both to reveal certain forms of injustice and to help formulate effective ways of resistance and awareness and strategies for the subversion of such injustices.

In a sense, it could be said that the term 'political turn' in analytic philosophy is deeply related to the activist sense of applied philosophy. Lippert-Rasmussen (2017) identifies seven conceptions of applied philosophy. According to the

---

14 See, e. g., Hornsby 1993; Langton and West 1999; Langton 2009; Maitra 2009; Mikkola 2011; and McGowan 2014.

15 See e. g., Fricker 2007; Dotson 2011 and 2012; Pohlhaus 2012; Medina 2012; McKinnon 2016; Tanesini 2016; and Anderson 2017.

16 See, e. g., Saul 2012 and 2018; Anderson, Haslanger, and Langton 2012; Tirrell 2012; Kukla 2014; Stanley 2015; Ayala-López 2016; Khoo 2017; Henderson and McCready 2019; Beaver and Stanley 2019; and Torices 2021.

activist conception, "Philosophy is applied if, and only if, it is motivated by an ambition of having a certain causal effect on the world" (Lippert-Rasmussen 2017, p. 10). Although having a certain impact on the world is undoubtedly very general, what counts as applied philosophy under this conception is something much more restricted. Proof of this is that, as an example of applied philosophy in this sense, Lippert-Rasmusen cites feminist epistemology and its liberatory socio-political aims. The political turn lies, thus, in bringing to the fore the agents' political and social status in the philosophical analysis of epistemic and linguistic practices, among others, not only to gain a better understanding of them but also, and mainly, to tackle the injustices and unfair power relations such as oppressions underlying such practices. This activist strategy is exemplified, for instance, in the ameliorative project in metaphysics, which attempts, among other things, to develop concepts that can help victims of hermeneutic injustice, who suffer damage to their ability to make sense of their own experiences due to the lack of collective conceptual resources (see Ayala-López 2022). Another example is the different forms of epistemic resistance (see Medina 2012; Berenstain et al. 2021; and Pinedo and Villanueva 2022) to subvert the different injustices suffered by epistemic agents or to undermine the power of those who inflict them.

To conclude, we are now better positioned to see the contrast between the political turn and the intersections between politics and analytic philosophy that we discussed in the previous section. First, the political turn manifests an interest in very concrete political phenomena such as epistemic or linguistic injustices toward disadvantaged groups in contrast with the more abstract political problems like social organization or justice that the analytic political philosophy focuses on. Second, this interest takes shape by using specific theories and conceptual tools from different areas of analytic philosophy ranging from speech act theory to the epistemology of testimony. As such, the political turn is a form of activism, because it puts these tools at the service of understanding and shedding light on different forms of injustices and oppressions to amend them, thus promoting social changes.[17]

---

17 Although it is beyond the scope of this chapter, it is worth noting that the turn, in its eagerness to do away with injustice in its many forms, breaks through the analytic/continental barrier influenced by the Critical Race Theory and the works of authors belonging to other philosophical traditions such as Marx, Foucault, Angela Davis, Judith Butler, and Simone de Beauvoir, to name just a few.

4 The Ideal/Non-Ideal Theory Distinction

The previous sections have been devoted to showing that there is something genuinely original in the way of doing philosophy that analytic feminism inaugurates and that differentiates it from what had been done up to that point in the analytic tradition. In what follows, we will elaborate on the notion of the political turn to illustrate more clearly what is distinctive about it. As far as we know, what the political turn is in analytic philosophy has never before been made explicit. But there have been attempts to explain what makes the traditional philosophy of language different from the political philosophy of language. Since this would be a particular instance of the political turn in analytic philosophy, we will take this debate as our starting point.

Recently, some authors have claimed to account for the difference between the traditional and the political philosophy of language by taking from political philosophy the distinction between ideal and non-ideal theory (see, e. g. Lackey 2018; Beaver and Stanley 2019; and Cappelen and Dever 2019). In the remainder of this section, we explore whether this distinction can be successfully applied to understand the difference between traditional analytic philosophy and the political turn in analytic philosophy. We will focus on two possible ways of interpreting the ideal/non-ideal theory distinction, the first one due to Rawls (1999) and the second one due to Charles Mills (2005 and 2017). We conclude, along with Cappelen and Dever (2021), that the ideal/non-ideal the distinction is not helpful for capturing differences beyond the debate in political philosophy. Their arguments for holding that this distinction does not work to set the political philosophy of language apart from the traditional project also lead to rejecting the claim that the ideal/non-ideal distinction accounts for the difference between traditional analytic philosophy and the political turn. However, we avoid their skeptical conclusion that there is no such a turn. On the contrary, we argue in the next section that there is another basis for characterizing the political turn as a genuine move within the Analytic tradition.

In the Rawlsian interpretation, ideal/non-ideal stands for strict/partial compliance. On the one hand, strict-compliance theory studies well-ordered societies, that is, those in which everyone accepts the same principles of justice and knows that others accept them as well (Rawls 1999, p. 4). On the other hand, partial-compliance theory concerns the principles that govern specific contexts of injustices such as war, punishment, and opposition to unjust regimes (Rawls 1999, p. 8). As Cappelen and Dever (2021, p. 94) point out, the strict or partial compliance with principles of justice does not make much sense when applied to certain debates in the philosophy of language, mind, epistemology, and metaphysics.[18] It does not seem that our linguistic, epistemic, and ontological practices, to name just a few, always rest on principles of justice, and there is nothing wrong with that. Thus, to make this distinction workable in other domains, it must be assumed that the type of principles we humans accept varies according to the activity we carry out. The idealization consists of assuming full compliance with the relevant principles that govern a given activity. But what could be, for instance, the linguistic principle that speakers and listeners are supposed to comply with in their conversational interactions? Following Beaver and Stanley (2019), such a principle could be what we can call Grice's Transparency of Intentions Principle (TIP, for short). Accordingly, the Gricean program is ideal philosophy of language as it relies on the assumption that the speaker's intentions are transparent to the addressee as if "no one has devious, hidden intentions" (Beaver, and Stanley 2019, p. 502). Griceans may argue, however, that their program aims to account for what speakers do in certain conversational contexts and, consequently, deny that TIP holds true in all possible conversational interactions. According to Cappelen and Dever, "there's no one in the Gricean tradition who ever assumed full compliance with this [Cooperative Principle] or related principles" (Cappelen and Dever 2021, p. 95). Be that as it may, regardless of whether the Gricean program is ideal or not, which is probably a matter of degree and comparison, the important point is that even if it were true that the Gricean program rested on detrimental idealizations, as Beaver and Stanley (2019) argue, it can by no means follow that such idealizations are at the basis of all the theories and proposals shaping the philosophy of language. In other words, it does not seem clear that we can find an idealized guiding principle that actually drives any inquiry in the philosophy of language and leads to the view that the philosophy of language as a whole is ideal theory. Thus, if the ideal/non-ideal theory distinction stands for strict/partial-compliance theory, the distinction is useless for differentiating traditional from political philosophy of language. Analogous counterexamples abound in other philosophical disciplines such as philosophy of language, mind, epistemology, and metaphysics. The Rawlsian ideal/non-ideal distinction, then, fails to set traditional analytic philosophy apart from the political turn. It can perhaps be applied *locally*, that is, to distinguish between theories within the same discipline, but it cannot be applied *globally*, that is, it cannot be applied to differentiate between two ways of con-

---

18 Cappelen and Dever focus exclusively on the philosophy of language and examine whether the ideal/non-ideal distinction can shed light on the distinction between traditional philosophy of language and political philosophy of language.

ducting the same discipline. Yet, it is precisely a global distinction that we need to account for the political turn.

The second interpretation, owed to Mills (2005 and 2017), is commonly adopted by those who have drawn on the ideal/non-ideal distinction to motivate the use of a different framework for addressing politically significant issues in the philosophy of language (Lackey 2018 and Beaver and Stanley 2019). According to Mills, the ideal/non-ideal theory distinction matches with the distinction ideal-as-idealized-model/ideal-as-descriptive-model. When theorizing about any social or natural phenomenon P, one starts with a model or ideal of that phenomenon, which is a representation of P. As Mills notes, this representation "purports to be descriptive of P's crucial aspects (its essential nature) and how it actually works (its basic dynamic). Call this descriptive modeling sense ideal-as-descriptive-model." (Mills 2005, p. 166).

A descriptive model of an actual P is an abstraction of its essential properties that leaves aside its superfluous features and imperfections:

> Since a model is not coincident with what it is modeling, of course, an ideal-as-descriptive-model necessarily has to abstract away from certain features of P. So one will make simplifying assumptions, based on what one takes the most important features of P to be, and include certain features while omitting others: this will produce a schematized picture of the actual workings and actual nature of P. (Mills 2005, p. 166)

The resulting picture is thus as accurate a representation as possible of the real phenomenon that theorists want to study. But as Mills points out, "for certain P (not all), it will also be possible to produce an idealized model, an exemplar, of what an ideal P *should* be like. Call this idealized model ideal-as-idealized-model." (Mills 2005, p. 166, emphasis added).

The descriptive model does not have to lead to an idealized model, but it can. When the idealized model and the descriptive one are too far apart, the former becomes a pointless model for explaining and understanding the actual world. The ideal theory works on idealized models that leave out crucial aspects of what is being modelled. The main problem with the ideal theory, according to Mills, is its over-optimism in its attribution of capacities and abilities to humans and its lack of interest in understanding "actual historic oppression and its legacy in the present, or current ongoing oppression" (Mills 2005, p. 168). Thus, the ideal theory works based on the attribution of *idealized capacities* to social agents and *the silencing of oppression*, among other flawed assumptions. The non-ideal theory, however, seeks to redress these methodological shortcomings. It explores how oppression to the extent that it is caused by unfairly asymmetrical power relations affects the social context and capabilities of the social agents involved. In other words, unlike ideal theory, non-ideal theory takes the perspective of the oppressed and does not rely on idealizations concerning human capabilities. Humans are understood as socially situated agents whose social reality shapes their epistemic and linguistic capacities, their preferences and concerns, and so on.

Understanding the ideal/non-ideal theory distinction this way, could one argue that the difference between traditional analytic philosophy and the political turn in analytic philosophy can be explained in such terms? Again, we think, along with Cappelen and Dever, that the answer is in the negative. It hardly makes sense to argue that analytic philosophy as a whole was an idealization in the negative sense of assuming over-idealized capacities or of silencing oppression. For example, is anything relevant to the dispute about the semantic nature of moral or aesthetic predicates left out of the debate by not considering gender, race, or class? Does all debate on the epistemology of disagreement rest on pernicious idealizations? Is the debate between externalists and internalists about the justification of knowledge or the debate among philosophers of science about whether and how induction provides genuine knowledge flawed by not considering the axes of oppression mentioned above?

To begin with, our answer to these questions is no. Perhaps if we look at these debates from a political point of view, we may find significant omissions or pernicious idealizations in debates where we did not think we would find them. Indeed, there are debates where omissions of gender, race, and class are theoretically relevant. For instance, ignoring the social status of epistemic agents in the analysis of our capacity as knowledge providers has negative theoretical and political consequences. However, it does not seem to make sense to accuse all the analytic tradition of being ideal theory in the sense of attributing idealized capacities to humans or of silencing oppression. The ideal/non-ideal theory distinction is thus helpful for pointing out theoretical deficiencies in particular debates but does not help to account for what the political turn in analytic philosophy is all about. As we said above, the ideal/non-ideal theory distinction is a locally useful distinction that allows us to differentiate theories from each other according to their assumptions and omissions, but it is not a globally useful distinction to differentiate traditional analytic philosophy from the political turn.

Having shown that the usefulness of the ideal/non-ideal distinction is quite restricted and does not explain what the political turn in analytic philosophy means, it is worth asking whether there is any distinction or criterion that would help us in this endeavor. Our answer is in the affirmative. In the following section, we reject the Cappelen and Dever conclusion that there is no such difference between analytic philosophy and the political turn, based on their rejection of the distinction between the traditional and the political philosophy of language, and provide what, we believe, is an illuminating distinction to shed light on that difference.

5 The Political Turn as the Transition from Non-Activist to Activist Theory

As we have seen, the political turn in analytic philosophy is the shift from traditional analytic philosophy to a politically engaged analytic philosophy, and the ideal/non-ideal theory distinction fails to explain it. However, the political turn is not merely a change of interest in the topics addressed by analytic philosophers. What we call political turn is not, pace Cappelen and Dever, "a more or less random collection of topics with no internal unity that for various hard to understand sociological reasons tend to be lumped together" (Cappelen and Dever 2021, p. 102). It seems to us that there is something substantive in a politically engaged analytic philosophy that explains the turn in how analytic philosophy is conducted and that confers unity to the issues it addresses. The political turn is committed, as we said above, to understanding and identifying particular forms of social injustice and intervening to eradicate them. The political turn consists in engaging in what we call, partially following Lippert-Rasmussen (2017), (philosophical) activism.

Activism takes many forms. It is characterized by the type of issues it addresses (e. g. economic activism and environmental activism) and how it is carried out (e. g. hacktivism and visual activism). Philosophical activism is making philosophy a straightforward and explicit contribution to activism. In other words, it is making philosophy a contribution to social and political change on particular and tangible issues concerning social injustice and unfair power relations.[19] Philosophical activism is far from new in the history of philosophy (the work of Angela Davis or Michel Foucault are some examples). What is original, what supports speaking of a turn, is philosophical activism building on analytic philosophy. Bearing this in mind, we argue that the political turn consists in moving from doing non-activist theory to activist theory in disciplines traditionally alien to political issues in the analytic tradition.

---

19 While activism is a set of socially transformative practices, philosophical activism is a theoretical enterprise closely linked to such practices. Philosophical activism can, for example, contribute to the development of useful tools to carry out different forms of activism such as linguistic activism or epistemic activism (see, e. g. Medina 2018). Of course, these forms of activism do not need philosophy to flourish, but philosophy can do its bit.

The distinction between non-activist and activist theory allows capturing the two ways of conducting analytic philosophy mentioned above: politically unengaged and politically engaged, respectively. Conducting activist theory in this context consists in placing at the center of philosophical inquiry those injustices that some disadvantaged groups experience as epistemic, linguistic, or intentional agents that are generated by unfair power relations and draw on the philosophical tools of analytic disciplines including the philosophy of language, epistemology, and metaphysics, to name just a few, to understand, identify, and resist those injustices. Thus, the activist theory's starting point is not to explain what deviations have taken place so that certain epistemic and communicative practices, for instance, have stopped going right as if going right were the norm. Rather, its starting point is the assumption that those practices often go wrong and that this has always been the case. As Fricker writes:

> The focus on justice creates an impression that justice is the norm and injustice the unfortunate aberration. But, obviously, this may be quite false. It also creates the impression that we should always understand injustice negatively by way of a prior grasp of justice. But, less obviously, the route to understanding may sometimes be the reverse. (Fricker 2007, p. viii)

Thus, the first novelty of philosophical activism, as we have said, is putting injustice in the middle of the philosophical inquiry.[20] There is a second novelty, though: philosophical activism involves a novel dimension assessing how successful a particular theory is. As Pinedo and Villanueva point out, this novel dimension is "the conviction that our philosophical instruments are as good as the role that they can play in the identification of social and economic inequalities and in supporting the project of intervening to eradicate or alleviate them" (Pinedo and Villanueva 2018, p. 123). This move brings the political turn and its philosophical activism closer to Critical Theory as defined, following Nancy Fraser, by Haslanger: "Critical social theory begins with a commitment to a political movement and its questions; *its concepts and theories are adequate only if they contribute to that movement*" (Haslanger 2012, p. 22, emphasis added). Engaging in non-activist theory, however, is not taking social injustice as an object of inquiry—its understanding, identification, and ways of challenging it—whether or not it is relevant to the issue at hand. One caveat, though: it is not our in-

---

20 Putting injustice at the center of the philosophical analysis is sometimes considered the defining move of the non-ideal theory. However, putting injustice at the center is not enough to speak of non-ideal theory. One can put sexism or racism at the center of philosophical analysis and do so on the basis of idealized assumptions. In other words, there can be ideal theory of injustice.

tention to moralize the distinction. One can work on highly abstract philosophical issues and be a very politically and socially engaged person. Nor is it our intention to argue against those who pursue non-activist theory. Many topics in philosophy are fascinating and valuable in their own right, even if they are not aimed at understanding and alleviating social injustices. It is also worth noting that many philosophical proposals and conceptual tools developed by non-activist theorists have turned out to have enormous activist potential (e.g., speech act theory). In short, while the ideal/non-ideal distinction is not merely descriptive but evaluative, the activist/non-activist distinction is meant to be merely descriptive. Engaging in ideal theory is theorizing based on flawed assumptions because of its idealizations and its omissions. Non-activist theory, however, can rest on flawed assumptions or not, and only when it does so, that is, only when it is not only non-activist but also ideal, is it doomed to failure.

In introducing the activist/non-activist theory distinction, our aim is just to take a snapshot of some of the many proposals recently advanced in analytic philosophy and find a common denominator. This shared concern among analytic philosophers engaged in activist theory, as we have said, is to put social injustice and unfair power relations in its manifold manifestations at the core of philosophical reflection. Thus, while the ideal/non-ideal theory distinction is too coarse to capture the difference between analytic philosophy and the political turn, we do find the non-activist/activist distinction useful for this purpose. It is not, therefore, a matter of substituting one distinction for an equivalent one. Both distinctions are not equivalent, but orthogonal. One can conduct ideal activist theory, non-ideal non-activist theory, ideal non-activist theory, and non-ideal activist theory. For instance, one could argue, following Crary (2018), that Fricker's work on epistemic injustice is an example of ideal activist theory, or at least that her proposal, while activist, has components that can be deemed as ideal theory. In particular, Crary argues that "Fricker unquestioningly —and incorrectly—takes for granted that ethical neutrality is a regulative ideal for all world-directed thought" (Crary 2018, p. 47), and ethical neutrality can be considered a common assumption of ideal theory.

Gettier's (1963) work on knowledge can be deemed as a non-ideal non-activist theory. On the one hand, Gettier challenges the idealized notion of knowledge according to which 'someone knows that p' is equivalent to 'someone has a justified true belief that p'. As such, one might say that his contribution to the philosophical debate is non-ideal, or at least less ideal than Chisholm's (1957) proposal, for instance, as it is an attempt to de-idealize the traditional notion of knowledge. On the other hand, Gettier's work is not political epistemology; it does not seek to identify, understand or resist forms of epistemic injustice or other politically relevant epistemic phenomena, that is, it is non-activist.

The Gricean model of communication (see Grice 1989) is a non-activist theory and let us assume for the sake of argument and following Lackey (2018) and Beaver and Stanley (2019), that it is also an instance of ideal theory. It is non-activist, as it does not attempt to explain how communication works in politically significant contexts: neither in those where the Cooperative Principle is one of the principles observed by the participants nor in those where it is not, for example, in cases of political manipulation. It is also an ideal theory, as Beaver and Stanley point out, since the assumption that speakers' intentions are transparent to listeners not only faces difficulties in explaining, for instance, cases of propaganda or dogwhistles but also in accounting for everyday communication—the kind of scenario it is meant to account for. As Beaver and Stanley point out, following Tirrell (2012), "the communicative effects of discourse can go well beyond what the speaker is willing to admit, and even well beyond what the speaker recognizes as the communicative effects of her discourse" (Beaver and Stanley 2019, p. 517).

The work of Quill Kukla (2014, written as Rebecca Kukla) is a case of non-ideal activist theory. Their proposal is activist because it aims to identify and explain a linguistic kind of injustice that they call discursive injustice. It is non-ideal because it takes a particular instance of injustice as a starting point and does not rely on assumptions that idealize the capacities and abilities of speakers and listeners. On the contrary, Kukla argues that speaker social identity such as gender plays an essential role in shaping the addressees' uptake, which, thus understood, de-idealizes the speaker-listener interaction in everyday conversations by bringing new factors to bear, beyond "the correct recognition of the speaker's intention," on the understanding of what it means to perform a speech act (Kukla 2014, p. 444).

Regardless of whether one agrees or disagrees with how some of the specific instances we have presented above have been listed, the crux of the matter is to show that the ideal/non-ideal and non-activist/activist distinctions can intersect between them and that it is the latter one that allows us to explain the two ways of conducting analytic philosophy mentioned above. What we call the political turn in analytic philosophy is accordingly to engage in activist theory. In other words, it is about embracing philosophical activism from an analytic lens.

6 Concluding Remarks

This chapter aimed to define what we have called the political turn in analytic philosophy. To speak of a turn is to speak of a change in the way things were being done up to a certain point in time. We have seen that although there are philosophers in the analytic tradition who have been committed to certain political ideals from the very beginning of analytic philosophy, this has not been the general trend, nor is the kind of political commitment that the political turn involves. We have also noticed that adopting an analytic perspective in traditional political philosophy does not correspond to the political turn either. As we have argued, the political turn takes its first steps with the rise of analytic feminism. The extension of analytic methods and concepts along with the use of particular theories from the philosophy of language and mind, epistemology or metaphysics to gender issues, on the one hand, and the adoption of a gender perspective to traditional problems of the analytic philosophy, on the other, constituted a significant shift within the analytic tradition.

Of course, this change in how analytic philosophy is conducted, initiated by analytic feminism, has taken time to spread. However, in the last two decades, there has been an increase in the number of philosophers who have adopted the political turn, and this has given rise to what we can call a proper political turn in analytic philosophy. The political turn is characterized by adopting philosophical activism, that is, an interest in different forms of injustice and unfair power relations experienced by disadvantaged groups. The ultimate aim of philosophical activism is to generate new conceptual tools or extend the application of the old ones to the political realm to identify otherwise invisible forms of injustice and unfair power relations and formulate effective ways of tackling them. Besides a change in the way we engage in analytic philosophy, as we have seen, the turn involves a new way of evaluating how successful our philosophical methods and tools are: we test them according to their potential and effectiveness in responding to and revealing particular instances of injustice. As a consequence, the research of a philosophical activist will be informed and guided by specific values, those directly or indirectly related to the injustice or oppression being fought. As Ayala-López notes:

> when philosophizing about other people's identities, I will have ethical principles guiding my inquiry, and I will use these principles as I reflect on my interest on the subject matter, the goals of my inquiry (e.g., Why am I interested in this? What do I want to attain with my research?), and the way I proceed in my research. This does not imply that my philosophizing will be spoiled or coerced towards specific positions. It means I will examine my subject matter not from an impossibly abstract and supposedly neutral armchair, but acknowledging and being transparent about the values that are guiding my inquiry in more or less implicit ways. (Ayala-López 2020, p. 13)

Being transparent about the ethical values that guide and motivate our academic work should not be a cause for concern. There is nothing wrong with recognizing that concrete values guide our philosophical practice. In fact, all research is guided by values, although they are not always made explicit.

Bibliography

Anderson, Luvell (2017): "Hermeneutical Impasses." In: *Philosophical Topics* 45. No. 2, pp. 1–19.

Anderson, Luvell, Sally Haslanger, and Rae Langton (2012): "Language and Race." In: Gillian Russell and Delia Graff Fara (Eds.): *The Routledge Companion to Philosophy of Language*. New York, London: Routledge, pp. 753–768.

Anderson, Luvell, and Ernie Lepore (2013): "Slurring Words." In: *Noûs* 47. No. 1, pp. 25–48.

Ayala-López, Saray (2016): "Speech Affordances: A structural take on how much we van do with our words." In: *European Journal of Philosophy* 24. No. 4, pp. 879–891.

Ayala-López, Saray. (2020): "(Philosophizing about) Gender Open Children." In: *American Philosophical Association Newsletter in Feminism and Philosophy* 19. No 2, pp. 45–49.

Ayala-López, Saray (2022): "Hermeneutical Injustice and Conceptual Landscaping: The Benefits and Responsibilities of Expanding Conceptual Landscaping Beyond Failure Reparation." In: David Bordonaba-Plou, Víctor Fernández-Castro, and José R. Torices (Eds.): *The Political Turn in Analytic Philosophy: Reflections on Social Injustice and Oppression*. Berlin: De Gruyter, pp. 211–228.

Baumgaertner, Bert (2014): "Yes, no, maybe so: a veritistic approach to echo chambers using a trichotomous belief model." In: *Synthese* 191. No. 11, pp. 2549–2569.

Beaney, Michael (2006): "Soames on Philosophical Analysis (Critical Notice of Scott Soames, Philosophical Analysis in the Twentieth Century)." In: *Philosophical Books* 7, pp. 255–271.

Beaney, Michael (2013): "What is Analytic Philosophy?" In: Michael Beaney (Ed.): *The Oxford Handbook on the History of Analytic Philosophy*. Oxford: Oxford University Press, pp. 3–29.

Beaver, David, and Jason Stanley (2019): "Toward a Non-Ideal Philosophy of Language." In: *Graduate Faculty Philosophy Journal* 39. No. 2, pp. 503–547.

Berenstain, Nora, Kristie Dotson, and Julieta Paredes et al. (2021): "Epistemic oppression, resistance, and resurgence." In: *Contemporary Political Theory* https://doi.org/10.1057/s41296-021-00483-z,_accessed on 22 September 2020.

Bordonaba, David (2019): "Polarización como impermeabilidad: cuando las razones ajenas no importan." In: *Cinta de Moebio* 66, pp. 295–309.

Bramson, Aaron, Patrick Grim, and Daniel J. Singer et al. (2017): "Understanding Polarization: Meanings, Measures, and Model Evaluation." In: *Philosophy of Science* 84. No. 1, pp. 115–159.

Bright, Liam K. (2017): "Logical empiricists on race." In: *Studies in History and Philosophy of Biological and Biomedical Sciences* 65, pp. 9–18.

Callahan, Joan (1996): "Symposium: A Roundtable on Feminism and Philosophy in the Mid-1990s: Taking Stock: Introduction." In *Metaphilosophy* 27. No. 1/2, pp. 184–188.

Cappelen, Herman, and Josh Dever (2019): *Bad Language*. Oxford: Oxford University Press.

Cappelen, Herman, and Josh Dever (2021): "On the Uselessness of the Distinction Between Ideal and Non-Ideal Theory (At Least in the Philosophy of Language)." In: Justin Khoo and Rachel Sterken (Eds.): *The Routledge Handbook of Social and Political Philosophy of Language*. New York, London: Routledge, pp. 91–105.

Carnap, Rudolf, Hans Hahn, and Otto Neurath (1973): "The Scientific Conception of the World: The Vienna Circle." In: Marie Neurath and Robert S. Cohen (Eds.): *Empiricism and Sociology*. Dordrecht: Reidel, pp. 298–318.

Chisholm, Roderick (1957): *Perceiving: A Philosophical Study*. New York: Cornell University Press.

Cohen, Gerald A. (1978): *Karl Marx's Theory of History: A Defence*. Oxford: Oxford University Press.

Crary, Alice (2018): "The methodological is political: What's the Matter with 'analytic feminism.'" In: *Radical Philosophy*. No. 2.02, pp. 47–60.

Cudd, Ann E. (1995): "Analytic Feminism: A Brief Introduction." In: *Hypatia* 10. No. 3, pp. 1–6.

Cudd, Ann E. (2005): "Analytic Feminism." In: Edward Craig (Ed.): *Encyclopedia of Philosophy*. New York: Routledge, pp. 157–159.

Cudd, Ann E. and Kathryn J. Norlock (2018): "The Society for Analytical Feminism: Our Founding Twenty-Five Years Ago." In: Pieranna Garavaso (Ed.): *The Bloomsbury Companion to Analytic Feminism*. London: Bloomsbury, pp. 37–43.

Dotson, Kristie (2011): "Tracking Epistemic Violence, Tracking Practices of Silencing." In: *Hypatia* 26. No. 2, pp. 236–257.

Dotson, Kristie (2012): "A Cautionary Tale: On Limiting Epistemic Oppression." In: *Frontiers: A Journal of Women Studies* 33. No. 1, pp. 24–47.

Elster, Jon (1985). *Making Sense of Marx*. Cambridge: Cambridge University Press.

Elzinga, Benjamin (2020): "Echo Chambers and Audio Signal Processing." In: *Episteme* (online first), pp. 1–21.

Flew, Antony (Ed.) (1951): *Essays in Logic and Language*. Oxford: Blackwell.

Fricker, Miranda (2007): *Epistemic Injustice: Power and the Ethics of Knowing*. Oxford: Oxford University Press.

Garavaso, Pieranna (2018): "What is Analytic Feminism?" In: Pieranna Garavaso (Ed.): *The Bloomsbury Companion to Analytic Feminism*. London: Bloomsbury Academics, pp. 3–16.

Garry, Ann (2018a): "Analytic Feminism." In: *The Stanford Encyclopedia of Philosophy* (Fall 2018 Edition), Edward N. Zalta (Ed.), URL = <https://plato.stanford.edu/archives/fall2018/entries/femapproach-analytic/>, accessed on 22 September 2021.

Garry, Ann (2018b): "Why Analytic Feminism?" In: Pieranna Garavaso (Ed.): *The Bloomsbury Companion to Analytic Feminism*. London: Bloomsbury Academics, pp. 17–36.

Gerrard, Steve (1997): "Desire and Desirability: Bradley, Russell and Moore versus Mill." In: William Tait (Ed.): *Early Analytic Philosophy: Frege, Russell, Wittgenstein*. Chicago: Open Court, pp. 37–74.

Gettier, Edmund (1963): "Is Justified True Belief Knowledge?" In: *Analysis* 23. No. 6, pp. 121–123.

Glock, Hans-Johann (2008): *What is Analytic Philosophy?* Cambridge: Cambridge University Press.

Grice, H. Paul (1989). *Studies in the Way of Words*. Cambridge: Harvard University Press.

Hacker, Peter M. S. (2007): "Analytic philosophy: beyond the linguistic turn and back again." In: Michael Beaney (Ed.): *The Analytic Turn: Analysis in Early Analytic Philosophy and Phenomenology*. London: Routledge, pp. 125–141.

Hacker, Peter M. S. (2013): "The Linguistic Turn in Analytic Philosophy." In: Michael Beaney (Ed.): *The Oxford Handbook on the History of Analytic Philosophy*. Oxford: Oxford University Press, pp. 926–947.

Haslanger, Sally (2000): "Feminism in Metaphysics: Negotiating the Natural." In: Miranda Fricker and Jennifer Hornsby (Eds.): *The Cambridge Companion to Feminism in Philosophy*. Cambridge: Cambridge University Press, pp. 107–126.

Haslanger, Sally (2012): *Resisting Reality: Social Construction and Social Critique*. New York: Oxford University Press.

Henderson, Robert, and Elin McCready (2019): "Dogwhistles and the At-Issue/Non-At-Issue Distinction." In: Daniel Gutzmann and Katharina Turgay (Eds.): *Secondary Content*. Leiden: Brill, pp. 222–245.

Hornsby, Jennifer (1993): "Speech acts and pornography." In: *Women's Philosophy Review* 10, pp. 38–45.

Howard, Don (2009): "Better Red than Dead—Putting an End to the Social Irrelevance of Postwar Philosophy of Science." In: *Science & Education* 18, pp. 199–220.

Irvine, Andrew David (2021): "Bertrand Russell." In: *The Stanford Encyclopedia of Philosophy* (Spring 2021 Edition), Edward N. Zalta (Ed.), URL = <https://plato.stanford.edu/archives/spr2021/entries/russell/>, accessed on 20 September 2021.

Jeske, Diane (2002): "Feminism, Friendship, and Philosophy." In: *Canadian Journal of Philosophy* 28 (Suppl.), pp. 63–82.

Kant, Immanuel (1948): *Groundwork of the Metaphysic of Morals*. Translated and analyzed by H. J. Paton. New York: Harper & Row.

Kelly, Thomas (2008): "Disagreement, Dogmatism, and Belief Polarization." In: *The Journal of Philosophy* 105. No. 10, pp. 611–633.

Kenyon, Tim (2014): "False polarization: debiasing as applied social epistemology." In: *Synthese* 191, pp. 2529–2547.

Khoo, Justin (2017): "Code Words in Political Discourse." In: *Philosophical Topics* 45. No. 2, pp. 33–64.

Kukla, Rebecca (2014): "Performative Force, Convention, and Discursive Injustice." In: *Hypatia* 29. No. 2, pp. 440–457.

Kukla, Rebecca (2018): "Slurs, Interpellation, and Ideology." In: *Southern Journal of Philosophy* 56. No. S1, pp. 7–32.

Lackey, Jennifer (2018): "Silence and Objecting." In: Cassey R. Johnson (Ed.): *Voicing Dissent: The Ethics and Epistemology of Making Disagreement Public*. Routledge, pp. 82–97.

Langton, Rae (2009): *Sexual Solipsism: Philosophical Essays on Pornography and Objectification*. New York: Oxford University Press.

Langton, Rae, and Caroline West (1999): "Scorekeeping in a Pornographic Language Game." In: *Australasian Journal of Philosophy* 77. No. 3, pp. 303–319.

Lippert-Rasmussen, Kasper (2017): "The Nature of Applied Philosophy." In: Kasper Lippert-Rasmussen, Kimberley Brownlee, and David Coady (Eds.): *A Companion to Applied Philosophy*. Oxford: Wiley-Blackwell, pp. 3–17.

Mackenzie, Catriona, and Natalie Stoljar (2000): "Introduction: autonomy refigured." In: Catriona Mackenzie and Natalie Stoljar (Eds.): *Relational Autonomy: Feminist*

Perspectives on Autonomy, Agency and the Social Self. New York: Oxford University Press, pp. 3–31.

Maitra, Ishani (2009): "Silencing Speech." In: *Canadian Journal of Philosophy* 39. No. 2, pp. 309–338.

Maitra, Ishani (2013): "Subordination and Objectification." In: *Journal of Moral Philosophy* 10. No. 1, pp. 87–100.

Manne, Kate (2017): *Down Girl: The Logic of Misogyny.* Oxford: Oxford University Press.

McGowan, Mary Kate (2014): "Sincerity Silencing." In: *Hypatia* 29. No. 2, pp. 458–473.

McKinnon, Rachel (2016): "Epistemic Injustice." In: *Philosophy Compass* 11. No. 8, pp. 437–446.

Medina, José (2012): *The Epistemology of Resistance: Gender and Racial Oppression, Epistemic Injustice, and Resistant Imaginations.* New York: Oxford University Press.

Medina, José (2018): "Resisting Racist Propaganda: Distorted Visual Communication and Epistemic Activism." In: *Southern Journal of Philosophy* 56. No. S1, pp. 50–75.

Mikkola, Mari (2011): "Illocution, Silencing and the Act of Refusal." In: *Pacific Philosophical Quarterly* 92. No. 3, pp. 415–437.

Miller, David (1998): "Political Philosophy." In: Edward Craig (Ed.): *Routledge Encyclopedia of Philosophy.* https://n9.cl/z0icz, accessed on 30 November 2020

Mills, Charles W. (2005): "'Ideal Theory' as Ideology." In: *Hypatia* 20. No. 3, pp. 165–184.

Mills, Charles W. (2017): *Black Rights/White Wrongs: The Critique of Racial Liberalism.* Oxford: Oxford University Press, .

Moseley, Alexander (2002): "Political Philosophy: Methodology." In: *The Internet Encyclopedia of Philosophy.* https://iep.utm.edu/polphil/, accessed on 10 October 2020.

Nussbaum, Martha (2006): *Frontiers of Justice: Disability, Nationality, Species Membership.* Cambridge: Harvard University Press.

Nguyen, C. Thi (2020a): "Echo Chambers and Epistemic Bubbles." In: *Episteme* 17. No. 2, pp. 141–161.

Nguyen, C. Thi (2020b): "Cognitive islands and runaway echo chambers: problems for epistemic dependence on experts." In: *Synthese* 197. No. 7, pp. 2803–2821.

Nozick, Robert (1974): *Anarchy, State and Utopia.* New York: Basic Books.

Pennebaker, Lee (2015): "The Epistemological Significance and Implications of Belief Polarization." In: *Res Cogitans* 6, pp. 93–101.

Pinedo, Manuel, and Neftalí Villanueva (2018): "Power to the People. The Indispensable Nature of the Normative Vocabulary and the Political Turn in Analytic Philosophy." In: Cristian Saborido, Sergi Oms, and Javier González de Prado (Eds.): *Proceedings of the IX Conference of the Spanish Society of Lógic, Methodology and Philosophy of Science.* Madrid, España, pp. 122–124.

Pinedo, Manuel, and Neftalí Villanueva (2022): "Epistemic De-Platforming." In: David Bordonaba-Plou, Víctor Fernández-Castro, and José R. Torices (Eds.): *The Political Turn in Analytic Philosophy: Reflections on Social Injustice and Oppression.* Berlín: De Gruyter, pp. 105–134.

Pohlhaus, Gaile (2012): "Relational Knowing and Epistemic Injustice: Toward a Theory of Willful Hermeneutical Ignorance." In: *Hypatia* 27. No. 4, pp. 715–735.

Rawls, John (1999): *A Theory of Justice.*, rev. ed. Cambridge: Harvard University Press.

Reisch, George A. (2005): *How the Cold War Transformed Philosophy of Science: To the Icy Slopes of Logic.* Cambridge: Cambridge University Press.

Reisch, George A. (2009): "Three Kinds of Political Engagement in Philosophy for Science." In: *Science & Education* 18, pp. 191–197.

Richardson, Sarah S. (2009a): "The Left Vienna Circle, Part 1. Carnap, Neurath, and the Left Vienna Circle Thesis." In: *Studies in History and Philosophy of Science* 40, pp. 14–24.

Richardson, Sarah S. (2009b): "The Left Vienna Circle, Part 2. The Left Vienna Circle, Disciplinary History, and Feminist Philosophy of Science." In: *Studies in History and Philosophy of Science* 40, pp. 167–174.

Robson, Jon (2014): "A social epistemology of aesthetics: belief polarization, echo chambers and aesthetic judgement." In: *Synthese* 191. No. 11, pp. 2513–2528.

Romizi, Donata (2012): "The Vienna Circle's 'Scientific World-Conception': Philosophy of Science in the Political Arena." In: *HOPOS: The Journal of the International Society for the History of Philosophy of Science* 2. No. 2, pp. 205–242.

Russell, Bertrand (1916): *Principles of Social Reconstruction*. London: George Allen and Unwin.

Russell, Bertrand (1926): *On Education, Especially in Early Childhood*. London: George Allen and Unwin.

Russell, Bertrand (1931): *The Scientific Outlook*. London: George Allen and Unwin.

Russell, Bertrand (1938): *Power: A New Social Analysis*. London: George Allen and Unwin.

Sandel, Michael (1982): *Liberalism and the Limits of Justice*. Cambridge: Cambridge University Press.

Saul, Jennifer (2006): "Pornography, Speech Acts and Context." In: *Proceedings of the Aristotelian Society* 106. No. 2, pp. 227–246.

Saul, Jennifer (2012): *Lying, Misleading, and What is Said: An Exploration in Philosophy of Language and in Ethics*. Oxford: Oxford University Press.

Saul, Jennifer (2018): "Dogwhistles, Political Manipulation and Philosophy of Language." In: Daniel Fogal, Daniel Harris, and Matt Moss (Eds.): *New Works on Speech Acts*. New York: Oxford University Press, pp. 360–383.

Schlosser, Markus (2019): "Agency." In: *The Stanford Encyclopedia of Philosophy* (Winter 2019 Edition), Edward N. Zalta (Ed.), URL = <https://plato.stanford.edu/archives/win2019/entries/agency/>, accessed 15 October 2020.

Sen, Amartya (2009): *The Idea of Justice*. London: Allen Lane.

Soames, Scott (2003a): *Philosophical Analysis in the Twentieth Century, Volume 1: The Dawn of Analysis*. Princeton: Princeton University Press.

Soames, Scott (2003b): *Philosophical Analysis in the Twentieth Century, Volume 2: The Age of Meaning*. Princeton: Princeton University Press

Stadler, Friedrich (1991). "Otto Neurath-Moritz Schlick: On the Philosophical and Political Antagonisms in the Vienna Circle." In: Thomas E. Uebel (Ed.): *Rediscovering the Forgotten Vienna Circle: Austrian Studies on Otto Neurath and the Vienna Circle*. Dordrecht: Kluwer Academic Publishers, pp. 159–168.

Stanley, Jason (2015): *How Propaganda Works*. New Jersey: Princeton University Press.

Stanley, Jason (2018): *How Fascism Works. The Politics of Us and Them*. New York: Random House.

Tanesini, Alessandra (2016): "'Calm Down, Dear': Intellectual Arrogance, Silencing and Ignorance." In: *Aristotelian Society Supplementary Volume* 90. No. 1, pp. 71–92.

Tirrell, Lynne (2012): "Genocidal Language Games." In: Ishani Maitra and Mary Kate McGowan (Eds.): *Speech and Harm: Controversies Over Free Speech*. Oxford: Oxford University Press. pp. 174–221.

Torices, José R. (2021): "Understanding dogwhistles politics." In: *Theoria* 36, vol. 3, pp. 321–339.

Uebel, Thomas (2005): "Political Philosophy of Science in Logical Empiricism: The Left Vienna Circle." In: *Studies in History and Philosophy of Science* 36, pp. 754–773.

Uebel, Thomas (2019): "Vienna Circle." In: *The Stanford Encyclopedia of Philosophy* (Spring 2019 Edition), Edward N. Zalta (Ed.), URL = <https://plato.stanford.edu/archives/spr2019/entries/vienna-circle/>. accessed on 10 October 2020.

Vellacott, Jo (1980): *Bertrand Russell and the Pacifists in the First World War*. Brighton, Sussex: Harvester Press.

Yap, Audrey (2022): "Conceptual Engineering and Neurath's Boat: A Return to the Political Roots of Logical Empiricism." In: David Bordonaba-Plou, Víctor Fernández-Castro, and José R. Torices (Eds.): *The Political Turn in Analytic Philosophy. Reflections on Social Injustice and Oppression*. Berlin: De Gruyter, pp. 31–51.

Wolff, Jonathan (2006): "Analytic Political Philosophy." In: Michael Beaney (Ed.): *The Oxford Handbook of The History of Analytic Philosophy*. Oxford: Oxford University Press, pp. 795–823.

Audrey Yap

Conceptual Engineering and Neurath's Boat: A Return to the Political Roots of Logical Empiricism

Abstract: Logical empiricism is not frequently associated with social and political philosophy, but several of the logical empiricists were politically active during their earlier careers and did consider the ways in which their scientific philosophy could provide tools for political engagement. The scientific worldview that they explicitly endorsed was intended to be allied with other modernist projects for the improvement of ordinary life. This chapter argues that some of the philosophical tools and frameworks they developed for those ends can be fruitfully reclaimed by contemporary analytic philosophers engaged in thinking about our social world. Rudolf Carnap's work, for example, is complementary to some contemporary work on conceptual engineering, such as Sally Haslanger's work on ameliorative concepts. Additionally, Otto Neurath's holism, and discussion of the work of auxiliary motives, can help us think through situations in which we have competing empirically adequate theories. I will demonstrate the potential benefits of these frameworks by showing how they might be brought to bear on contemporary issues such as misogyny, transphobia, and scientific racism. I do not claim that the logical empiricists can solve those issues for us, but rather that their work is friendly to the work of philosophers trying to deploy analytic philosophy for liberatory ends. In other words, I argue that there are potential benefits for our viewing the contemporary political turn in analytic philosophy as a return to the political in scientific philosophy.

1 Introduction: The Political Vienna Circle

Logical empiricism is not frequently associated with social and political philosophy, but several of the logical empiricists were in fact very politically active during their earlier careers. Otto Neurath, for example, was involved in socialist politics in the Bavarian Republic in 1918–1919 and was in fact imprisoned and tried for treason as a result (Cartwright et al. 1996, pp. 43–53). Rudolf Carnap himself notes in his *Intellectual Autobiography* that many members of the Vienna Circle were interested in social and political progress and were in fact socialists. Though with the exception of Neurath, they preferred to keep their philosophical work mostly separated from their political aims (Carnap 1963, p. 23). Carnap was

https://doi.org/10.1515/9783110612318-004

particularly insistent on the idea that their logical and scientific work be politically neutral, but, following Reisch (2005), Howard (2003) and, to some extent, Roberts (2007), I will raise some questions in later sections about the extent to which he could do so entirely. For instance, after the completion of his *Logical Structure of the World*, Carnap gave a lecture at the Dessau Bauhaus entitled "Science and Life", in which he told the audience, 'I work in science, and you in visible forms; the two are only different sides of a single life' (Galison 1990, pp. 709 – 710). There was a strong sense that the scientific world-conception, for instance as expressed in their *Wissenschaftliche Weltauffassung: Der Wiener Kreis* would be of general social benefit. In other words, the scientific world-conception, with its anti-metaphysical stance, was intended to be allied with other modernist endeavours that hoped to reform the practices of ordinary life.

Other projects undertaken by Carnap, Neurath and other logical empiricists such as Charles Morris and Philipp Frank in the 1930's, such as their involvement in the Unity of Science movement and the *International Encyclopedia of Unified Science* (though Carnap withdrew his name from the *Encyclopedia* project at a late stage) had an explicitly socially progressive orientation. Though there were tensions between the Carnap and Neurath at various stages of their collaborations, they did share a commitment to various ways in which science could become more unified, for instance by improving means of communication between scientists across international and disciplinary borders. Their vision of unified science was not a reductionist plan, but a social one, for a scientific community that had a common language with which to communicate ideas and results. This, together with Vienna Circle's opposition to many elements of Nazi ideology, such as the idea of a transcendent German essence (Galison 1990, p. 744), had clear political implications.

In this sense, the philosophy of science of many of the logical empiricists was absolutely socially engaged, though not politically engaged in the sense of party politics. So when Carnap and others claim not to be engaging in politics, they are nevertheless working with the awareness that scientific philosophy has social implications. Neurath in particular, as a social scientist, was fully aware that his work dealt with issues of broad social consequence. For instance, he describes some of his work in social planning as economic-technical but non-political, where by non-political he seems to mean something like "non-partisan". In his own description of that work, still notes that he is considering how various social measures will influence conditions of life. What he refrains from discussing (the political matters) are the ways in which those measures are to be implemented through distributions of power (Neurath 1996, p. 19). So even non-political projects are undertaken with social consequences in mind.

In later sections, I will talk more specifically about what Carnap and Neurath, in particular, can bring to contemporary social philosophy, through Carnap's work on conceptual engineering in section 2 and Neurath's embedding of scientific philosophy in ordinary life in section 3. As such, and in thinking of the political turn in contemporary analytic philosophy as a kind of *return*, I argue that we can reclaim some helpful tools for thinking about social change. While I will not spend much time on the issue of how logical empiricism as it has been received in North America became so depoliticized, much of the story has to do with the careers of key figures of the Vienna Circle after emigrating due to World War II. Social factors such as tenuous status in an America concerned about the threat of communism did a great deal to shape their activities after the 1930's. Frank and Carnap, who had both immigrated to the US, were investigated by the FBI as 'potential subversives' (Reisch 2005, p. 115). In fact, several people involved with the Unity of Science movement – Morris, Frank, and Carnap – faced a great deal of anti-communist pressure, though they were shielded in part by the prestige of their institutional employers, Chicago and Harvard (Reisch 2005, pp. 259–282). Because their political work was explicitly suppressed, we cannot know the extent to which their philosophical work would have developed along political lines had they been given the freedom for it. As such, especially in Carnap's case, we will focus much more on the extent to which logical empiricist philosophical work *can* be useful for social and political ends, rather than the extent to which it was.

2 Conceptual Engineering and Ameliorative Projects

Carnap's *Logical Syntax of Language* sets the stage by stating his *Principle of Tolerance:* 'It is not our business to set up prohibitions, but to arrive at conventions' (Carnap and Smeaton 2002, p. 51). He clarifies this pluralist statement on the following page, writing, 'Everyone is at liberty to build up his own logic, i.e. his own form of language, as he wishes. All that is required of him is that, if he wishes to discuss it, he must state his methods clearly, and give syntactical rules instead of philosophical arguments' (Carnap and Smeaton 2002, p. 52). This rejection of philosophical arguments in favour of syntactical, or logical, rules, might seem to run counter to the idea of the political turn, but we can understand this in terms of a distinction in Carnap's thought between the theoretical and the practical (Carnap 1934). Alan Richardson interprets the distinction between theoretical questions and practical decisions as follows:

The former is a matter for scientific investigation of nature; the latter is a matter of the need, within the realm of action, for decision. While theoretical knowledge might help one make a proper decision, the insistence upon the practical is, for Carnap, an insistence upon the need for decision in the realm of activity and an insistence that no amount of knowledge is the very same thing as the decision to act upon it. (Richardson 2007, p. 299)

Read in light of such distinctions, I think that we can understand Carnap's Principle of Tolerance, not as prohibiting our taking interest in the practical, but as distinguishing between the practical decision to use a linguistic framework and the theoretical knowledge that we might obtain using it. We make mistakes when we attempt to frame questions as theoretical when they are in fact practical.

A running example in *Logical Syntax* is the ontological status of numbers in the philosophy of mathematics, where we might contrast a Platonist view with a nominalist one. Under the former view, numbers are genuine existent, but non-physical entities. Under the latter, we would reject the real existence of abstract objects of any kind, including numbers. As such, we might find a Platonist about mathematical objects claiming that the number five is a real thing, where a nominalist might disagree, claiming instead that it is simply a number. Carnap does not take sides in this debate, but claims instead that the way in which it is framed and conducted is fundamentally mistaken because of the ways in which it misuses language. More specifically, Carnap believes that the debate is not really (or should not really be) about the number five itself but is in fact about how the word "five" ought to be used. This means that instead of asking whether five is a thing or a number, we should ask instead whether the word "five" is meant to refer to things or is used in some other way (Carnap and Smeaton 2002, pp. 285–287). But it then becomes clear that we can only answer such a question relative to particular languages and systems of rules. Then the question of which language we ought to use is not a theoretical question, but seems to be a practical one. The theoretical uses to which such a language can be put may be informative with respect to our practical decision, but the decision to use one language over another is still a practical matter (Carnap 1950, pp. 35–36).

The task, then, for the philosopher (or anyone constructing logical frameworks) is a kind of conceptual engineering in the service of science (Richardson 2007, pp. 304–305). Given that feminist philosophers of science have long since debunked the value-free ideal of science (see, e. g., Longino 1990; Douglas 2000; and Wylie and Nelson 2007) and socially engaged philosophy of science continues to gain in popularity (Fehr and Plaisance 2010 and Cartieri and Potochnik 2014), we might want to think of how to revise Carnap's stated commitment to political neutrality. I have argued elsewhere that his general ideas about the reframing of debates are basically compatible with the value-ladenness of science.

So even though his belief in the value-free ideal might have been misplaced, this arguably only means that he was wrong about what our practical decisions about the choice of frameworks might look like (Yap 2010). In fact, some contemporary feminist projects can be seen as conceptual engineering work that is very much in line with these Carnapian ideas (Dutilh Novaes 2020). The best example of this is Sally Haslanger's ameliorative approach to answering "what is *X?*" questions. In describing her approaches to philosophy of race and gender, Haslanger seems to be writing exactly in the spirit of Carnapian conceptual engineering:

> I've cast my inquiry as an analytical – or what I call here an *ameliorative* – project that seeks to identify what legitimate purposes we might have (if any) in categorizing people on the basis of race or gender, and to develop concepts that would help us achieve these ends. I believe that we should adopt a constructionist account not because it provides an analysis of our ordinary discourse, but because it offers numerous political and theoretical advantages. (Haslanger 2012, p. 366)

While Haslanger starts with the practical rather than the theoretical question, she still makes use of a distinction between the ways in which people might use a race or gender term in ordinary speech and a proposal that we adopt a particular definition for a race or gender term. The former project is descriptive, and sensitive to different kinds of considerations than the latter ameliorative kind of project. Haslanger's own proposed definitions for gender and race terms are correspondingly offered as revisionary proposals (Haslanger 2012, pp. 234 – 240). We likely still want our race and gender terms to overlap somewhat with common usage – otherwise it is difficult to see how they would remain useful. But their differing from the ways in which they are used in some idiolects does not really speak against them.

What Haslanger provides is a theoretical framework for such terms that we can decide to use if we decide that it best suits our purposes. In that vein, someone might raise problems with Haslanger's definition if they believe that it is not the best definition for certain shared purposes. But it would not be genuine engagement to argue against Haslanger on the grounds that many people do not use the word "man" as she defines it. Along the former lines, Katharine Jenkins argues that Haslanger's definition is not the best one for working against gendered injustice, since it would deny the gender identity of at least some trans people. And since trans people face many kinds of gendered injustice, a transexclusionary definition of gendered terms would be inadequate and ought to be revised as well.

Whether Jenkins' revised definition is to be preferred over Haslanger's original ameliorative proposal is orthogonal to my argument; what is important for my purposes is the *way* in which they disagree. In describing her argumentative strategy, and before providing an ameliorative proposal of her own, Jenkins writes:

> Focusing on trans women, I show that according to Haslanger's definition of woman, some trans women would not count as women. Given that a target concept is a normative proposal for how feminists ought to use the concept woman, this is an unacceptable result: the adoption of the concept would exacerbate the existing (and illegitimate) marginalization of trans women within feminist discourse. (Jenkins 2016, p. 396)

The disagreement, then, is firmly in the realm of the practical. As such, a transphobe or other anti-feminist would likely find neither Jenkins' nor Haslanger's proposals acceptable, since they would presumably share very different goals for an analysis of gender terms. For them, trans-exclusionary definitions of gender terms might be preferable, since they would have quite different political aims. In a later section, I will return to this issue and discuss how a logical empiricist-style scientific world-conception might be able to help us deal with such disagreements.

Other feminist philosophers have used ameliorative analyses of other concepts. Kate Manne (2018) offers an ameliorative definition of misogyny, in order to distinguish it from sexism. As background for her work, she considers various incidents of violence such as the Isla Vista Killings, as well as the high-profile treatment of relatively powerful women such as Hillary Clinton's election campaign in 2016. In discussing the media discourse around these events, Manne notes that there is considerable imprecision in our ordinary use of the term "misogyny". In common usage, people often use the term misogyny to talk about a hatred of women. This encourages us to think primarily in terms of individual misogynists and their particular hatred of women. The problem with this is that it fails to help us link acts of violence explicitly targeting women to an overall patriarchal system. As such, Manne calls the use of misogyny to refer to the hatred of women the 'naive conception'(Manne 2018, p. 18) of misogyny and suggests an ameliorative definition instead.

As we saw in Jenkins' and Haslanger's arguments above, the reason for offering an ameliorative definition is not to codify how a term or concept is in fact used. Indeed, in this particular case, Manne explicitly argues that we should alter our use of the word misogyny. Her arguments against the naive conception are that it has serious limitations if we want to try and identify the causes of misogynist violence in order to stop them. For instance, if misogyny is essentially matter of personal feeling, any charge of misogyny whatsoever can be denied by the accused, on the grounds that they are the only ones with first-person access to their feelings. Manne also argues that the definition fails to capture many cases, in part due to problems with scope – do misogynists hate *all* women? Surely that would mean there are almost no misogynists. Is there a threshold? Again, this leaves us with the issue that there would be virtually no misogynists under such a definition (Manne 2018, pp. 42–48). And defining misogynists out of existence does not help us (and arguably hinders us) in achieving social goals such as the reduction of gendered violence and greater equality overall.

Instead, Manne's ameliorative definition of misogyny contrasts it with sexism, such that the two ideas each play a distinct role in reinforcing patriarchy. For Manne, sexism plays a primarily *justificatory* function, such that it provides a background ideology for the rationalization of patriarchal social relationships and gendered inequality. In contrast, the function of misogyny is to *enforce* that patriarchal social order. Sexist ideology, then, provides us with background assumptions, stereotypes, and cultural tropes that reinforce and valorise patriarchal social arrangements. It provides us with reasons to think that, say, wives should submit to their husbands, or that women are "naturally" less capable at carrying out an assortment of high-status tasks (Manne 2018, pp. 78–79). The complementary role of misogyny is to ensure that those who threaten such arrangements are put back in their place. Its mechanisms are those through which women are sanctioned for failing to provide feminine-coded goods such as love and affection, or for competing with men for masculine-coded goods such as money and power (Manne 2018, p. 117). The strategy underlying Manne's analysis is to expose and pull apart the various mechanisms that contribute to gendered oppression, and the concepts she outlines in her framework are the ones she sees as best suited to help us understand those mechanisms.

Having outlined the affinities between Carnap's ideas about conceptual engineering and contemporary ameliorative projects in feminist philosophy, I will suggest that other logical empiricist ideas can help to supplement them. Though Carnap and Neurath disagreed on several issues, such as the extent to which we could separate the theoretical and the practical, I nevertheless argue that a Neurathian holism can help address some natural objections to conceptual engineering projects. More specifically, one natural objection one might have to a Haslanger-style approach to gender is that it does not go far enough. Trans women simply *are* women, and trans men simply *are* men, and to allow this to be treated as a disagreement between competing frameworks gives too much legitimacy to transphobia.

The sentiment behind this objection – that we should not legitimize transphobic views – is reasonable but has two problems. First, in our current unjust world, transphobia already enjoys a great deal of legitimacy. As feminists, we might not want to grant legitimacy to transphobic worldviews, but they are already entrenched in many institutions, such as the legal and health care systems. Second, simply insisting the truth of trans-inclusive claims this to someone already committed to an anti-feminist view will likely not advance the aim of mitigating injustice. Addressing this objection is one place in which I think that the scientific world-conception broadly endorsed by members of the Vienna Circle can be helpful, particularly Neurath's holistic views about the connections between theoretical and practical. The following section will outline some aspects of that world-conception and draw some points of connection with other analytic feminist philosophers.

3 The Scientific World-Conception and Neurath's Boat

The general scientific outlook of the Vienna Circle was set out in their *Wissenschaftliche Weltauffassung* (Scientific World-Conception), collectively written by Carnap, Neurath and the mathematician Hans Hahn. The manifesto, as it is sometimes called, closes with the following claim: 'Die wissenschaftliche Weltauffassung dient dem Leben und das Leben nimmt sie auf' (Carnap, Hahn and Neurath 1929, p. 315). This can be translated as 'The scientific world-conception serves life, and life receives it.' Carnap echoed this sentiment in his Dessau Bauhaus lecture, and in the Preface to the 1928 edition of his *Logische Aufbau der Welt* (Friedman 2000, p. 17). Though they disagreed on many of the philosophical details, Carnap and Neurath shared a common world view, political goals and overall stance. The consistent sense that we get from their writing at this time is that scientific philosophy was intended to be allied with other modernist endeavours that hoped to reform the practices of ordinary life for greater social benefit.

The Vienna Circle world-conception, though, did not consist in the dogmatic prescription of a particular set of scientific theses, however, but rather in a particular attitude to inquiry (Carnap, Hahn and Neurath 1929, p. 305). This attitude was empiricist and physicalist, committed to logical analysis (Carnap, Hahn and Neurath 1929, p. 307), but with the goal of clarity and the avoidance of a particular kind of transcendent metaphysics (Uebel 2004, p. 47 and Friedman 2000, pp. 13–15). The widely known logical empiricist rejection of metaphysics is best seen, then, as continuous with their scientific world-conception, stemming from a version of neo-Kantianism that seeks to dissolve rather than resolve metaphysical disputes (Richardson 2007, p. 298).

Logical empiricism has not gained widespread traction among feminist empiricists, however. Many important feminist empiricists use Quinean naturalized epistemology or a fact-value holism as a starting point for feminist philosophy of science (Anderson 2004; Antony 1993; Clough 2004; and Nelson 1990). The benefit of this kind of fact-value holism, or what Miriam Solomon (2012) has called the web of valief, is that we might find ourselves able to confirm some feminist values that are built into empirically successful theories (Anderson 2004). Or we might be able to call racist claims, not just morally wrong, but empirically false (Clough and Loges 2008). But one concern that has been raised about this naturalized approach to epistemology as a resource for feminism is that it might too easily accommodate social conservativism. That is to say, those committed to racist worldviews, for example, can and do often find ways to claim that their beliefs are well supported by evidence. The worry is that naturalized epistemology does not give us sufficient resources to work against things like well-entrenched scientific racism (Yap 2016). But it may be that what Neurath has called his antiphilosophy, exemplified by his boat metaphor, may be helpful in that respect.

This metaphor, later adopted by Quine, appears in the context of a discussion about the impossibility of eliminating imprecision from our language. Our scientific languages are generally formulated in terms of our ordinary languages, and there is considerable overlap between the two. In Neurath's view, imprecise verbal clusters, *Ballungen*, will always be part of our language, precisely because there is no blank slate from which we are starting. In this context, Neurath then writes, 'We are like sailors who have to rebuild their ship on the open sea, without ever being able to dismantle it in drydock and reconstruct it from the best components' (Neurath 1932, p. 92).

The naturalism exemplified by the boat metaphor is a philosophical point on which both Quine and Neurath diverge from Carnap. Unlike Carnap, Neurath saw the theoretical and the practical as standing in a reciprocal relationship, rather than as being separable. This represents a more thoroughgoing holism than the Duhemian holism that Carnap and other Vienna Circle members already endorsed (Cartwright et al. 1996, pp. 92–93). This means that Neurath, like Quine, would not endorse a clear distinction between the questions we might ask within a theory and the questions we might ask about whether to adopt a theory. But Cartwright and Uebel give us some reason to believe that Neurath's version of holism would not be vulnerable to some of the same issues as Quine's. In discussing the boat metaphor, they write:

> In and of itself an attitude that seeks to make do with holistic considerations in the absence of secure foundations for knowledge may well tend towards conservatism. (Quine's adoption of Neurath's Boat may serve as an example.) But it is clear that this was not Neurath's way. Neurath was a follower of the Enlightenment. He was fired by an idea simple in conception, yet difficult to realise: to develop and employ a conception of *knowledge as an instrument of emancipation*. (Cartwright,and Uebel 1996, p. 43)

For Cartwright and Uebel, what saves Neurath from conservatism are his anti-foundationalist commitments. Neurath was opposed to philosophy that starts from first principles, as well as to attempts to place our thinking on secure and unrevisable foundations. For Neurath, given the impossibility of First Philosophy, we should expect the stock of human knowledge to be changing indefinitely. We will never eliminate all *Ballungen*, or imprecision, from our language, but doing so is not our goal as scientific philosophers or scientists. Rather, we will build and rebuild our boat (on the open sea) in ways that we hope will help us with the navigational challenges that arise.

One consequence of Neurath's holism and anti-foundationalism is his idea of auxiliary motives. Given even a Duhemian holism, theories are underdetermined by evidence; but if decisions between theories are urgent, perhaps for immediate reasons of public policy, such decisions will then rely on motives beyond evidence. And in such cases, when reason does not single out a correct decision, but provides several equally likely alternatives, it would be intellectually dishonest to pretend that we can determine the truth of things any better through appeals to evidence than we could by drawing lots or flipping a coin (Neurath 1913). Acknowledgement of this is not a denial of rationality, but a recognition of its limits.

The role of extra-theoretical considerations in science is, as mentioned above, something many feminist philosophers of science have pointed out, in debunking the myth of value-free science. And Neurath – much more than other members of the Vienna Circle – saw science as always embedded in a particular social context. Kathleen Okruhlik points out the affinities between Neurath's auxiliary motive and feminist philosophy of science as follows:

> Just as feminist philosophers of science have emphasized the need to articulate and make visible the auxiliary *assumptions* that shape hypothesis development and testing, the suggestion here is that the same thing must be done for auxiliary *motives*. Even if the empirical evidence underdetermines certain decisions, there may nonetheless be good (non-evidential, non-empirical) reasons for preferring one decision to another. (Okruhlik 2004, p. 60)

Auxiliary motives can then be responsible for many of our choices between scientific frameworks. These are clear places in which the theoretical and the practical cannot be separated, because choices about theoretical frameworks need to be made on extra-theoretical grounds. Of course, the mere acknowledgement of auxiliary motives is not yet a tool for social progress. After all, nothing about the fact that we do make such choices tells us that we *ought* to choose in ways that makes society better or promotes the welfare of others. Neurath's own discussions of socialism make it clear that he was well aware of this. He was certainly committed to social planning and social arrangements that would promote human well-being. His son notes that, 'In later years one sentence occurred repeatedly in Otto Neurath's writings: "The sumtotal of human happiness is too small. It must be made bigger" ("Die Glücksumme in der Welt ist zu klein. Sie muß vergrößert warden")' (P. Neurath 1996, p. 16). But despite these commitments, to socialism and to the improvement of human life, he remained relatively guarded about the extent to which he exhorted others to adhere to such views as well. Rather than advocating for socialism in categorical terms, as the ideal social arrangement, he tended to phrase his endorsements as hypotheticals, for instance framing it as the only way to victory for those who love the working class (Cartwright and Uebel 1996, pp. 41–42). Such statements made his own views extremely clear but fell short of making claims he would see as unwarranted, namely, that everyone should be a socialist. After all, there do seem to be many people who do not love the working class.

What makes auxiliary motives important for us, and distinct from the conceptual resources of feminist naturalized epistemology, is that they expose points of decision. Even if we cannot cleanly separate these decisions from the theoretical matters, as Carnap would have hoped, such choices between scientific frameworks are not *determined* by those frameworks, nor by the data we collect. In cases in which our decisions are not mandated by evidence, they might be made by deferring to chance, or by other things that we value, which is entirely consistent with feminist analyses of scientific practice (Longino 1983). And as Howard also argues, there is no reason why such decisions should be hidden from public view or public discourse. Neurath, Howard argues, would go so far as to advise us to explicitly direct these decisions in ways that serve progressive social ends (Howard 2003, pp. 43–44). The following section will focus on these points of decision and on the auxiliary motives that might be brought to bear in such cases. I will also discuss how Neurath's views about the entanglements between theory and practice might help circumvent some of the problems raised earlier for ameliorative projects at these kinds of decision points.

4 Ameliorative Projects, Decisions and Values

One reason why we might prefer a feminist empiricist view like Elizabeth Anderson's is that it allows for the confirmation of our values through scientific means, allowing us to say that our feminist values simply *are* the ones best confirmed by scientific theories (Anderson 2004). This is what Clough and Loges (2008) draw on explicitly when they criticize scientific racism as being empirically false. Their targets are scientists like Philippe Rushton and Arthur Jensen who take intelligence to be something that varies across people of different races. Clough and Loges point out many scientific problems with a Rushton and Jensen-style framework, such as the reliance on a suspect concept like IQ and significant concerns with the extent to which we can even treat race as a biological category.

But scientists like Rushton and Jensen do take their views to be well-justified by scientific evidence. For instance, in response to their critics who point out the social constructedness of race, they fall back on the claim that what they are primarily interested in is heredity and genetics (Rushton and Jensen 2005a, pp. 237–238). Then even though race is arguably socially constructed, there are still able to fall back on something more scientifically well-founded, like heredity. So, thinking strategically, it may be that arguments grounded primarily in the language of scientific methods are not the best solutions for dislodging scientific racism, as a naturalized epistemology approach might suggest. Instead, I will propose a Neurathian approach to discussing theory choice, which involves two different kinds of commitments. First, it requires a commitment to acknowledging the entanglement between theories, social values and data. Second, this approach involves a commitment to practices of clear communication, which for Neurath were exemplified by his anti-metaphysical stance. I will outline each of these commitments in some detail below.

Acknowledging the three-way entanglement among our scientific theories, the ways in which those theories are responsive to various kinds of data, and the society in which those theories are formed, is part of both Neurath's thought and the views of many feminist philosophers of science. Both of these ways of thinking acknowledge that all of our scientific thinking begins with antecedent concepts that are first formed through our ordinary life experiences. This is the basis of much of Neurath's anti-foundationalism, such that the groundings of the structures of science are

> wobbly, not firm. Nor do the data determine what is built upon them – the same ground can support different structures. Those with different cultural backgrounds, with different instruments, with different methods, with different kinds of imagination or different problems to solve, those who think mathematically versus those who want to feel a stress before they understand the concept "force", those who favour statistics versus those who look for the Golden Event, all these will choose different places to start and will take different things for granted in their constructions. The empirical base and the theoretical structure alike are historically conditioned. (Cartwright and Uebel 1996, pp. 46–47)

In that case, a variety of factors can lead a person to construct a theory or interpret data in a particular way. Among this list of factors, we see both constitutive and contextual values in Longino's sense, where the former are the more purely scientific values, such as our preferences for simpler theories or those that use certain mathematical methods over others, and the latter values are those that result primarily from the social context in which we live (Longino 1990). Taking these ideas into account, we can understand why researchers like Philippe Rushton might have come up with scientific theories that linked race and intelligence, with particularly negative consequences for African Americans. After all, North American society is rife with, and arguably built on, anti-black racism and other oppressions (Mills 1997). It is unsurprising, then, that science emerging from a racist society would arrive at racist conclusions. But then we should probably not expect those same scientific mechanisms from that same racist social context to disrupt those conclusions. Yet these are precisely the places at which we might need to make decisions that are not determined by our scientific evidence. The basis for such decisions involves the second Neurathian commitment, namely, to clear communication.

One of the best-known accounts of Neurath's hatred of metaphysics involves the ways in which he carefully tried to guard against it in Vienna Circle discussions and would call out "Metaphysics!" whenever he believed someone had made an empty claim. As the story goes, when they discussed Wittgenstein's *Tractatus*, Neurath interrupted continuously, such that someone suggested that he ought to hum "M-m-m-m" instead. His counter to that suggestion was that he might instead simply say "not-M" when he though they were not being misled into metaphysics (Cartwright and Uebel 1996, p. 40). Neurath's opposition to metaphysical discussion was not irrational antipathy, however, but the reflection of a deep commitment to free and open communication, which he believed was only hindered by metaphysical discussion. These are the primary reasons motivating his endorsement of physicalism and the rejection of the a priori in science.

Rejecting the transcendent in favour of the concrete means taking actions like replacing the guidance of religion with that of social planning and understanding the ways in which science fails to be prescriptive. Rather than telling us which social arrangements are best – in an undefined sense of "best" – we are instead to understand which social conditions lead to which effects. So rather than telling us what we ought to do, science will instead tell us that A and B result in C (Neurath 1931, p. 622). Whether C is something we want to bring about is a further question, and a matter on which we must decide. But as we know from the boat metaphor, we cannot expect a language entirely free from *Ballungen*, or imprecise clusters; the most we can strive for is the elimination of metaphysics from our ordinary physicalist language and the building up of a universal jargon for science (Neurath 1932, pp. 91–92).

With this in mind, we will return to the two problematic cases of theory choice that we have considered so far in this paper: scientific racism and definitions of gender terms. I will argue that a Neurathian approach can help us think through alternative argumentative strategies that we can employ. Rushton and Jensen long held to their hereditarian views in the face of criticism from within the scientific community. Such criticisms took a variety of forms, such as arguing against the use of IQ tests as measuring intelligence or providing alternative environmental explanations of how living in a racist society might be the cause of low test scores among members of particular ethic groups. But another way to talk about the choice we might have between a hereditarian framework and an environmental one to explain difference in test scores would be in terms of their consequences. This would require cooperation across the social sciences and humanities, between psychologists, historians and sociologists, just as Neurath would have endorsed. For instance, we could draw on historical resources that we have about the ways in which racial essentialism is built into our culture and into the theoretical frameworks that emerge from it. This applies to relatively specialized disciplines like academic philosophy (Zack 2014), but also to education in our public-school systems. Basic Canadian geography textbooks used in the early half of the twentieth century explicitly endorsed racialized views of character and intelligence that coincided extremely well with federal policies, portraying:

> 'the White Race' as 'the most active, enterprising, and intelligent race in the world,' thus establishing 'whites' as the positive norm against which the other 'races' could be evaluated. It represented Asians as the opposite of this norm. It told students that 'the Yellow Race' included 'some of the most backward tribes of the world and, as a rule, are not progressive.' It then described Africans and Aboriginal peoples in the Americas as in need of the paternal guidance of whites. 'The Red Race' was 'but little civilized, although a few are beginning to develop industries, such as basketry, pottery, and a little farming,' whereas 'the Black Race' was 'somewhat indolent, like other peoples whose homes are in tropical countries. They are often impulsive in their actions, but they are faithful and affectionate to any one for whom they care'. (Stanley 2011, p. 108)

Federal policies that were developed in this kind of background context resulted in atrocities such as the residential school system that was responsible for widespread systematic abuse of Indigenous children and separation from their families.

In this light, we could make a case about the likely outcomes of supporting hereditarian frameworks about intelligence and allowing them to (continue to) shape our institutions. The logical empiricists themselves arguably held views about race and racial explanations that were compatible with this idea, namely, that racial explanations are empirically meaningful, yet should be avoided on moral and political grounds (Bright 2017, p. 16). Reviewing the social scientific literature that would be required to make this claim about hereditarian frameworks in our contemporary social context is beyond the scope of this paper, but I will at least outline what might be involved. We might look into the connections between the effects of naturalizing intelligence and linking it to heredity and decreased educational opportunities. This claim would entail that students who are perceived as belonging to racial groups believed to have statistically lower IQs would likely receive less teacher attention and be offered fewer enrichment opportunities. This decrease in educational opportunity might then be linked to lower socio-economic status, all of which are reductions in measures currently acknowledged to be social determinants of health. While the current state of science does not seem to have a universal jargon, the idea that there are social components to health is at least relatively widespread.

These considerations might not be persuasive for scientists currently working within a scientific racist paradigm, but they might help influence the amount of uptake such paradigms receive. After all, even Rushton and Jensen consider their work to be embedded within a social context – just not in the ways we would like if we care about anti-racism. Indeed, they take their research in intelligence to counteract anti-racist efforts, writing, 'we will never make progress in race relations if we operate on the belief that one segment of society is responsible for the plight of another segment and that belief is false' (Rushton and Jensen 2005b, p. 334). In other words, they are offering a competing explanation for the obvious racialized social inequalities, namely, that it can be attributed, at least in part, to inherited racial characteristics rather than to white supremacy. Still, if we accept the described outcomes as the likely result of entrenching hereditarian paradigms of intelligence, then we might note that we are at a point of decision.

If we do not think that environmental explanations of IQ differences are clearly better supported than hereditarian ones (or vice versa), then our decisions can be informed by the likely outcomes described above. To generalize, Neurath suggests that we take (social) science to be providing us with conditional statements about the consequences of adopting certain social arrangements. The science does not dictate which social arrangements ought to be adopted, or which psychological frameworks ought to be favoured. But it can give us some insight into which scientific frameworks embody which kinds of values. In which case we need to acknowledge that the decisions we make between frameworks are part of what constitute our institutions' values. For a society built on maintaining white supremacy, educational arrangements that are entwined with racial essentialism could be of value, since they would help perpetuate various racist social inequalities. But otherwise, we might have an auxiliary motive to favor alternative accounts of test score differences, if they lead to outcomes that we would find preferable.

But just as we can see ways in which the practical might helpfully be brought to bear on discussions of the theoretical, we might also bring in theoretical considerations on matters that seem primarily ethical or practical. This is not quite the same as the feminist empiricist argument that certain values are better confirmed by empirical evidence. Rather, it is the claim that certain approaches to questions, such as the best definitions of gender terms, make better sense of the empirical phenomena than others. This does not necessarily help us determine whether we might prefer Jenkins's approach to Haslanger's but can at least tell us that either of these approaches is preferable to those that deny people's gender claims through naive appeals to biology. More specifically, we can lean on a scientific worldview to support the idea that gender terms like "woman" and "man" are not straightforward biological concepts.

Even if we disagree on values, or the desired outcomes of particular social arrangements, we might nevertheless agree on some things we want out of our scientific frameworks. While it is virtually inevitable that scientific theories will idealize to some extent, we might nevertheless prefer those theories that can account for more of the world's complexities. If this is the case, then adopting the scientific world-conception might give us a reason to reject simplistic biological definitions of gender terms, that equate gender categories with, say, XX or XY chromosomes. After all, biologists such as Anne Fausto-Sterling (2012) and Joan Roughgarden (2004) give arguments about why there is no neat mapping between gender identity and biological classifications. They do not provide the same positive proposals, but if our current understanding of biology and the social world indicates that there is no neat one-to-one correspondence between, say, the type of gametes that a person produces and their gender expression, then linking the two by fiat does not sound like a particularly scientific stance. We might have a reason to adopt such a framework if our values dictated that a person's gender identity ought to be determined by their reproductive capacities, but otherwise it is unclear why we would want to do so. In Neurath's view, the biologist ought to refrain from categorically advocating a particular way to connect physiology to gender terms, but can at least point out the consequences of a stipulated correspondence between XX chromosomes and womanhood.

Again, Neurath's holism does not suppose that we can cleanly separate the theoretical questions from the practical ones but acknowledges that the two are thoroughly entangled. After all, some decisions we might make about what counts as evidence in the first place require prior theoretical commitments, which might themselves have been motivated by social values. For instance, accepting various aspects of Rushton and Jensen's hereditarian results requires us to accept the validity of IQ as a measure of intelligence, which is at the very least controversial (Evans and Waits 1981). We might also need to make decisions about the extent to which we accept arguments by philosophers like Talia Bettcher (2009) that we should treat people as though they have first-person authority over their gender. Accepting Bettcher's account of ethical first-person authority and the ways in which trans people are typically denied that authority by mainstream cis-normative society has wide-ranging implications about the social context we would need to be in to properly understand people's avowals of their gender. For instance, we need to be in a context in which claims such as "I am a woman" or "I am a man" are not understood as communicating something about the speaker's genitals. In other words, the claim "I am a man" might be taken to convey different evidence in different contexts: in a trans-inclusive context, it is simply an avowal of the speaker's gender, whereas in a trans-exclusive context, it is also taken to imply that the speaker has a penis. So, all these supposedly theoretical claims about which views are best supported by evidence are nevertheless made in terms of concepts and methods that are developed in a particular cultural and historical context.

In general, then, Neurathian holism can help us think through how to argue in situations in which we have competing scientific theories. Though he does not separate the theoretical and the practical as Carnap does, Neurath's idea of auxiliary motives allows us to highlight ways in which decisions between competing frameworks are informed by something other than purely evidentiary considerations (Okruhlik 2004, p. 60). While we could leave such decisions to chance, they are frequently informed by what we value or by the background culture in which we are embedded. As such, the overall picture is not radically different from the one initially presented as Carnap's view of scientific philosophy in terms of conceptual engineering. Where Neurath is able to go beyond Carnap is in his greater understanding of the interconnections between the theoretical and practical, such that practical decisions are always intertwined with theoretical issues and theoretical frameworks begin in social contexts informed by the practical and social.

5 Conclusion: Scientific Philosophy for Liberatory Purposes

This paper has argued that scientific philosophy, at least as it was conceived of by some members of the Vienna Circle, has more to offer contemporary social philosophy than we might think based on its popular reputation. I have not claimed to have provided an exhaustive list of ways in which Vienna Circle members engaged in social activism, nor have I even provided a complete set of their tools that might be useful for contemporary socially-engaged analytic philosophy. But at the very least, I have pointed out that there are clear connections between a Carnapian conceptual engineering view and contemporary ameliorative projects in feminist philosophy. And I have also indicated several ways in which a Neurathian holism might be better for feminist empiricism than naturalized epistemology.

Admittedly, scientific philosophy for liberatory purposes seems to promise less than some versions of feminist empiricism. We likely will not find our values confirmed by science, though we might find that some of our preferred scientific frameworks are those that promote certain social values. But if we are anti-foundationalists in Neurath's sense, we will recognize that where we are scientifically is as people constantly building and rebuilding the languages and frameworks that have been handed down to us. We were never at any point starting from scratch. As such, we can expect some imprecision in our language, and in the stock of concepts available to us. This, among other things, means that there are many points at which we can simply try to determine scientifically what the results will be of various social arrangements, but realize that adopting such arrangements is a decision that must be made which is underdetermined by the scientific evidence.

The members of the Left Vienna Circle considered here have not provided a means through which we can ensure scientifically that progressive values will be adopted. But it is not clear that any philosophical theory could do so. Scientific racists will likely continue to insist that their views about race and intelligence are well-confirmed by evidence. It is also likely that transphobes will similarly continue to insist that we should deny an assortment of people's claims about their own gender. However, what the conceptual engineering approach and Neurathian holism can provide us with are some strategies for countering arguments that purport to rest on science. Using such strategies, understanding that concepts can be redefined, we can argue for ways in which doing so can help us make better sense of the phenomena in the world in which we live, and ameliorate the conditions of life for various people in that world. Still, analytic philosophy, even in the service of social ends, cannot compel us through reason to care about other people. But as socially minded analytic philosophers have pointed out before, it can provide us with conceptual tools that we can use in the service of life, in the hopes that life will also receive it.

Bibliography

Anderson, Elizabeth (2004): "Uses of Value Judgments in Science: A General Argument, with Lessons from a Case Study of Feminist Research on Divorce". In: *Hypatia* 19. No. 1, pp. 1–24.

Antony, Louise. (1993): "Quine as Feminist: The Radical Import of Naturalized Epistemology". In: Louise Antony and Charlotte Witt (Eds.): *A Mind of One's Own: Feminist Essays on Reason and Objectivity.* Boulder: Westview Press, pp. 185–226.

Bettcher, Talia M. (2009): "Trans Identities and First-Person Authority". In: Laurie Shrage (Ed.): *You've Changed: Sex Reassignment and Personal Identity.* New York: Oxford University Press, pp. 98–120.

Bright, Liam K. (2017): "Logical Empiricists on Race". In: *Studies in History and Philosophy of Biological and Biomedical Sciences* 65, pp. 9–18.

Carnap, Rudolf (1934): "Theoretische Fragen Und Praktische Entscheidungen". In: *Natur Und Geist* 2, pp. 257–260.

Carnap, Rudolf (1950): "Empiricism, Semantics, and Ontology". In: *Revue Internationale de Philosophie* 4. No. 11, pp. 20–40.

Carnap, Rudolf (1963): "Intellectual Autobiography". In: Paul A. Schilpp (Ed.): *The Philosophy of Rudolf Carnap.* La Salle: Open Court, pp. 3–84.

Carnap, Rudolf, Hans Hahn, and Otto Neurath (1929): "Wissenschaftliche Weltauffassung: Der Wiener Kreis". In: Arthur Wolf (Ed.): *Veröffentlichungen Des Vereines Ernst Mach.* Wien: Verlag, pp. 301–317.

Carnap, Rudolf, and Amethe Smeaton (2002): *Logical Syntax of Language.* Peru: Open Court.

Cartieri, Francis, and Angela Potochnik (2014): "Toward Philosophy of Science's Social Engagement". In: *Erkenntnis* 79, pp. 901–916.

Cartwright, Nancy, Jordi Cat, Lola Fleck, and Thomas E. Uebel (1996): *Otto Neurath: Philosophy Between Science and Politics.* Cambridge: Cambridge University Press.

Cartwright, Nancy, and Thomas E. Uebel (1996): "Philosophy in the Earthly Plane". In: Elisabeth Nemeth and Friedrich Stadler (Eds.): *Encyclopedia and Utopia: The Life and Work of Otto Neurath (1882–1945).* Dordrecht: Kluwer Academic Publishers, pp. 39–52.

Clough, Sharyn (2004): "Having It All: Naturalized Normativity in Feminist Science Studies". In: *Hypatia* 19. No. 1, pp. 102–118.

Clough, Sharyn, and William Loges (2008.): "Racist Value Judgments as Objectively False Beliefs: A Philosophical and Social-Psychological Analysis". In: *Journal of Social Philosophy* 39. No. 1, pp. 77–95.

Douglas, Heather (2000): "Inductive Risk and Values in Science". In: *Philosophy of Science* 67, pp. 559–579.

Dutilh Novaes, Catarina (2020): "Carnapian Explication and Ameliorative Analysis: A Systematic Comparison". In: *Synthese* 197, pp. 1011–1134.

Evans, Brian, and Bernard Waits (1981): *IQ and Mental Testing: An Unnatural Science and Its Social History.* London: Macmillan Press.

Fausto-Sterling, Anne (2012): *Sex/Gender: Biology in a Social World.* New York: Routledge.

Fehr, Carla, and Kathryn S. Plaisance (2010): "Socially Relevant Philosophy of Science: An Introduction". In: *Synthese* 177, pp. 301–316.

Friedman, Michael (2000): *A Parting of the Ways: Carnap, Cassirer, and Heidegger.* Chicago: Open Court.

Galison, Peter (1990): "Aufbau/Bauhaus: Logical Positivism and Architectural Modernism". In: *Critical Inquiry* 16. No. 4, pp. 709–752.

Haslanger, Sally (2012): *Resisting Reality: Social Construction and Social Critique*. New York: Oxford University Press.

Howard, Don (2003): "Two Left Turns Make a Right: On the Curious Political Career of North American Philosophy of Science at Midcentury". In: Gary L. Hardcastle and Alan W. Richardson (Eds.): *Logical Empiricism in North America*. Minneapolis: University of Minnesota Press, pp. 25–93.

Jenkins, Katharine (2016): "Amelioration and Inclusion: Gender Identity and the Concept of Woman". In: *Ethics* 126, pp. 394–421.

Longino, Helen (1983): "Beyond 'Bad Science': Skeptical Reflections on the Value-Freedom of Scientific Inquiry". In: *Science, Technology, & Human Values* 8. No. 1, pp. 7–17.

Longino, Helen (1990): *Science as Social Knowledge: Values and Objectivity in Scientific Inquiry*. Princeton: Princeton University Press.

Manne, Kate (2018): *Down Girl: The Logic of Misogyny*. New York: Oxford University Press.

Mills, Charles W. (1997): *The Racial Contract*. Ithaca: Cornell University Press.

Nelson, Lynn H. (1990): *Who Knows: From Quine to a Feminist Empiricism*. Philadelphia: Temple Press.

Neurath, Otto (1913): "The Lost Wanderers of Descartes and the Auxiliary Motive". In: Robert S. Cohen and Marie Neurath (Eds.): *Philosophical Papers 1913–1946*. Dordrecht: Kluwer Academic Publishers, pp. 1–12.

Neurath, Otto (1931): "Physicalism: The Philosophy of the Vienna Circle". *The Monist* 41. No. 4, pp. 618–623.

Neurath, Otto (1932): "Protocol Statements". In: Robert S. Cohen and Marie Neurath (Eds.): *Philosophical Papers 1913–1946*. Dordrecht: Kluwer Academic Publishers, pp. 91–99.

Neurath, Paul (1996): "Otto Neurath (1882–1945) – Life and Work". In: Elisabeth Nemeth and Friedrich Stadler (Eds.): *Encyclopedia and Utopia: The Life and Work of Otto Neurath (1882–1945)*. Dordrecht: Kluwer Academic Publishers, pp. 15–28.

Okruhlik, Kathleen (2004): "Logical Empiricism, Feminism, and Neurath's Auxiliary Motive". In: *Hypatia* 19. No. 1, pp. 48–72.

Reisch, George (2005): *How the Cold War Transformed Philosophy of Science: To the Icy Slopes of Logic*. New York: Cambridge University Press.

Richardson, Alan (2007): "Carnapian Pragmatism". In: Michael Friedman and Richard Creath (Eds.): *The Cambridge Companion to Carnap*. Cambridge: Cambridge University Press, pp. 295–315.

Roberts, John T. (2007): "Is Logical Empiricism Committed to the Ideal of Value-Free Science?" In: Harold Kincaid, John Dupré and Alison Wylie (Eds.): *Value-Free Science: Ideal or Illusion?* New York: Oxford University Press, pp. 143–163.

Roughgarden, Joan (2004): *Evolution's Rainbow: Diversity, Gender, and Sexuality in Nature and People*. University of California Press.

Rushton, J. Philippe, and Arthur R. Jensen (2005a): "Thirty Years of Research on Race Differences in Cognitive Ability". In: *Psychology, Public Policy, and Law* 11. No. 2, pp. 235–294.

Rushton, J. Philippe, and Arthur R. Jensen (2005b): "Wanted: More Race Realism, Less Moralistic Fallacy". In: *Psychology, Public Policy, and Law* 11. No. 2, pp. 328–336.

Solomon, Miriam (2012): "The Web of Valief. An Assessment of Feminist Radical Empiricism". In: Sharon L. Crasnow and Anita M. Superson (Eds.): *Out from the Shadows: Analytical Feminist Contributions to Traditional Philosophy*. New York: Oxford University Press, pp. 435–450.

Stanley, Timothy J. (2011): *Contesting White Supremacy: School Segregation, Anti-Racism, and the Making of Chinese Canadians*. Vancouver: UBC Press.

Uebel, Thomas E. (2004): "Education, Enlightenment and Positivism: The Vienna Circle's Scientific World-Conception Revisited". In: *Science & Education* 13, pp. 41–66.

Wylie, Alison, and Lynn H. Nelson (2007): "Coming to Terms with the Values of Science: Insights from Feminist Science Studies Scholarship". In: Harold Kincaid, John Dupré and Alison Wylie (Eds.): *Value-Free Science: Ideal or Illusion?* New York: Oxford University Press, pp. 58–86.

Yap, Audrey (2010): "Feminism and Carnap's Principle of Tolerance". In: *Hypatia* 25. No. 2, pp. 437–454.

Yap, Audrey (2016): "Feminist Radical Empiricism, Values, and Evidence". In: *Hypatia* 31. No. 1, pp. 58–73.

Zack, Naomi (2014): "The Philosophical Roots of Racial Essentialism and Its Legacy". In: *Confluence: Online Journal of World Philosophies* 1, pp. 85–98.

Part II: **Mind, Knowledge, and the Social World**

José Medina
Political Epistemology

Abstract: This essay elucidates the recent shift from social epistemology to political epistemology, focusing specifically on the shift in the literature of epistemic injustice from an ethics of knowing to a politics of knowing. I give an analysis of how normative issues concerning epistemic oppression and marginalization need to go beyond the epistemic normativity inscribed in interpersonal dynamics and engage with the normative side of epistemic group dynamics and with the epistemic life of institutions and structures of public discourse. Drawing from recent discussions of the epistemic virtues and vices of groups and institutions, the essay pays particular attention to epistemic injustices within carceral institutions and the responsibility of institutions and publics to eradicate these injustices. Focusing on recent discussions in the literature on epistemic injustice, in the first section I elucidate some aspects of the shift from an ethics of knowing to a politics of knowing to highlight the role of institutions and group dynamics in epistemic practices and their dysfunctions. In section 2, I discuss how to normatively assess epistemic agency and epistemic responsibility at the institutional and collective level. In section 3, I address how institutional and collective epistemic injustices should be resisted through what I call epistemic advocacy and epistemic activism.

1 From an Ethics of Knowing to a Politics of Knowing

The subtitle of Miranda Fricker's pioneering monograph *Epistemic Injustice* (2007) is *Power and the Ethics of Knowing*. As this subtitle makes clear, Fricker focuses the discussions of epistemic injustice around the ethics of knowing, that is, the normative issues concerning responsibility and fair and unfair treatment in knowledge-producing and knowledge-sharing practices. Recent discussions of epistemic injustice—also including Fricker's more recent contributions (2010, 2012, and 2013)—have expanded this focus to also include the politics of knowing, that is, the normative issues concerning collective responsibility,

---

Note: I am grateful for the detailed feedback I received from the editors and anonymous reviewers on prior versions of this chapter.

https://doi.org/10.1515/9783110612318-005

group dynamics, and institutional arrangements as they pertain to fair and unfair treatment in knowledge-producing and knowledge-sharing practices[1].

In the ethics of knowing outlined by Fricker (2007), epistemic injustice is something that happens in epistemic transactions among individuals, and it takes two forms: testimonial and hermeneutical. In testimonial injustice, according to the ethics of knowing approach, the epistemic mistreatment occurs when a speaker is unfairly denied the *credibility* she deserves by her interlocutors; and, in hermeneutical injustice, the epistemic mistreatment occurs when the *intelligibility* of a speaker's attempt to communicate her experience is unfairly assessed by her interlocutors because of a lack of language or a scarcity of shared expressive and interpretative resources. As we shall see in what follows, a political approach to epistemic injustice broadens the phenomena that are considered under the rubrics of testimonial and hermeneutical injustice so as to include not only unfair epistemic transactions among individuals, but also unfair epistemic patterns and dynamics displayed by groups and institutions. The politics of knowing approach emphasizes that the epistemic subjects engaging in unfair credibility and intelligibility assessments and therefore perpetrating testimonial or hermeneutical injustice can be groups (as we shall see with epistemic bubbles in 2.2) or institutions (as we shall see through the analysis of epistemic institutional vices in 2.1). This shift from the ethics of knowing to the politics of knowing transforms how we think about epistemic agency: it shifts the focus from the individual agency of particular communicators to the collective agency of groups and institutions; and it invites us to think of speakers and hearers as members of collectives and officers of institutions. Even when we zero in on very specific and individualized instances of epistemic injustice, the politics of knowing approach calls attention to the social and institutional conditions of the epistemic transaction in question and of the epistemic agency of the individual participants, and to the systemic and structural aspects of the exchange.

Although in her original account there are already key elements that underscored the political side of epistemic injustice (collective and structural elements that were especially salient in the hermeneutical cases), Fricker (2007) conceptualizes the issues of epistemic responsibility and epistemic agency concerning unfair treatment in testimonial and hermeneutical dynamics primarily in personal and interpersonal ways and in the realm of ethics. More recently, however, unfair epistemic treatment—whether testimonial or hermeneutical—has been discussed in the context of the epistemic responsibility of groups and institutions. In fact, some have argued for the priority of the collective and struc-

1 See also Alessandra Tanesini's chapter in this volume.

tural aspects of epistemic injustice since the epistemic interactions of individuals (and their responsibility and agency therein) are embedded in group dynamics and institutional arrangements. Hence the priority of the politics of knowing over the ethics of knowing underscored by the early criticisms that Fricker's account of epistemic injustice received from Linda Alcoff (2010) and Elizabeth Anderson (2012), among others.[2]

Alcoff (2010) calls into question Fricker's proposal that testimonial injustice should be neutralized and overcome by exercising critical reflection on one's testimonial sensibility and through the subject's efforts to become aware of her biases and to adjust her credibility assessments accordingly. Alcoff argues that conscious and volitional practices are doomed to be ineffective because "identity prejudice operates via a collective imaginary, as Fricker suggests, through associated images and relatively unconscious connotations" (Alcoff 2010, p. 132). Alcoff emphasizes that the implicit bias literature has provided ample evidence that prejudice operates below the level of conscious awareness and it can operate "quite effectively even when it runs counter to a person's own consciously held values and commitments" (Alcoff 2010, p. 132). Fricker suggests that subjects can exploit the cognitive dissonance between their unconscious, prejudicial appraisals of credibility and their self-avowed beliefs to trigger critical reflection and to adjust their credibility assessments accordingly. But, as Alcoff observes, unconscious prejudices and self-avowed beliefs are typically insulated from each other in such a way that subjects are rarely aware of any dissonance in their mental states.

Building on Alcoff's work, Anderson (2012) argues that testimonial injustice and epistemic injustice more generally are neither exclusively nor primarily an interpersonal phenomenon but a structural and systemic one that calls for structural and systemic remedies at the level of group dynamics and structural background conditions, rather than purely personal and psychological or volitional remedies. As Fricker herself emphasizes, hermeneutical injustice goes beyond the personal and transactional and is always structural because it is produced by the scarcity of *collective* expressive and interpretive resources.[3] But Anderson points out that at least some forms of testimonial injustice, as described by Fricker, are also structural: "Testimonial exclusion becomes structural when institutions are set up to exclude people without anyone having to decide to do so" (An-

---

2 See also Medina (2012) and Pohlhaus (2012).
3 Mason (2011), Pohlhaus (2012), and I (2012) have argued that hermeneutical injustice also includes transactional elements and personal failings, namely, the failure to learn and use the hermeneutical resources available to members of other groups. This is well illustrated by the phenomenon of willful hermeneutical ignorance analyzed by Pohlhaus (2012).

derson 2012, p. 166). Fricker illustrates these forms of structural testimonial in-justice through what she calls "pre-emptive testimonial injustice" (Fricker 2007, p. 130), which includes, for example, cases in which the courts did not allow people of color to give testimony.

Anderson argues that "structural injustices call for structural remedies" (Anderson 2012, p. 171); and, thus, for example, if a structural testimonial injustice results from an institution excluding a group of people from testimonial dynamics (e. g. the justice system excluding people of color from giving testimony), then an institutional reform is needed to guarantee the testimonial inclusion of the members of the previously excluded group.[4] But Anderson also argues that "structural remedies need to be stressed even when the injustices at issue are transactional" (Anderson 2012, p. 168), since we need to improve the epistemic environments and climates[5] in which the epistemically defective transactions take place and the group dynamics in which those transactions are embedded. Anderson urges us to think of the structural and the transactional as interpene-trating each other, rather than as separate domains; and, accordingly, she sug-gests that we think of structural remedies for epistemic injustice as facilitating (rather than competing with) virtue-based remedies: "Many structural remedies are put in place to enable individual virtue to work, by giving it favorable con-ditions" (Anderson 2012, p. 168).

Anderson emphasizes that "structural remedies may be viewed as virtue-based remedies for collective agents" (Anderson 2012, p. 168). She encourages us to think of the virtue of epistemic justice as a virtue of institutions and collec-tives as much as a virtue of individual knowers: "Just as individuals are account-able for how each acts independently, we are accountable for how we act collec-tively. Epistemic virtue is needed at both individual and structural scales" (Anderson 2012, p. 171). At the structural level, issues of epistemic justice call for normative assessments of epistemic agency and epistemic responsibility at the institutional and collective level. In section 2, I focus on normative failures at the level of institutional and collective epistemic agency that a politics of

---

4 Recent discussions in vice epistemology have also addressed the issue of individual vs. struc-tural remedies for epistemic vices. See esp. Tanesini's chapter in this volume and Medina 2021a and 2021b.
5 Lorraine Code forcefully argues for this point in *Rhetorical Spaces: Essays on Gendered Loca-tions* (1995), where she anticipates the notion of epistemic injustice by calling attention to op-pressive communicative contexts and climates that silence women and handicap their equal and fair participation in epistemic practices. Code's notion of rhetorical spaces calls attention to the issue of epistemic agency as facilitated or impaired by communicative contexts and struc-tures.

knowledge should address. I also address, in section 3, how institutional and collective epistemic injustices should be resisted through what I call epistemic activism, which involves epistemic advocacy and the empowerment of marginalized epistemic communities.

2 The Epistemic Life of Institutions and Public Discourse

In this section, I will discuss, first, how to normatively assess the epistemic life of institutions and, second, how to normatively assess the epistemic life of groups in order to detect and combat epistemic injustices at the structural and political level.

2.1 The Epistemic Life of Institutions

In "Capital Vices, Institutional Failures, and Epistemic Neglect in a County Jail" (2021a), I analyze what I describe as the phenomenon of epistemic neglect embedded in a fatal case of medical neglect in a county jail: the death of Matthew McCain, a detained subject at the Durham County Detention Facility (hereafter DCDF). Mr. McCain died unattended on January 19th, 2016, after suffering a seizure while detainees in his pod repeatedly pressed emergency call buttons. The guards on duty during McCain's death displayed a complete lack of concern for his well-being, which is primarily a moral failing; but they also displayed a systematic epistemic failing: the guards repeatedly contended that they do not take seriously the detainees' emergency calls because they do not *believe* them. As I argue in my full analysis of this case (2021a), the epistemic vice of testimonial insensitivity occurs both as a personal and as an institutional epistemic failing. Not only the individual correctional officers on duty but also the county jail, DCDF, as an institution are to be blamed. We are dealing with a vitiated institutional space that, far from guaranteeing that the voices of detained subjects are properly heard, stacks the decks against their voices receiving proper uptake, that is, an institutional space that encourages testimonial insensitivity against detained subjects.[6]

---

6 This fits well what Quill Kukla (writing as Rebecca) calls discursive injustice. As they put it, discursive injustice occurs "when members of a disadvantaged group face a systematic inability to produce a specific kind of speech act that they are entitled to perform—and in particular when

After Mr. McCain's death by medical and epistemic neglect came to light, under the pressure of the activist organization Inside/Outside Alliance, it was uncovered that there had been multiple similar cases at DCDF in the preceding months and many more in the preceding years. Once we notice that these are not isolated events, we cannot simply attribute the epistemic failings *only* to the individual epistemic vices of the particular guards on duty in each case. There is in fact a *pattern* of epistemic neglect at DCDF. In light of this pattern, we can (and should) move the analysis of epistemic vices from the personal to the institutional level. But how do we establish the presence of an epistemic vice in an institution? What counts as institutional epistemic behavior? One way of proceeding is by paying attention to what the institution *should have been doing* to guarantee the fair epistemic treatment of detained subjects and to guarantee, as much as possible, that tragic cases like the epistemic neglect of emergency calls do not happen.[7] In this way, we can detect the epistemic negligence of the institution itself as setting the stage for specific instances of epistemic neglect in the interaction between detained subjects and detention officers.

There are three key criteria we can use to assess the epistemic behavior of an institution and to hold it accountable to standards of fair epistemic treatment: (1) by assessing the epistemic protocols established by the institution that lay out the paths that the individual behavior of the officers of the institution will take; (2) by assessing the epistemic side of training procedures that instill particular habits, attitudes, and patterns of communicative behavior in the officers of the institution, teaching them to interpret and apply epistemic protocols in particular contexts in particular ways; and (3) by assessing the accountability procedures that makes sure that the officers follow the epistemic protocols and the training they have received in applying those protocols, guaranteeing that failures to do so will not happen with impunity and thus demonstrating the institutional values and commitments to the epistemic dignity of all subjects within the institution.[8]

---

their attempts result in their actually producing a different kind of speech act that further compromises their social position and agency" (Kukla 2014, p. 440). The structural testimonial injustice I have described also fits well Saray Ayala's account of how social structures unfairly restrict the range of speech affordances allowed for certain subjects in certain social positions (Ayala 2016).

7 See Goldberg (2017) for an account of "should have known" that sheds light on how absence of epistemic standing may be epistemically detrimental to oneself and others.

8 Of course, there are more general structural factors such as the overcrowding of the prison system that goes well beyond the epistemic institutional design I have sketched here, and these factors will undoubtedly impact the treatment that incarcerated subjects receive. Institutional responsibility for fair treatment (epistemic and otherwise) clearly falls not only on the

First, a carceral institution should have protocols that make it impossible for guards to arbitrarily disregard emergency calls, so that as soon as an emergency button is pushed and/or a cry for help is heard, they are required to check on the alleged emergency immediately and cease all other activities until this requirement is fulfilled. And in order to prevent testimonial insensitivity against detained subjects in carceral life, institutional protocols that implement a communicative-epistemic policy of taking the expressive behavior of detained subjects seriously are also needed in non-emergency situations. In particular, protocols are needed for guaranteeing that detainees' complaints and grievances are properly handled. At DCDF, detained subjects do have opportunities to file grievances, which is crucial in order to ensure that they have epistemic subjectivity and agency to give testimony about their problems and concerns in carceral life. The problem is that such grievances are rarely properly processed, responded to, and acted on. Protocols and proper procedures are needed here to ensure that detainees' voices are respected, and that their utterances and pronouncements are taken seriously and as deserving of trust. Unfortunately, that is not the case at DCDF. As Medina and Whitt's (2021) analysis of testimonial dysfunctions within DCDF underscores, hundreds of detainee letters collected by Inside-Outside Alliance offer hundreds of examples of jail staff verbally disregarding, minimizing, or dismissing detainee complaints and requests out of hand.

Second, in order to ensure that the epistemic dignity of detained subjects is respected, and sufficient levels of trust are maintained, a carceral institution should provide the adequate training to officers and jail staff to guarantee proper epistemic functioning and fair treatment of the detained subjects. As Medina and Whitt's analysis of testimonial dysfunctions at DCDF shows, "correctional officers are explicitly and implicitly trained to not trust, fraternize with, or empathize with incarcerated persons" (Medina and Whitt 2021). The institution lacks training processes that foster empathy, unprejudiced attitudes, and good listening habits. Carceral institutions should work with psychologists and educational experts who can help them to design training practices aimed at mitigating or bypassing the influence of prejudices. But, more importantly, they should also consult with formerly incarcerated subjects in designing training practices

---

county and state levels of the prison system, but also on different higher levels of the judicial and penal system. It is beyond the scope of this paper to elucidate how institutional responsibility of this sort concerns multiple institutions and at different levels, but it is important to keep in mind that what I am offering here is only one element in the complex task of (re-)designing institutional structures for more fair epistemic treatment. The discussion of protocols, training, and accountability procedures that follows has been adapted from Medina 2021a.

so that their perspectives and vulnerabilities as subjects of knowledge under the care and protection of the institution are considered and properly addressed.

Finally, the best protocols and training procedures would be ineffective if officers and jail staff are not held accountable when they violate them; and the best training for instilling healthy epistemic attitudes and habits would be ineffective if officers and jail staff who do not act in consequence with their training face no consequences. So, third, a carceral institution needs to implement the use of procedures of accountability so that officers and jail staff are held accountable for epistemic mistreatment, that is, so that they face consequences if they unfairly distrust detainees or treat them in epistemically undignified and disrespectful ways. A carceral institutional commitment to a communicative-epistemic policy of taking detainees' words and expressive behavior seriously requires clear accountability procedures that are made known to officers and jail staff and are conscientiously followed to guarantee that infringing the institution's expectations about epistemic respect for detainees and proper epistemic treatment cannot happen with impunity. Unfortunately, DCDF allows correctional officers to demonstrate epistemic insensitivity toward detainees and to neglect them epistemically without facing any consequences.

DCDF's record of testimonial insensitivity against detained subjects is clearly tied to the epistemic failures of the institution, and not only to the individual failings of particular guards. And we can clearly identify these failures by assessing the institutions' epistemic protocols, training procedures, and accountability procedures. On all these fronts, the institution shows clear deficiencies. The three institutional deficiencies I have highlighted—the lack of adequate protocols, of proper training, and of accountability procedures—clearly demonstrate that the institution has failed to take proactive steps to guarantee that the detainees' expressive behavior is taken seriously and trusted as a matter of communicative-epistemic policy. The institution exhibits the epistemic vice of testimonial insensitivity with respect to detainees since it has not instituted the default attitude of trusting detainees unless evidence to the contrary surfaces. And the epistemic vice can be ascribed to the institution based on these failures alone and independently of actual patterns of distrust exhibited by officers on the grounds. Note that the epistemic vice can be attributed to the institution even when there is no pattern of epistemic mistreatment of detainees that can be detected over time by the accumulation of cases. Think of the following two kinds of possible cases. On the one hand, an institution can be epistemically vicious even if, out of sheer luck, it happens to be staffed by epistemically virtuous officers: epistemically virtuous guards at DCDF would be acting virtuously out of personal commitment and independently achieved skills, but not because of the training received at DCDF or the protocols and accountability procedures to which they are subject. On the other hand, consider the following counterfactual scenario different from DCDF and other crowded county jails: think of a jail with so few detainees and with such a quick turn-around that detainees are processed in and out of the facilities without there being much of a chance for a pattern of epistemic neglect to be formed and clearly exhibited. Still, in this scenario the epistemic vice can be counterfactually attributed to the jail as a carceral institution given its lack of protocols, training, and accountability procedures to prevent epistemic neglect from occurring. A single instance of epistemic neglect in this scenario is likely to be attributed to the flawed detention officers on the grounds, but, in fact, it is already a *systemic* problem because of the absence of adequate protocols, training, and accountability procedures.

2.2 The Epistemic Life of Groups and Public Discourse

Besides paying attention to the epistemic behavior of institutions, in order to combat epistemic injustices at the supra-individual level, we need to normatively assess the epistemic agency and epistemic responsibility of collectives, that is, we need to identify and resist the normative failures of epistemic group dynamics. For it is not just institutions that may misbehave epistemically, it is not just institutional frameworks and conditions that we may need to dismantle in order to eradicate a pattern of epistemic injustice; in some cases, it is groups that misbehave, and it is collective epistemic attitudes, habits, and arrangements that need to be eradicated or radically transformed.

As Medina and Whitt (2021) and Medina (2021a) have emphasized, epistemic injustices within carceral institutions (such as the systemic testimonial insensitivity against detained subjects at DCDF discussed in the previous sub-section) are not only internal problems, but are tolerated and supported by the complicity of multiple publics in active and passive ways and are thus part of wider epistemic pathologies of public life. These epistemic injustices are passively supported by the indifference and apathy of the general public about what goes on within carceral institutions. But they are also supported actively by the circulation of stigmatizing stereotypes about detainees who are depicted as criminals who are not deserving of any trust—and it is important to note that the stigmatization of criminality applies even to those awaiting trial in county jails who have not been convicted of any crime and are legally required to be viewed and treated as innocent; and such stigmatization is even extended to the advocates of detainees who are depicted as defenders of criminals.

One may naively think that epistemic advocacy about epistemic injustices in carceral institutions consists simply in making these injustices visible by entering the relevant information about them in public discourse and educating the general public so that people would disrupt their complicity by making carceral institutions and their officers accountable to standards of fair epistemic treatment. However, it is not that easy because large parts of the public may be communicatively insulated from this information even if it is widely circulated; and for some publics, even if the information reaches them, they may assimilate it in distorted ways, so that the evidence about mistreatment is discounted or reinterpreted and justified as fair treatment. This is what happens when there are publics that operate as epistemic bubbles within which the stigmas of criminality function as informational filters that in some cases prevent information about carceral life from coming in and in other cases reinterpret and distort information as it comes in. For Anderson, epistemic bubbles are self-segregated networks for the circulation of ideas that are resistant to belief correction. There are different models for explaining how epistemic bubbles are formed and maintained. I will briefly consider the psychological models and then focus on the social cognition model and the discursive model proposed by Anderson in more detail.[9]

Two prominent psychological models of how epistemic bubbles work can be found in Cass Sunstein's group polarization theory and in Dan Kahan's cultural cognition theory. Sunstein's (2017) group polarization theory explains how epistemic bubbles function through cognitive biases—e.g. assimilation biases—that prompt individuals to radicalize their beliefs and to make them resistant to correction. Individuals entrapped in epistemic bubbles assimilate information through biased processes that include a confirmation bias that leads them to seek and believe evidence that confirms their beliefs, and a disconfirmation bias that leads them to repudiate evidence that disconfirms their beliefs. As Anderson explains, "Sunstein's theory predicts group polarization entirely through in-group processes: each relatively segregated group is separately driven to extreme beliefs on opposite sides of a particular claim" (Anderson 2021, p. 15). By contrast, Kahan's (2012) cultural cognition theory explains the communicative insularity of epistemic bubbles and the reluctance to learn and correct beliefs of their members in terms of hostile intergroup processes. Cultural cognition theory explains group polarization as an intergroup phenomenon produced by culturally antagonistic memes (e.g. "the enemy of the people") that foster hostility between members of the groups in question: the members of one group

---

9 In the rest of this section, I expand on the discussion of epistemic bubbles contained in section 2 of my 2021b.

stigmatize the members of the other group as epistemic adversaries[10] who are not worthy of trust and have to be excluded from epistemic cooperation and collective learning altogether.

Whether it is through assimilation biases or through antagonistic biases, the psychological models of epistemic bubbles explain the dysfunctions of these groups in terms of the distorting cognitive attitudes of their individual members. These psychological models help us understand the dysfunctional epistemic functioning of individuals within epistemic bubbles, but they do not explain how their cognitive biases are formed or get activated by epistemic group norms and discursive norms. If these psychological explanations were the whole story, it would be impossible to understand why some groups are more vulnerable to becoming epistemic bubbles than others and why individuals exhibit cognitive biases and dysfunctional epistemic functioning while operating within some groups but not while operating in other groups or by themselves, as ample evidence in social psychology strongly suggests.[11] So, for example, the psychological models would explain that viewers of Fox News are resistant to learning from information about epistemic injustice in carceral institutions whereas viewers of MSNBC are not because the former exhibit assimilation and antagonistic biases in this domain, whereas the latter do not. Although this is definitely part of the story, it does not seem to be the whole story. This individualistic explanation fails to account for how these TV networks constitute their viewership through the epistemic group functioning that they foster and the discursive norms they use to frame and package information. I will elucidate how additional, supra-individual factors can explain the kind of epistemic bubble that would be resistant to learning about epistemic injustices in carceral institutions through the collective and discursive models of epistemic bubbles proposed by Anderson.

Underscoring the explanatory limitations of the psychological models that conceptualize epistemic bubbles in individualistic terms, Anderson argues that the analysis of epistemic bubbles needs to get more social and more political by modeling cognitive biases not as operating inside individuals' heads, but as operating externally in the group behavior of collectives as they interact (or

---

10 This is my own expression, which I will use and explain through an example below.

11 Anderson (2021, pp. 15 – 18) reviews the psychological and social scientific evidence relevant here. She emphasizes that while the laboratory evidence produced by the psychological models suggests cognitive symmetry among individuals across ideological lines (e. g., between Republicans and Democrats in the US), "the field evidence suggests a substantial partisan asymmetry in entrapment" (Anderson 2021, p. 17)—an asymmetry that, as Anderson argues, only the different epistemic and discursive norms of these groups can explain.

fail to interact) in the public sphere according to particular group epistemic and discursive norms. Anderson proposes two additional models to supplement the psychological models: a group cognition model and a discursive model. According to the group cognition model, all that is needed to explain entrapment in epistemic bubbles is participation in a group that enacts biased social norms for information processing, independently of whatever cognitive attitudes and habits the individuals in question happen to have outside the group. The group cognition model distinguishes norms of assertion from norms for individual belief. Individual members of a group may vary in their cognitive attachments and commitments outside the group, but all that is needed for their participation in the collective cognitive activities of the group is that they follow the epistemic norms of the group when acting and thinking together. So, if, for example, a group is committed to remaining ignorant with respect to epistemic injustices in carceral institutions, all the group needs to do is to enforce the epistemic group norm that evidence to that effect be filtered out or automatically rejected.

Finally, Anderson also offers a discursive model of epistemic bubbles that explains how publics become entrapped by discursive norms in such a way that utterances and other communicative contributions become devoid of assertoric force and content, and become identity-expressive communicative moves, thus transforming, for example, an assertion that expresses a criticism or a complaint into an attack or an insult. Discursive entrapment occurs when "a group adopts discursive norms to treat certain ostensibly empirical assertions as identity-expressive" (Anderson 2021, p. 23). What characterizes discursive "epistemic" bubbles is identity-expressive discourse[12] that affirms and celebrates ingroup members (or their creed) while rejecting and denigrating out-group members (or their discursive contributions).[13] Discursive epistemic bubbles transform empirically informed and learning-oriented discourse into ad hominem discourse. Therefore, these bubbles are not, strictly speaking, epistemic, but their functional equivalent, since epistemic force and content disappear as a result of the discursive distortions. As Anderson puts it, what is discursively produced is "the functional equivalent of epistemic bubbles in public discourse, by replac-

---

12 As Anderson puts it, "Identity-expressive discourse expresses the speaker's group identity and positions the speaker in relation to people with the same or other identities. It may signal whose side one is on, who is the enemy, or doesn't belong, who is illegitimate, who is superior to whom" (Anderson 2021, p. 23).

13 Note that what results from this is affective polarization, that is, polarizing emotional attitudes such as anger, animus, arrogance, hostility, outrage, etc. See the essays on affective polarization in Lynch and Tanesini (2020). See also Archer and Mills (2019).

ing empirical discussion with trolling and insults" (Anderson 2021, p. 20). Kahan's culturally antagonistic memes function similarly as a vehicle of symbolic positional competition between rival identity groups. But, as Anderson points out, the difference is that while for Kahan the problem lies with processes going on inside people's heads, for Anderson, "identity-expressive discourse should be modeled instead in terms of rules of a language game." (Anderson 2021, p. 24) It is a game of insults and pokes in the eye, rather than a game of collective learning.

The group cognition model and the discursive model that Anderson offers can overlap, since epistemic and discursive entrapment within bubbles can converge. The example of bubbles shaped by stigmatizing stereotypes of criminality can illustrate this. The venue of public discourse that is Fox News and the group constituted by its viewership can be characterized both as a discursive bubble and as an epistemic bubble in Anderson's sense. When spokespersons of prison education programs and prison activist groups give information about epistemic injustices in the US criminal justice system, networks such as Fox News often depict them as the defenders of criminals and the attackers of the protectors of public order. They do this, for example, by associating members of Black Lives Matter with criminal life and attacking them with the opposing slogan Blue Lives Matter. In this discursive frame, activists are stigmatized in such a way that everything they say is depicted as baseless insults and attacks on so-called blue lives, that is, on the defenders of public order. The stigmatization of activists leaves viewers no option but to regard them, not as delivering information at all, but as indulging in dangerous identity-expressive statements. This stigmatization discursively positions viewers in such a way that the only uptake that they are encouraged to give to the activists' utterances is dismissal and name-calling since their criticisms of and challenges to the criminal justice system have to be understood as concealed attacks on and insults to blue lives. This is a good example of discursive entrapment in a bubble accomplished by discursive mechanisms, that is, through the deployment of narrative and rhetorical frames that stigmatize or glorify through the polarizing opposition Black Lives Matter versus Blue Lives Matter. What the stigmatization of activist organizations such as Black Lives Matter accomplishes is the systematic discrediting of its members and spokespersons, so that they are considered dangerous to talk to, intrinsically *untrustworthy*, and are thus excluded from epistemic transactions. And note here the convergence of discursive entrapment with epistemic entrapment: the discursive activation of stigmatizing stereotypes provides an excuse or alibi for epistemic exclusion; the discursive framing creates an epistemic environment in which the activists' evidence is filtered out or automatically rejected and presented as intrinsically flawed.

Both the discursive frames deployed by Fox News and the epistemic group functioning of its viewers cast out-group interlocutors as adversaries with whom no epistemic cooperation is possible: in one case the adversarial relation is forged discursively and non-epistemically, and in the other case epistemically. But discursive adversaries are typically also epistemic adversaries, and neither discursive adversaries nor epistemic adversaries are considered eligible for engaging in fruitful epistemic interaction, cooperation, and collective learning.

As Anderson emphasizes, "if epistemic bubbles are a problem, we need to figure out how to burst them;" and different models recommend "different strategies for popping epistemic bubbles" (Anderson 2021, p. 12). The different dimensions of epistemic bubbles highlighted by the different models clue us in as to the different kinds of remedies that can be mobilized against them. At the psychological level, the goal is to eradicate the cognitive biases of individuals (overcoming assimilation biases and antagonistic biases); but at the level of group cognition and structures of public discourse, the goal is to suspend dysfunctional epistemic and discursive norms and replace them with better ones. At the discursive level, we need to unmask and denounce ad hominem discourse, and promote the introduction of norms against deploying discursive frames that activate stigmatizing stereotypes. At the epistemic level, we need to speak against the filtering-out and automatic rejection of information and promote norms against dismissing speakers and bodies of evidence out of hand.[14] Of course, there will be discursive-epistemic environments and publics resistant to admit these changes,[15] but the very attempt to introduce them can spark critical discussions that can facilitate the achievement of meta-lucidity[16] about the epistemic and discursive norms that those environments and publics are following, typically without any awareness of how these norms work or of their implications. Mitigating degrees of self-denial about epistemic and discursive entrapment in bubbles can create opportunities for epistemic melioration. We need collective and concerted efforts to fight against forms of epistemic marginalization and oppression such as the ones that epistemic bubbles protect or produce. In the final section, I

---

14 Of course, some filtering of information is necessary given that our attention and capacity to process information are always limited and selective, but what is crucial is that such filtering be brought under rational scrutiny and open to be justified and corrected if need be.

15 Recent discussions of filter bubbles and echo chambers have highlighted and elucidated the recalcitrant nature of these epistemically dysfunctional groups and how difficult it is to change them. See Pariser (2011) and Nguyen (2020).

16 See the discussions of my notion of meta-lucidity in *The Epistemology of Resistance* (2013), esp. pp. 186–206.

will provide a preliminary sketch of how to think about collective epistemic resistance and what I term epistemic activism.

3 Collective Epistemic Agency and Epistemic Activism

Just as in order to fully appreciate the breadth of the phenomena of epistemic injustice we need to go beyond the interpersonal level and the ethics of knowing, in order to fully appreciate the forms of agency and the variety of resources available in the fight against epistemic injustices—i. e. in epistemic resistance (Medina 2013), we need to go to the supra-individual level and look at forms of collective agency that exert resistance against institutional and group failures in epistemic dynamics. As Code has pointed out, addressing questions about epistemic responsibility requires engaging with epistemic subjectivities *and* epistemic communities: "It is about the ethics and politics of knowledge, and indeed about epistemic subjectivity in its multiple instantiations" (Code 2013, p. 90). As Code and others have emphasized, epistemic responsibility concerns the accountability and responsivity of individual knowers, but also of groups or publics and of institutions (see also Anderson 2021). In her discussions of the politics of epistemic location (2006), Code has argued that epistemic advocacy is required to unmask our complicity with ongoing injustices and to mobilize publics to fight against them. Code's concept of advocacy is a key component of what I have called epistemic activism, namely, the kind of activism that can mobilize differently situated subjects and publics in order to resist epistemic injustices (see Medina 2019; Medina 2021a; and Medina and Whitt 2021).

As Code (2006) suggests, we need to engage in epistemic advocacy that denounces epistemic oppression and demands the overcoming of dysfunctional epistemic dynamics and the removal of epistemic obstacles that marginalized subjects face. But what is needed is not only speaking up against epistemic marginalization. The eradication of epistemic injustice requires something more than a purely communicative move; it requires epistemic resistance not only in word but in deed, that is, epistemic resistance in action, what I call epistemic activism. Epistemic activism is the kind of activism that aims at epistemic self-empowerment and at transformation or restructuration of epistemic environments. These two goals of epistemic activism bring two kinds of activist moves and techniques associated with them that I will explore in the two subsections that follow: first, epistemic activism that aims at epistemic self-empowerment, that is, at gaining or augmenting epistemic agency for those who have been marginalized or disempowered in epistemic dynamics; and second, epistemic activism that aims at epistemic structural transformation, that is, the restructuration of conditions, climates, and institutional frameworks that mediate epistemic interactions. Both subsections will use the case study of institutional epistemic injustice at DCDF discussed above, elucidating the forms of epistemic activism mobilized to fight against structural epistemic injustices in carceral contexts.

3.1 Epistemic Self-Empowerment: Gaining Epistemic Agency through Activism

As explained by Medina and Whitt, we can think of epistemic activism as concerted efforts and interventions in epistemic practices that aim to "augment the epistemic agency of unfairly disadvantaged subjects, amplifying their voices and facilitating the development and exercise of their epistemic capacities" (Medina and Whitt 2021, p. 309). Epistemic activism can take many different shapes and forms. Its strategies and tactics will be dictated by who engages in it, in what contexts, and against what patterns of interaction and institutional frameworks. Differently situated subjects, both oppressed and non-oppressed subjects, can become epistemic activists. "Oppressed subjects can become epistemic activists—sometimes by necessity if not by choice—when they actively fight against their epistemic marginalization and work towards forms of self-empowerment that can achieve the epistemic agency they are unjustly denied" (Medina and Whitt 2021, p. 312). Within carceral contexts, incarcerated subjects themselves can (and often do) become epistemic activists by denouncing and trying to resist unfair patterns of epistemic neglect, and by expressing epistemic solidarity by backing up one another's testimonies, so that they mitigate the harmful consequences that individual acts of protest typically encounter. A good example of epistemic activism cultivated by detained subjects at DCDF is provided by Medina and Whitt:

> [At DCDF] an unknown number of detainees recently organized the First Five Grieving Committee, a "non-violent" and "non-gang affiliated" cooperative that anonymizes and amplifies the grievances of individual detainees. By working together, the members of the Committee have successfully directed their concerns to the Durham County Sheriff, whereas individual grievances are typically heard—if they are heard at all—by subordinate staff members. This is an instance of epistemic activism, within the context of the jail, starting to ameliorate the testimonial disadvantage that detainees face (Medina and Whitt 2021, p. 312).

Note that the formation of the First Five Grieving Committee exemplifies how detained subjects, through concerted efforts at epistemic activism, form a collective voice—a collective epistemic agent, in fact—by pooling their agential resources and coordinating their actions. Detained subjects organize their voices and interventions in such a way that their individual epistemic actions become mutually supportive and protective of each other, thus contributing to the exercise of collective epistemic agency. This way of gaining and protecting epistemic agency through epistemic activism illustrates well how epistemic self-empowerment can take place even in the most adverse conditions.

We can also find ways of exercising collective epistemic agency to resist epistemic injustices within carceral institutions on the outside, with family members of incarcerated subjects, allies, and activists pooling their resources and coordinating their actions to facilitate the epistemic empowerment of detained subjects. In cases of epistemic neglect within carceral contexts, "activists, scholars and journalists, family members, political leaders, social media participants, and in short the general public can join forces with jail detainees to help ensure that their voices are heard and their concerns are addressed" (Medina and Whitt 2021, p. 313). I want to pay special attention here to the work of epistemic resistance of the Inside-Outside Alliance (hereafter IOA), a local activist organization that describes itself as a group of people trying to support the struggles of those inside (or formerly inside) Durham County jail, and their families and friends. OA members—friends and family of incarcerated subjects, formerly incarcerated subjects, and activists—engage in epistemic interventions, programs, and initiatives, which they subsume under the heading Amplify Voices Inside. There are two kinds of interventions in the epistemic activism of IOA that I want to highlight: those that try to reach *inside* the jail and those that try to have an impact on public discourse *outside* the jail.

First, IOA members use their voices and epistemic agency *within* the jail itself to echo, support, and empower the neglected voices of detained subjects and to put pressure on DCDF and its workers to meliorate their dynamics and policies. Think of cases of epistemic neglect of detainees' grievances. Detainees often complain that their unaddressed grievances disappear in the system, and when these grievances have been especially urgent, they have worked with IOA members to put external pressure on jail administrators.

In November 2014, the activist group organized call-ins to overwhelm administrators' phone lines when evening meals were reduced to two cold sandwiches. They adopted similar tactics in September 2015, when jail staff would not grant emergency medical transfer to a detainee in severe pain. In these actions, phone calls from diverse community members— many of whom do not consider themselves to be 'activists'—*echoed detainee grievances in ways that made them more difficult to disregard or disbelieve.* Additionally, the phone calls reminded jail staff of their accountability to the local community. In these ways, the actions temporarily disrupted typical patterns of interaction inside and outside of the jail, and indicated the possibility for alternative, less dysfunctional patterns. For alternative patterns to take hold, however, it may take repeated activist interventions. (Medina and Whitt 2021, p. 313)

What this epistemic activism can achieve is the (at least temporary) interruption of dysfunctional dynamics of epistemic neglect, making it difficult (at least temporarily) for detention officers and jail staff to act on epistemic vices (such as testimonial insensitivity). The epistemic resistance of IOA members by itself will not create fair epistemic dynamics, but it can help trigger a process of amelioration of testimonial dynamics and it puts pressure on individuals and institutions to discontinue epistemic vices and improve testimonial sensibilities.

Second, IOA's epistemic activism also takes place *outside* the county jail in order to procure epistemic standing for detainees' perspectives and some degrees of epistemic agency for their voices in the outside world. As Medina and Whitt put it, "detainees' voices rarely reach places of political authority without being distorted, translated into other idioms or discourses, or ventriloquized by others. For this reason, it is important to have forms of epistemic activism in which outside allies lend their voices as instruments or extensions of the detainees' own, without interpreting or translating them." (Medina and Whitt 2021, pp. 313–314). IOA members do this "by reading detainees' letters in City Council meetings and County Commissioner meetings, disrupting 'business as usual' with the testimonies of individuals who have been excluded from the sites of official power." (Medina and Whitt 2021, p. 314) Other ways in which IOA members seek to amplify detainee voices include: "publishing their letters verbatim, usually without context or commentary, on the website *Amplify Voices Inside*;" and publishing "detainee letters and artwork in a print magazine called *Feedback*." (Medina and Whitt 2021, p. 314.) In these different ways IOA members try to ensure that the voices of detainees are heard in the outside world and their stories, problems, and concerns neglected inside the jail can reach other institutions and authorities as well as the general public. These are examples of epistemic empowerment of marginalized groups and ways of gaining epistemic agency through epistemic activism.

3.2 Restructuring Epistemic Environments through Activism

In carceral contexts, the work of epistemic resistance needs to happen both at the level of interpersonal dynamics, targeting the testimonial insensitivity of carceral workers, and at the level of the institution itself, targeting the epistemic deficiencies of the institutional framework in its protocols and procedures or lack thereof. The activist interventions of IOA show ways in which not only the epistemic attitudes and habits of guards and jail staff, but also the institution itself and its protocols and procedures are put under pressure to change and improve by becoming accountable to outside publics and authorities. The epistemic activism mobilized against the institutional failures of DCDF discussed above focuses on three areas of institutional design and practice. Let me distinguish between the short-term and long-term epistemic goals of this activism in each of these three areas. First, activism against epistemic neglect in carceral contexts should aim at epistemically adequate protocols, but also at assessment processes for the evaluation and revision of these protocols so that there are sustained efforts toward creating an institutional culture in which detainees' voices and perspectives are respected and treated fairly. Second, epistemic activists should demand from a carceral institution such as DCDF not only adequate training, but also constant re-training, which is crucial not only to address the epistemic flaws and failings of workers who are already in the institution, as opposed to newcomers, but also for revisiting and continuing the epistemic melioration of attitudes, habits and dynamics initiated in the training process, since no training practice will be able to guarantee once and for all that epistemic vices will not set in. Third, the epistemic restructuring of an institution such as DCDF should include not only *having* adequate accountability procedures, but also *enforcing* them conscientiously so as to ensure that protocols are followed, that the received training and re-training is incorporated in epistemic interactions, and that the institutional culture and values operate properly in the epistemic sensibility of the officers of the institutions in their daily activities.[17]

---

[17] Epistemic melioration at the institutional level should be coupled with the pursuit of epistemic virtue at the personal level. Indeed, we should always expect and demand epistemically responsible behavior both from institutions and from individual subjects. When an institution fails to meet its epistemic responsibilities, this failure does not excuse the personal and professional epistemic responsibilities of its officers (e.g. the guards of DCDF). As any (general or professional) ethics of knowing would tell us, officers of an institution have to live up to standards of fair epistemic treatment of the subjects they serve even if they are not expected or encouraged to do so by their institution.

4 Concluding Remarks

In this essay, I have analyzed the recent shift from the ethics of knowing to the politics of knowing in social epistemology and argued that in order to properly address epistemic injustices we need to attend not only to interpersonal communicative interactions, but also to how these interactions are embedded in group dynamics and in large-scale communicative patterns. The target of our normative epistemic evaluations should be not only how individuals communicate with each other, but also how and why they fail to communicate with others, and also how they (often unconsciously and non-deliberately) exercise selective attention to certain sources of information and certain agencies, platforms, and networks of interpretation and assessments, short-circuiting paths of testimonial information and paths of interpretation and assessment. Elucidating the epistemic failures of a carceral institution and the phenomenon of epistemic bubbles, I have tried to shed light on how to normatively assess epistemic agency and epistemic responsibility at the institutional and collective level, and I have also provided suggestions for engaging in epistemic resistance against institutional and collective epistemic injustices through what I call epistemic activism.

Bibliography

Anderson, Elisabeth (2012): "Epistemic Justice as a Virtue of Social Institutions." In: *Social Epistemology* 26. No. 2, pp. 163–173.

Anderson, Elisabeth (2021): "Epistemic Bubbles and Authoritarian Politics." In: Michael Hannon and Elizabeth Edenberg (Eds.): *Political Epistemology.* Oxford, New York: Oxford University Press.

Archer, Alfred, and Georgina Mills (2021): "Anger, Affective Injustice and Emotion Regulation." In: *Philosophical Topics* 47, issue 2, pp. 75–94.

Ayala, Saray (2016): "Speech affordances: A structural take on how much we can do with our words." In: *European Journal of Philosophy* 24. No. 4, pp. 879–891.

Code, Lorraine (1995): *Rhetorical Spaces: Essays on Gendered Locations.* New York: Routledge.

Code, Lorraine (2006): *Ecological Thinking: The Politics of Epistemic Location.* London, New York: Oxford University Press.

Code, Lorraine (2014): "Culpable Ignorance?" In: *Hypatia* 29. No. 3, pp. 670–676.

Dotson, Kristie (2014): "Conceptualizing epistemic oppression." In: *Social Epistemology 28.* No. 2, pp. 115–138.

Fricker, Miranda (2007): *Epistemic Injustice: Power and the Ethics of Knowing.* Oxford, New York: Oxford University Press.

Fricker, Miranda (2010): "Can There Be Institutional Virtues?" In: Tamar S. Gendler and John Hawthorne (Eds.): *Oxford Studies in Epistemology*. Oxford: Oxford University Press, pp. 235–252.

Fricker, Miranda (2012): "Group Testimony? The Making of a Collective Informant." In: *Philosophy and Phenomenological Research* 84. No. 2, pp. 249–276.

Fricker, Miranda (2013): "Epistemic Justice as a Condition of Political Freedom?" In: *Synthese* 190. No. 7, pp. 1317–1332.

Goldberg, Sanford (2017): "Should have known." In: *Synthese* 194. No. 8, pp. 2863–2894.

Kahan, Dan (2012): "Cultural Cognition as a Conception of the Cultural Theory of Risk." In: Sabine Roeser (Ed.): *Handbook of Risk Theory: Epistemology, Decision Theory, Ethics, and Social Implications of Risk*. Dordrecht: Springer, pp. 725–759.

Kukla, Rebecca (2014): "Performative Force, Convention, and Discursive Injustice." *Hypatia* 29. No. 2, pp. 440–457.

Lynch, Michael, and Alessandra Tanesini (Eds.) (2020): *Polarisation, Arrogance, and Dogmatism: Philosophical Perspectives*. New York: Routledge.

Mason, Rebecca (2011): "Two Kinds of Unknowing." In: *Hypatia* 26. No. 2, pp. 294–307.

Medina, José (2012): "Hermeneutical Injustice and Polyphonic Contextualism: Social Silences and Shared Hermeneutical Responsibilities." In: *Social Epistemology* 26. No. 2, pp. 201–220.

Medina, José (2013): *The Epistemology of Resistance: Gender and Racial Oppression, Epistemic Injustice, and Resistant Imaginations*. Oxford, New York: Oxford University Press.

Medina, José (2019): "Racial Violence, Affective Resistance, and Epistemic Activism." In: *Angelaki: Journal of the Theoretical Humanities* 24. No. 4, pp. 22–37.

Medina, José (2021a): "Capital Vices, Institutional Failures, and Epistemic Neglect in a County Jail." In: Ian J. Kidd, Heather Battaly, and Quassim Cassam (Eds): *Vice Epistemology: Theory and Practice*. New York: Routledge.

Medina, José (2021b): "Vices of the Privileged and Virtues of the Oppressed in Epistemic Group Dynamics." In: Michael Hannon and Jeroen de Ridder (Eds.): *Routledge Handbook of Political Epistemology*. New York: Routledge.

Medina, José, and Matt Whitt (2021): "Epistemic Activism and the Politics of Credibility: Testimonial Injustice Inside/Outside a North Carolina Jail." In: Nancy McHugh and Heidi Grasswick (Eds.): *Making the Case*. Albany: SUNY Press.

Nguyen, C. Thi (2020): "Echo Chambers and Epistemic Bubbles." In: *Episteme* 17. No 2, pp. 141–161.

Pariser, Eli (2011): *The Filter Bubble: What the Internet Is Hiding from You*. New York: Penguin Press.

Pohlhaus Jr., Gaile (2012): "Relational Knowing and Epistemic Injustice: Toward a Theory of Willful Hermeneutical Ignorance." In: *Hypatia* 27. No. 4, pp. 715–735.

Sunstein, Cass R. (2017): *#Republic: Divided Democracy in the Age of Social Media*. New Jersey: Princeton University Press.

Alessandra Tanesini

Intellectual Vices in Conditions of Oppression: The Turn to the Political in Virtue Epistemology

Abstract: This chapter details five forms of epistemic oppression and explains some of the epistemic injustices caused by these oppressive relations. It describes the ways in which oppression contributes to the creation and maintenance of the intellectual vices of superiority (arrogance and narcissism) and inferiority (servility and timidity). The chapter concludes by highlighting the roles played by vices in reproducing oppression and by mentioning some potential ameliorative measures.

1 Introduction

Oppressive structures are widespread in liberal Western democracies. In this chapter, first, I argue that some of these structures have epistemic dimensions since they involve unfair distributions of epistemic goods (such as information) and resources (like education) and/or failures to recognise fully the epistemic abilities of epistemic subjects belonging to some underprivileged social groups. Following Iris Young's (1990) model, I describe five faces of epistemic oppression and explain how they disadvantage and wrong members of some social groups (section 2). Second, I examine the psychological effects of each form of epistemic oppression on those who benefit from it and those who are burdened and wronged by it, respectively. I argue that being the recipient of epistemic privileges creates expectations, incentives and misleading evidence that conspire to promote vices of superiority in the privileged, such as intellectual arrogance and narcissism. I also show that being subjected to epistemic oppressions generates a different set of expectations, incentives and misleading evidence that conspire to facilitate vices of inferiority, like intellectual servility and timidity, in the oppressed. These diverse expectations, incentives and misleading bodies of evidence make the oppressive nature of oppressive structures invisible to agents. Nevertheless, the oppressed are also likely to experience some aspects of their situation as unfair. Therefore, they are more able than privileged people to access evidence that reveals the true nature of oppression (section 3). I conclude this chapter with some brief remarks about the adoption of a vice-theoretical frame-

https://doi.org/10.1515/9783110612318-006

work to understand the nature of epistemic oppression and to develop strategies for its eradication (section 4).

2 The Five Faces of Epistemic Oppression

In her influential work, Young (1990, ch. 2) identifies oppression as a heterogeneous family of structural relations between social groups. These relations are unjust because they result in distributive unfairness and/or because they create injustices of recognition.[1] That is, these structural relations often produce unfair distributions of goods and resources and/or generate contexts in which individuals belonging to some groups are not fully recognised as agents.[2] Young characterises five aspects or faces of oppression that are prevalent even in democratic regimes that are not tyrannical. These are: exploitation, marginalisation, powerlessness, subordination (or cultural imperialism) and violence.[3]

Members of some social group are exploited when they are systematically expected to offer services and produce goods primarily for the benefit of others, leading to a progressive diminution of useful resources accessible to the exploited individuals themselves. People are marginalised when, because of their group membership, they are excluded from the production and distribution of goods and resources. Agents are rendered powerless when they have little or no power over the structure of their daily lives. Individuals are subjected to cultural imperialism, and therefore subordinated, when they are viewed as inferior and deviant compared to members of the dominant groups. Finally, people are subjected to violence when they are intimidated, humiliated, mocked or assaulted or made to feel especially vulnerable to these treatments because of their prevalence and of widespread complacency about, or complicity with, such conduct.

Whilst Young focuses her discussion on a broad range of social and economic goods, her account of the nature and the variety of oppression is readily applicable to the epistemic domain. This approach lends itself to developing a capacious, but theoretically coherent, account of epistemic injustice as injustice

---

1 Young thinks that matters of equitable or fair distribution and equal recognition are intertwined and should not be thought as separate (Young 1997).
2 Mocking, humiliating and insulting behaviours are examples of treatments that fail to recognise the equal status of all agents. This conduct is an affront to the kind of dignity that all agents have in equal measure irrespective of status and ability.
3 Sometimes 'subordination' is used as a catch-all term synonymous with oppression. In this chapter, I use the term exclusively to refer to the kind of oppression that is due to cultural imperialism.

that is the result of epistemic oppression.[4] In turn, epistemic oppression is understood as circumscribing structural relations among social groups producing unfair distributions of epistemic goods and resources and/or generating contexts in which individuals belonging to some groups are not fully recognised as epistemic agents. The account of epistemic injustices as primarily concerning injustices that flow from epistemic oppression is, in one regard, much broader than Miranda Fricker's (2007) original account which is restricted to discriminatory injustice.[5] In another, however, it is narrower in so far as it ignores injustices due to idiosyncratic prejudices that do not align with differences in social power.[6] One advantage of the broader characterisation is that it avoids individualism since it acknowledges the contribution of structural features even to those injustices committed by a single perpetrator, but also because it can account for the existence of purely structural epistemic injustices that do not involve prejudicial attitudes on the part of any of the actors involved (cf., Anderson, 2012).[7]

The idea that epistemic injustice is the product of oppressive relations concerning the maldistribution of epistemic goods and/or the failure to recognise properly the epistemic agency of oppressed individuals has been developed in different directions by, among others, José Medina (2013), Kristie Dotson (2011 and 2014), Nora Berenstain (2016), Ian Kidd (2020), and Fricker (1998) herself. To my knowledge, however, these theorists do not explicitly invoke Young's multifaceted account of oppression. In what follows, I lend further support for their views by interpreting them through the lenses of Young's framework.

Before I outline the five faces of epistemic oppression, two qualifications are in order. First, I do not wish to suggest here that all the forms of epistemic oppression I discuss are distinctively epistemic in a sense that makes them irreducible to other kinds of social and political oppression. I am inclined to think that each of the five faces includes instances of oppression that are irreducibly epistemic; but this is not the focus of my discussion. For the purposes of this paper,

---

4 Oppression is thus broader than those forms of social control that for Fricker are an essential part of the stereotyping and prejudice that is in her opinion at the root of epistemic injustice (2007, see esp. ch. 1).

5 Following Coady (2010 and 2017), Fricker (2017) is now willing to countenance the existence of two varieties of epistemic injustice: distributive and discriminatory.

6 For instance, a doctor might suffer a testimonial injustice at the hands of a hearer who is prejudiced against the medical profession. I accept that epistemic injustice can take this form. However, like Fricker, I do not take cases of this sort to be paradigmatic. Thanks to Mona Simion and Katherine Jenkins for pressing this point.

7 For example, qualified members of an oppressed group might never be consulted by an official body not because the officials are prejudiced but because they are working with an out of date list that reflects past prejudices.

one can think of them as merely the manifestations of social and political oppression in the epistemic realm. Second, I do not wish to suggest that these five faces are exhaustive or mutually exclusive. There might well be forms of epistemic oppression not covered in this taxonomy. There certainly exist oppressive social relations that instantiate more than one face. The classification is only meant to illuminate the distinctive features of partly overlapping forms of epistemic oppression.

2.1 Epistemic Exploitation

Following Young, I define exploitation as the product of social relations where members of one group prevalently produce and transfer goods and resources primarily for the benefit of members of other groups.[8] Over time, the exploited group is deprived of goods whilst other groups are able to accumulate them. Deprivation occurs because members of the exploited group have to focus on producing goods that are not a priority for them at the detriment of having the time and energy to produce those things that would satisfy their needs. It also occurs because they often transfer these items to others without receiving an equal amount of goods in return.

Epistemically exploitative relations are widespread in current Western liberal democracies. These relations concern the production and dissemination of epistemic goods such as knowledge and information within an epistemic community as well as the access to those epistemic resources such as libraries, formal education, ICT equipment that facilitate the production and circulation of epistemic goods. Epistemic exploitation exists when these relations are exploitative. Berenstain (2016) has highlighted some aspects of this phenomenon. In her view, 'Epistemic exploitation occurs when privileged persons compel marginalized persons to produce an education or explanation about the nature of the oppression they face' (Berenstain 2016, p. 570). But, as her account also indicates, the true scope of epistemically exploitative relations is broader than this along several dimensions.

First, exploitation can concern the supply of information on many topics rather than exclusively about the discrimination suffered by members of the exploited group. Second, compulsion is not necessary for exploitation, other forms

---

8 I do not want to make a hard and fast distinction between goods and resources. Broadly speaking, I think of goods as items that are constitutively or instrumentally valuable and of resources as items that are of instrumental value in the pursuit of further goods. Resources so understood are a kind of good.

of manipulation or trickery are also exploitative. Thus, for instance, epistemic oppression can occur if the exploited individual is emotionally manipulated to provide the information because they are made to feel sorry for the privileged person. Third, epistemic exploitation can also take place in the absence of any one exploiting individual.[9] For example, some people, because they are members of a social group, might end up disproportionally shouldering the burden of epistemic labour that mostly serves the epistemic interests of members of groups other than their own (Olson 2016).

Finally, exploitative relations are not limited to burdens regarding information but can also concern the legitimisation of ignorance. Sometimes privileged people do not demand to be educated about the oppression of the exploited but demand to be excused for their ignorance. These requests for emotional support, when it is extracted through white tears and expressions of white guilt, are also exploitative since they are manipulative.[10]

In conclusion, epistemic exploitation characterises a range of relations between members of social groups where individuals belonging to one group systematically (1) carry out epistemic labour to produce and transmit knowledge and (2) perform emotional labour to legitimise ignorance whenever these activities primarily serve the epistemic interests of another group and result in the depletion of cognitive resources of members of the exploited group. Epistemic exploitation is thus an injustice in the distributive sense because it produces an unfair distribution of epistemic goods and resources. It is also an injustice of recognition because it involves the systematic objectification of some individuals who are treated as mere means at the service of the interests of others.[11] Being subjected to this kind of treatment is disrespectful. It is an affront to dignity that impacts on one's ability to develop and sustain self-respect.

2.2 Epistemic Marginalisation

Marginalisation occurs when individuals are unfairly cut out from the processes of production and distribution of goods. In some sense marginalisation is worse than exploitation because it involves being excluded from the community as opposed to being exploited within its boundaries. Epistemic marginalisation is the product of being shunned from an epistemic community and its practices when

---

9 It is a feature of Young systemic notion of oppression that it is possible to have oppressed groups even though no particular group does the oppressing.
10 On white tears and fragility, see Applebaum (2017).
11 See Nussbaum (1995) on instrumentalization as one form that objectification can take.

one is entitled to inclusion. It also involves unfair barriers to accessing epistemic goods and resources.

Examples of epistemic marginalisation include the existence of unfair and systemic obstacles in the way of accessing epistemic resources such as good quality formal education.[12] Individuals are also marginalised when they are prevented from contributing to inquiries and conversations. Exclusions occur when people are not admitted to the spaces where these activities are carried out because they are not invited, informed or elected. They can occur when the contributions of some people are systematically and wrongly ignored, dismissed, or not taken seriously. There are many instances of epistemic marginalisation. The marginalisation of the scientific knowledge developed by non-Western people is a case in point (Schiebinger 2008);[13] another is the tendency to view some social groups as suitable object of study by sociologists and anthropologists while assuming that these individuals cannot also actively contribute to this research (Vaditya 2018).[14]

In addition, Dotson's (2014) account of epistemic oppression focuses primarily on the epistemic marginalisation of oppressed groups. Her discussion addresses the exclusions from inquiry of members of these groups due to hermeneutical injustices. These injustices can be traced back to the nature of existing conceptual frameworks that are among the epistemic resources shared by the community at large. Because these resources are unserviceable for, or unsuited to, the purposes of making intelligible the experiences of members of oppressed groups, they are unfairly rendered unable to participate fully in the life of the epistemic community and to influence its agenda about the topics and issues to prioritise (see also Medina 2013, ch. 4).[15]

In conclusion, epistemic marginalisation is a complex phenomenon, partially overlapping with subordination or cultural imperialism, that involves distributive injustices such as school exclusions, and failures of recognition, as when some people are denied participation in the life of the epistemic community simply because they are members of a given social group. These treatments are af-

---

12 In the UK, Black Caribbean, and Mixed White and Black Caribbean students were in 2018 nearly three times as likely to be permanently excluded from school than British White pupils. https://www.ethnicity-facts-figures.service.gov.uk/education-skills-and-training/absence-and-ex clusions/pupil-exclusions/latest accessed 15 October 2020.
13 For an attempt to redress some of these issues see, Ludwig (2018) and the researchers active in the decolonising project (Mignolo 2011).
14 See also several contributions to Harding (1993b). Harding (1998) is an early discussion of how to address these issues.
15 On hermeneutical injustices, see also Fricker (2007).

fronts to people's dignity since it is insulting to be ignored when one is entitled to be heard.

2.3 Epistemic Powerlessness

In Young's framework, powerlessness indicates the kind of oppression that occurs when people are not in charge even of the most trivial decisions concerning their everyday lives. Powerlessness so conceived characterises the working routines of the non-professional classes on production lines, in the dispatch centres of on-line retailers or in call-centres. Their work is dictated by rhythms that are not of their own making and that they have no power to influence. In the epistemic realm, powerlessness occurs primarily when members of some social groups are denied autonomy and control over their contributions to epistemic practices, or epistemic authority over matters about which they legitimately expect to be treated as authoritative.

Examples of epistemic powerlessness include being tricked, coerced or manipulated into testifying. Individuals who have their speech extracted by the police suffer from this form of oppression if they are subjected to forms of interrogation that undermine or by-pass their agency (Lackey 2020). This kind of treatment is exploitative and causes powerlessness, since those who are subjected to it have little control over their testimonies.

Further, powerlessness is the outcome of having one's epistemic authority undercut. For instance, patients can feel powerless when their authoritative experiential knowledge is discounted or dismissed (Freeman 2015 and Kidd and Carel 2017). A similar dynamic is at play when individuals suffer from what Medina has labelled 'ego skepticism' (Medina 2013, p. 42). This is a tendency to self-questioning and self-doubting that undermines self-trust and self-esteem. As a result, an individual might even question her own sanity. This phenomenon has many causes. They include: persistent dismissive incomprehension, when others dismiss a person's testimony as being unintelligible, irrational or a rant (Cull 2019); and gaslighting, when an agent is manipulated into becoming complicit in her epistemic undoing (Abramson 2014).

The injustice of being rendered powerless is primarily an injustice of recognition since it is tantamount to the diminution or even privation of autonomy that is one of the essential characteristics of epistemic agenthood. This form of oppression can cut especially deep when it recruits its victim into contributing to her own oppression through the erosion of self-trust. Nevertheless, there is also a distributive dimension to this form of epistemic oppression. As a consequence of the loss of autonomy, individuals' access to epistemic goods such as knowledge is restricted. Those who are prone to self-doubting are unlikely to have the confidence required to enjoy a variety of epistemic goods that would be of significance in their lives.

2.4 Epistemic Subordination (Cultural Imperialism)

Young characterises subordination or cultural imperialism as the kind of oppression that results in members of some social groups being represented as the Other. Shared social meanings, norms, metaphors and customs play a central role in these processes. In some societies, including our own, the features, experiences and characteristics of individuals belonging to some social groups are identified as defining of humanity. These properties are the reference point by which all people are judged. Thus, for instance, the lives and habits of affluent white males are in contemporary Western democracies often taken as the norm and yardstick by which to measure the behaviour of others.

Those people who are not the norm (both in the sense of being a standard and of being normal), are said to be the Other. To be thought of as the Other is to be taken as inferior to, deviant and derivative from the dominant person rather than merely different from him in some respects. Subordinated individuals are seen as inferior because they do not match what is taken to be the cultural standard understood as a norm to aspire to. Further, to the extent to which members of these groups are different from the so-called normality set by cultural norms, they are treated as deviant. Finally, subordinated individuals are treated as derivative because their worth is judged by the inverse proportion to their distance from a norm that they have not set for themselves.

Cultural imperialism primarily moulds the concepts, images and stereotypes that are available to people in a given community in their attempts to understand themselves as members of a social group and others who belong to groups different, at least in part, from one's own. In other words, this structure of oppression shapes people's social cognition; that is, their processing and sharing of information about themselves and others, social groups and social situations. Therefore, cultural imperialism is mainly about epistemic resources. It concerns the range of notions, metaphors, stereotypes, social beliefs, norms and assumptions that are prevalent within a community or at least available to individuals within it. These resources enable, and constrain, people's thinking about social reality and their ability to conceive alternatives to it (Medina 2013).

The study of these phenomena has generated a vast literature. It is usually traced back to Du Bois' (1990) influential account of the double consciousness of Black Americans who, out of necessity, learn to look at themselves and their own experiences simultaneously from two perspectives. One is their own, but the other is the dominant viewpoint that characterises them as inferior, deviant and pitiable. Since De Bois' pioneering work, the notion of double consciousness has been extensively explored by feminists and critical race theorists (Collins 2000 and Medina 2013).

More recently, Fricker's discussion of hermeneutical injustice has prompted the exploration of similar themes. This kind of injustice, in Fricker's view, occurs when 'a gap in collective interpretive resources puts someone at an unfair disadvantage when it comes to making sense of their social experiences' (Fricker 2007, p. 1). This characterisation has subsequently been modified and refined by Medina (2013, 2017), who has convincingly argued that interpretative resources might be differentially shared between social groups. Hence, for example, whilst dominant individuals often lack the resources to understand subordinated others, members of oppressed groups might have developed the concepts and images necessary to make some sense of their experiences. That said, it is also possible for some people to find themselves in a situation in which they are unable to find the means to make sense of themselves.[16]

More often, subordinated individuals have the conceptual resources to make sense of their experiences but, out of necessity, silence their own voices. Dotson (2011) who first identified this phenomenon refers to it as testimonial smothering. It is a form of self-silencing due to a reasonable assumption that one's audience is unable or unwilling to hear what one has to say.[17] In this instance, one is able to make sense of one's experience but one is not able to communicate such understanding to others because they (i.e., the others) lack the resources to comprehend it or because they wilfully refuse to avail themselves of these, while preferring to use distorting images and perspectives (Pohlhaus 2011). Some might even take a further step in this direction, and positively attempt to suppress available epistemic resources so that others are prevented from accessing them. Arguably, this dynamic plays a role in the general invisibility of white privilege to white people. Some individuals in the community produce and distribute stories about hard work, merit and colour blindness that obfuscate other people's (and especially other white people's) ability to understand

---

16 Medina (2017) describe these circumstances as cases of hermeneutical death.

17 Dotson thinks of testimonial smothering as a kind of epistemic violence because it involves a refusal to reciprocate communicatively in a linguistic exchange. As I argue below, I agree with Dotson that testimonial smothering increases the communicative vulnerability of speakers. However, it is also important to note that, in addition, testimonial smothering is a form of epistemic subordination that makes the points of view of non-dominant groups invisible to dominant audience who, because of their ignorance, are unable to hear properly subordinated voices.

social reality. In this way, what Mills (2007) and others have called white igno-rance is a phenomenon that is contiguous with hermeneutical injustice.

The injustice of cultural imperialism or epistemic subordination is primarily an injustice of recognition because it consists in the creation and maintenance of conceptual frameworks that construe members of subordinated groups as inferi-or and deviant epistemic agents. Some forms of epistemic subordination are ef-fective in atrophying the ability of oppressed people to make sense of them-selves; others cause oppressed individuals to be widely construed as unintelligible or stupid by the dominant culture;[18] others still force them to bite their tongues and play an inferior and diminished role in communicative ex-changes. Epistemic subordination, however, brings unfairness of distribution in its trail. These occur, for instance, when some epistemic resources are sup-pressed by some so others who would find them useful are unable to access them.

2.5 Epistemic Violence

The idea that there might be something that is properly called epistemic violence will undoubtedly sound preposterous to some. It is tempting to think that vio-lence is limited to physical assault. This definition is too restrictive, since people subjected to coercive and controlling behaviours are naturally thought as victims of violence even when they have not been physically assaulted. We think of them in these terms because, like victims of assault, they live in constant and real fear of violence. It is, thus, not wholly implausible to define as violent those actions that, like verbal assault, insult, mockery, intimidation, are either intended to, or systematically result in, their targets constantly feeling afraid and vulnerable. A consequence of these oppressive relations is that their victims are likely to re-spond to the oppressive behaviour by retreat. In this way, the mere threat of vio-lence, is an instrument of control.

Once violence has been reconceived in this extended sense, the idea that this concept can be applied to relations of epistemic dependence between agents is no longer outlandish. Individuals are often dependent on other agents in their epistemic activities. Much of what we know we have learnt by relying on the tes-timony of other people. Further, as providers of testimony, we rely on our audi-ence's willingness to take seriously our claims in order to be full participants

---

18 Dotson (2011) calls this testimonial quieting that consists in discounting the testimony of subordinated individuals.

within the epistemic community. More generally, many of our epistemic aims can only be fulfilled with the cooperation of other epistemic agents by means of distributions of epistemic labour, generous dissemination of information or charitable engagement with others in discussion. It is because epistemic agents are never wholly self-reliant that they are epistemically vulnerable. Our vulnerabilities are especially exposed when others on whom we need to rely to achieve our epistemic aims withdraw their cooperation.

Epistemic violence could then be thought as an appropriate label for those relationships that result in unjust differential epistemic vulnerabilities. Thus, for example, because of their group membership, some individuals are routinely mocked or intimidated when they attempt to contribute to epistemic practices. Being subjected to this treatment is bound to make one feel that one cannot depend on the good will of others. One might react to being quietened by others by smothering one's own speech (Dotson 2011). This reaction is a withdrawal from epistemic practices that is consonant with the kind of retreating response typical of those who are afraid.

Epistemic practices that create unjust distributions of vulnerabilities, caused by the withdrawal of good will and cooperation, also erode some agents' ability to rely on themselves. What I have in mind here is the effect of epistemic violence on a kind of fundamental self-trust that has been defined by Jones (2004) as basal security. This kind of self-trust is an orientation to risk and vulnerability. Those who feel secure have some confidence that they will not be harmed. However, those who have experienced great harms, or who are frequently the targets of harmful behaviours, develop an anxious attitude to risk and feel highly vulnerable. When their basal security is shattered, they become distrustful and paralysed. In this manner, all their epistemic capacities might in the longer term become atrophied.[19]

The injustice of epistemic violence is both a matter of the recognition of epistemic agents and of the distribution of epistemic goods. To the extent to which violence quietens and erodes the self-trust of its victims, it denies them recognition as full participants within epistemic practices. These failures of recognition impinge on the distribution of epistemic goods such as knowledge and resources like education. People who feel vulnerable and insecure are unlikely to avail themselves of opportunities if these are perceived to be associated with risks. As a consequence, they refrain from accessing epistemic resources and thus end up with fewer epistemic goods than would otherwise be available to

---

19 The development and exercise of every intellectual ability ultimately depends on possessing a healthy dose of intellectual self-trust.

them. Whilst this pattern of distribution might seem the outcome of the free choices made by vulnerable individuals, it is unfair because it is the result of lowering aspirations caused by the expectation that one might be subjected to threatening behaviour if one dared to claim a bigger share for oneself.

To summarise, Young's analysis of the oppressive structural relationships is readily applicable to the epistemic domain. In this section I have described cases of epistemic exploitation that involve the unfair distribution of cognitive labour. I have highlighted epistemic marginalisation where individuals because of their membership to a social group are excluded from epistemic communities. I have identified instances of epistemic powerlessness when the testimonies of some are extracted from them by means of interventions that undercut or manipulate their autonomy. I have explained how hermeneutical injustices are readily understood as examples of epistemic cultural imperialism. Finally, I have detailed cases of epistemic violence which result in the erosion of trust and also self-trust.

3 Vices of Superiority and Inferiority

In this section, I detail some of the effects of social relations of epistemic oppression on the minds of those that are privileged and disadvantaged by them, respectively. That is, I offer descriptions of some features of the psychologies of privilege and of oppression. My ultimate aim is to show how social relations of oppression produce characteristic psychological signatures that include the formation of vices of superiority, prevalent among the privileged, and of inferiority, mostly afflicting the oppressed. I focus primarily on two vices of superiority – arrogance and narcissism – and two vices of inferiority – timidity and servility.

I adopt here and elsewhere a capacious notion of epistemic or intellectual vice. In my view, these vices are clusters of psychological features of individuals that range from sensibilities (such as being inattentive or unobservant), to thinking styles (like wishful thinking) and character traits (arrogance, for example).[20] These psychological qualities are generally stable and tend to be manifested in varied situations. Intellectual vices systematically have bad epistemic effects. They are features that, as Medina (2013) and Quassim Cassam (2019) have argued, get in the way of knowledge, but also truth and understanding.

---

20 For a detailed account, see Tanesini (2021), esp. ch. 2. Both Medina (2013) and Cassam (2019) emphasise the heterogeneous nature of epistemic vices.

In my view, intellectual vices also have a motivational dimension. They have either been acquired under the continuous influence of epistemically bad motivations or include these among their components. Thus, for example, some individuals often fail to notice episodes of harassment in the workplace because they have cultivated an inattentive sensibility out of a desire to turn away from inconvenient truths. Similarly, intellectually arrogant people dismiss views contrary to their own, and derogate their critics because they are driven by the need to defend the self against any real or alleged threats to self-esteem (Tanesini 2018a and 2019). This behaviour is characteristic of what psychologists have described as motivated cognition. It involves thinking and reasoning that is not driven by the epistemic goal of accuracy but by goals of a different sort such as enhancing self-esteem or being socially accepted (Kahan 2013 and Kunda 1990). The pursuit of these goals results in biased searches and evaluations of the available information because the desires for self-enhancement or for social acceptance motivate one to discount or ignore the truth. To sum up, epistemic vices are sensibilities, styles of thinking and character traits with systematic bad epistemic effects and that are either the product of motivated cognition or have among their components motivations that systematically cause motivated reasoning.

Before I detail some aspects of the psychologies of privilege and of oppression, two qualifications are in order to avoid possible misunderstandings. First, individuals can be beneficiaries of the oppression of people belonging to other social groups without playing themselves the role of oppressors. Some injustices are wholly structural. In other cases, privileged individuals bear more responsibility for the injustice because of complacency or complicity, even though they do not themselves actively engage in oppressive behaviour.

Second, I do not wish to indicate that everyone in a social group is psychologically affected in the same way by their place in oppressive relations. Individuals belong to multiple social groups distinguished by gender, ethnicity, age and other identities. All of these contribute to complex patterns of relative advantages and disadvantages in different contexts.[21] People also have different experiences because of sheer environmental luck or misfortune. Serendipity and coincidences can make a big difference to the kind of person one eventually becomes. Finally, some temperamental tendencies might be innate (Veselka et al. 2010). That said, those who occupy a given social location are more likely because of

---

[21] Hence, even within an oppressed group there can be individuals that are doubly subordinated because their other identities make them different from the group's more privileged members. For instance, black women find themselves in this position regarding both their racial and gender identities.

their social position to develop some psychological tendencies. When these dispositions are epistemic (or moral) vices it makes sense to classify them as the vices that are characteristic of members of privileged or oppressed groups.

One might be tempted to think that social and economic privileges can only bring epistemic advantages in their trails. Indeed, some social and economic benefits translate into epistemic assets because of increased access to epistemic goods and resources such as high quality education, libraries, parental tuition, theatres and museums.[22] Conversely, oppression is unjust partly because it denies some people their fair share of epistemic goods and resources. There are, however, also epistemic downsides to being socially privileged and upsides to struggling against oppression. These reversals have long been highlighted and explored by feminist standpoint epistemologist and critical race theorists.[23]

Below I focus almost exclusively on the negative consequences of epistemic oppression on the characters of the privileged and the disadvantaged. I consider how oppressive structures create new social expectations, new incentives and produce misleading evidence.[24] These three broad structural features, I argue, impact on the psychologies of the privileged and of the oppressed in ways that facilitates the formation of characteristic intellectual vices.

3.1 Epistemic Exploitation

When one is the beneficiary of the epistemic exploitation, one's epistemic needs are serviced by others who instead have to sacrifice their priorities in the interest of fulfilling the wants of those who belong to the privileged group. In societies where epistemic exploitation is widespread, privileged individuals expect (in the sense of predict) that their epistemic needs to know and to be protected from upsetting knowledge are met by other people. Conversely, exploited individuals predict that privileged individuals have expectations about their epistemic needs being serviced by them. Expectations about others' behaviours naturally influence decisions; when these concern customary or habitual actions, they generate

---

22 I mention these because in Western societies they contribute to so-called cultural capital or are among the means of acquiring it. Other epistemic goods might be equally or even more important in different cultures and contexts. Thanks to the editors for pressing this point.
23 This literature is vast. The most prominent contributions include Collins (2000); Haraway (1991); Harding (1993a); Hartsock (1983); Intemann (2010); Du Bois (1990); and Medina (2013).
24 Therefore, although I discuss how oppression produces injustices of recognition, the psychology of oppression discussed here differs in some from accounts that foreground the desire for recognition. For a critical discussion of diverse psychologies of oppression, see Cudd (2006).

habits.[25] Thus, in exploitative societies, privileged individuals learn to behave habitually in ways that presuppose (or take for granted) that others do the legwork for them, answer their questions and shield them from uncomfortable truths (Sullivan 2006). Those who develop these habits become complicit with, or complacent about, exploitative structures.[26]

Often when people become habituated to something convenient or pleasurable, they become resentful when that source of pleasure is taken away, even though they are not entitled to it. Those who develop these reactive attitudes towards others whom they expect to serve their epistemic needs[27] have acquired new normative expectations. They no longer merely expect in the sense of predict that others service their needs, but they also demand this preferential treatment. Once these normative expectations are in place, privileged individuals arrogate their entitlements, develop a sense of self-importance and feelings of superiority. They might also lash out in anger in response to perceived attacks to their own inflated self-conception. These behaviours follow from the rationalisation of the belief (conscious or not) that others ought to serve their epistemic needs.

The prevalence of these predictive and normative expectations about the behaviour of privileged and exploited individuals creates a whole gamut of novel incentives. Privileged individuals have incentives to preserve their privilege since this is advantageous. But they also have incentives to see themselves as genuinely superior so that their entitlements would be merited rather than arrogated and/or to deny the existence of preferential treatment.[28] Either way, privileged individuals acquire incentives to engage in motivated cognition out of a desire to protect one's self-esteem as a good and competent person. Defensive reasoning leads these individuals to discount evidence of their limitations and promotes agentic attributional styles according to which successes are due to merit whilst failures are caused by bad luck or other features of the situation that are outside the agent's control.[29]

---

25 If I expect people to drive on the right, since I want to avoid accidents, I will also drive on the right. In countries where this is the norm, everyone habitually drives on the right.

26 Privileged individuals are not complacent or complicit with structures of exploitations merely because they have true expectations about others' behaviours. Complacency and complicity set in when people develop habits that depend on the continuation of that conduct, rather than trying to disrupt the unjust structures.

27 The locus classicus on resentment as a reactive attitude is Strawson (2008).

28 For both of these strategies, see Phillips and Lowery (2018).

29 Another self-esteem preserving response to challenges is to avoid applying oneself. This strategy is known as self-handicapping (Lupien et al. 2010). When it is adopted, success is attributed to natural talent, failure to lack of application, rather than lack of ability. This is perhaps

These are defensive strategies for the preservation of an inflated self-conception that is dependent on feeling special in order to feel good about oneself. They are strategies that feed a need not to know the truth about oneself, one's intellectual abilities and one's complicity in structures of injustice. Often these strategies also include tactics that rely on avoiding unwelcome evidence altogether rather than merely dismissing its importance. These involve attempts to isolate oneself from facts indicating that one is not so special. One might do this preemptively by not putting oneself in situations where one might face these facts or by developing patterns of inattentiveness that help one not to notice any evidence that might become available.

Whilst it is easier to appreciate that privileged individuals have strong incentives to ignore the fact of their privilege, one might wonder why exploited individuals do not rebel. In response, one might point to the range of instruments at the disposal of the privileged to induce others to service their epistemic needs. These involve monetary incentives but also the use of threatening behaviours to induce members of oppressed groups to sacrifice themselves for the benefit of others. These incentives and the knowledge of the expectation harboured by the privileged give reasons to oppressed people to comply out of the necessity with the demands imposed on them. Unfortunately, what might begin as strategic compliance with unreasonable demands in the knowledge of their unreasonableness, sometimes develops into something else.

One form of epistemic exploitation, I have argued, involves expecting others to carry out the emotional work of shielding the privileged from unwelcome information. Those who have incentives to tending to the emotional needs of others often progress from faking concern to actually feeling it. For example, it is a job requirement of many service industries that workers appear happy when assisting clients. As clients are adept at spotting fake happiness, service workers have powerful incentives to be happy to serve (Hochschild 2012). Similar considerations give exploited individuals reasons to become subservient, rather than merely pretending. In addition, fake subservience would require that one suppresses anger at one's treatment. But bottling emotions and resentment is often hard to bear. Hence, oppressed individuals have further powerful incentives to tell themselves that their treatment is not unfair or undignified.[30] In

---

why, among the British upper classes, seemingly not applying oneself was even more important than succeeding.

30 The main incentive is the pressure of cognitive dissonance. Humans have a need to think of themselves as honest people, rather than as deceivers. Thus, there is some tendency to conform beliefs to behaviour when there is a dissonance between the two. On cognitive dissonance, see Cooper (2008).

these cases, exploited individuals for the sake of social acceptance or in order to protect the self from threats, come to see themselves as deserving of their inferior roles. They might become deferential, ready to answer the beckon call of the privileged, ready to self-sacrifice and adopt other subservient behaviours. They might also feel ashamed because of their alleged shortcomings.

Exploited individuals might also be tricked or manipulated into accepting servitude. In such cases, individuals can be made to become complicit in their own exploitation by presenting them with false or misleading evidence that they are best suited to playing inferior and servile roles. As a matter of fact, one of the insidious features of exploitation is that it constitutes a cognitive environmental niche that is biased to indicate to privileged children that they are capable and gifted and to underprivileged children that they are stupid and lacking in talent. Because the social environment is so structured to impose excessive burdens onto children who belong to exploited groups, they will find it harder than their privileged counterparts to succeed. If they take these failures to be evidence of lack of talent, they are likely to conclude that they deserve their inferior status. When this happens, they have been tricked into being complicit in their own oppression. By contrast, the social environment is structured to ease the epistemic burdens of privileged children. These young people might, as a result, develop a mistaken sense of confidence in their own abilities because epistemic obstacles are smoothed out of the way by others. This sense of ease can also be taken as evidence of one's epistemic superiority since the obstacles that prevent others' epistemic successes are hidden from view.

To sum up, on the one hand, privileged individuals who benefit from exploitative relations are likely to develop dispositions to arrogate epistemic entitlements, and to reason defensively. They are at risk of having excessive confidence of their abilities, developing a sense of self-esteem that is dependent on feeling superior to other people. These are some of the principal features of intellectual arrogance which is therefore best thought as an epistemic vice of superiority (Tanesini 2018a and 2019). On the other hand, exploited individuals are at risk of developing deferential and subservient behaviours which are designed to gain the approval of dominant individuals. They might feel inferior, ashamed and doubt their own abilities. These are characteristics of the epistemic vice of intellectual servility that is one of the main epistemic vices of inferiority (Tanesini 2018b).

3.2 Epistemic Marginalisation

Being the beneficiary of other people's epistemic marginalisation also puts privileged individuals at a higher risk of developing features that are characteristic of intellectual arrogance. The primary effect of marginalisation on the marginalised is that it makes them susceptible to another vice of inferiority: intellectual timidity (Tanesini 2018b).

The main structural effect of marginalisation on the privileged is the creation of a biased cognitive environment that fails to include the contributions of some members of society. When other people are marginalised from the epistemic community, privileged individuals are cut out from the contributions that the marginalised members would make if they were present. They are thus unable, but often also unwilling, to hear criticism of their position and to be exposed to alternative points of view. This situation means that those whose views are left unchallenged because dissenting voices are marginalised find themselves in a situation where they might end up believing that they know it all.[31] They might reach this conclusion because they fail to notice that the absence of disagreement in their milieu is not due to the non-existence of alternatives but to the exclusion of dissenting viewpoints.[32] Thus, since being a know-it-all is another trademark of arrogance, structural relation of marginalisation, like those of exploitation, facilitate the formation of arrogant attitudes in privileged individuals.[33]

The effects of marginalisation on the marginalised are more complex. Being cut out from the epistemic community could result in the epistemic impoverishment of the marginalised. But this effect is likely to be found more acutely only among individuals unable to join counter-communities. Those who are truly isolated are denied the means to learn from others. Their epistemic capacities would thus be atrophied. Hence, it would not be surprising if they would become fatalistic about their alleged lack of ability, and timid in their contributions. In addition, the exclusion from the mainstream is bound to create a loss of self-confidence and a tendency to remain quiet in order to avoid the intimidating experience of being shut out. When this happens, individuals first bite their tongue. Subsequently, however, they might rationalise their behaviour and come to believe that they have nothing to say (Tanesini 2016a). When this occurs, oppressed individuals are at risk of becoming intellectually timid.

---

31 Medina (2013) describes this phenomenon metaphorically as lack of friction.
32 This failure is a result of defensive reasoning that is characteristic of arrogant people.
33 See Lynch (2019) for the view that arrogant people think they know it all.

Some marginalised groups have a long and proud history of resistance. Members of these groups have created alternative and lively epistemic communities in which they have access to each other's contributions. Further, since there is no escaping the perspective of the dominant, they will be able to compare different viewpoints.[34] Because of the mutual support they gain from community membership, these individuals are not at an enhanced risk of timidity. On the contrary, their ability to see things from multiple perspectives brings to them the distinctive epistemic advantages of double consciousness.[35]

3.3 Epistemic Powerlessness

People become beneficiaries of others' powerlessness when they gain from others' loss of intellectual autonomy and authoritativeness. These gains differ depending on if one is directly instrumental in others' powerlessness because one exercises control over their epistemic lives or whether one gains indirectly from, and is implicitly complicit with, this state of affairs. In the first case, privileged individuals experience a boost to their sense of powerfulness, autonomy and authoritativeness that comes from the ability to order others around.[36] This increase is not merely apparent; power structures genuinely increase the authority of their beneficiaries without necessarily conferring legitimacy to that authority.

Power creates a novel set of expectations in those who are powerful. They expect, in the sense of predict, that others follow their orders, but they are also likely to have normative expectations that others should act as they want them to. Thus, powerful individuals take others to be accountable to them. However, these power relations are not symmetric. Powerful individuals predict that they are not accountable to those to whom they issue orders. As these expectations turn into habits, they are likely to become normative. Powerful individuals thus behave as if they were wholly unaccountable. These same expectations are also likely to engender a sense of invulnerability, of complete self-reliance and of total autonomy, provided no one else has the power to tell them what to do.

Unequal power relations are also likely to generate incentives in the powerful to protect their power. The motivation to defend one's powerful position

---

34 Some of these themes have been explored by Medina (2013).

35 See Medina (2013) for arguments in favour of, whenever possible, bringing into dialogue multiple perspectives within oneself.

36 The benefits to those who are indirect beneficiary of others' powerlessness are very similar to those gained by being privileged in relations of exploitation and marginalisation.

would seem to be a consequence of the pleasure one gains from the ability to do as one pleases. Hence, these power relations are likely to reinforce the tendency of the privileged to engage in defensive motivated cognition. Finally, these same relations bias the cognitive environment by supplying misleading evidence that might inflate privileged self-evaluation of one's own abilities.

One psychological consequence of this kind of power is the deepening of intellectual arrogance into its hubristic version. This is the arrogance that is characteristic of those who feel invulnerable and wholly self-reliant. It is typical of those who take themselves to be so unaccountable that they are under the spell of the illusion that their views are correct simply because they believe them. I compare this position elsewhere to that of the umpire who announces that the batter is out because *he* (the umpire) calls him out (Tanesini 2016a).

By contrast, the experience of powerlessness in the long run creates expectations of failure, disappointment and frustration. These are highly aversive (unpleasant) experiences. Predicting that one is repeatedly subjected to them is utterly demoralising. Thus, powerlessness like marginalisation can engender an attitude of fatalism or resignation. It supplies incentives not to avail oneself of opportunities out of fear that one might fail. Thus, feelings of powerlessness might be a cause of self-silencing (London et al. 2012). These are among the hallmarks of intellectual timidity. It is not surprising that marginalisation and powerlessness have similar consequences on the psychology of the oppressed since these are structural relations that deny their targets agentic status within the epistemic community.

3.4 Epistemic Subordination (Cultural Imperialism)

Members of dominant social groups grow up learning that they are the standard by which others are to be judged. This point is aptly put by Elizabeth Spelman when she writes that in the US white children were told 'by well-meaning white adults that Black people were just like [them]' but they never heard it said that '[they] were just like Blacks' (Spelman 1990, p. 12). She labels this phenomenon 'boomerang vision' because when members of dominant groups look at others all that they see are (usually defective) versions of themselves. As Medina (2013) further observes, this inability to understand other people in their own terms is epistemically debilitating. It is also instrumental in fostering narcissism as well as arrogance in members of dominant groups.

Dominant individuals are at risk of developing narcissistic tendencies because they alone are positively affirmed in the cultural stereotypes circulating in their communities. That is, they see that the images of the kind of person that their societies value only include people like themselves. Dominant children, thus, learn to normatively expect to be the standard of comparison by which the worth of others is to be judged. Hence, they are educated into the narcissist posture of thinking of some feature as admirable because they possess it (rather than thinking of themselves as admirable only to the extent to which they possess some admirable feature). This training promotes motivated cognition when making judgments about one's abilities based on social comparisons.[37] Children who belong to dominant groups, are not habituated to compare themselves accurately to others, but to judge other people in ways that take the dominant's dominance for granted. This tendency to think of oneself as the yardstick of what is valuable is one of the defining characteristics of narcissism as a character vice (Tanesini 2021, ch. 6).

The effects of subordination on the psychology of the oppressed are varied but they can be devastating. Writing early in the twentieth century, Aime Césaire (1972) and Frantz Fanon (1986) have described the effects of colonialism on those it subordinated as causing an 'inferiority complex' (Fanon 1986, pp. 13, 42, 213–216). The characteristics of this complex are those that define intellectual servility including abasement, servility and despair (Césaire 1972, p. 22). Fanon (1986, p. 100) also details the desire to become white and to be approved of, and accepted by, white society.

The structures of cultural imperialism create a cognitive environment that is biased to indicate the superiority of some groups and the inferiority of others that are portrayed as the Other. Those oppressed individuals who do not have the double vision invoked by W. E. B. Du Bois (1990) only measure their worth by the tape of the dominant stereotypes and images. They thus see themselves as second rate and crave the approval of dominant society. The desire to be accepted in turn promotes motivated cognition that further exacerbates one's self-assessment as inferior. In this manner, cultural imperialism puts subordinated people at a heightened risk of losing self-respect.

3.5 Epistemic Violence

I have argued above that one of the effects of epistemic violence on those upon whom it is inflicted is the development of an enhanced sense of vulnerability and a loss of self-trust. By contrast, those who benefit from others' insecurities are able to impose their points of view without having to face challenges.

---

37 See Corcoran et al. (2011) for an overview of the motivations that inform these comparisons.

An epistemic environment in which some are subjected to widespread intimidation because they are shouted down, mocked or dismissed, gives vulnerable individuals an incentive to keep quiet and self-silence. It also promotes in them a legitimate fear of being subjected to similar treatment in future. Further, if the threat of violence appears to be widely tolerated, the realisation that the dominant majority is complicit in the maintenance of the status quo is likely to engender a sense of hopelessness that further exacerbates a propensity to self-silence and to keep one's head down. Thus, epistemic violence gives to vulnerable individuals incentives to become timid. Further, the widespread acceptance of the harsh behaviour that is directed at them could even induce some to believe that such treatment is warranted. In these cases, individuals' confidence in their own abilities might be severely eroded.

Not all privileged individuals are likely to engage in the kind of bullying behaviour that creates a threatening and hostile environment. Nevertheless, because of its existence, these individuals face less dissent and competition for positions of prominence in epistemic communities. Since this state of affairs is advantageous, it provides incentives to ignore that one's gains are unfairly obtained. Out of a desire to think well of oneself, privileged individuals are thus at heightened risk of engaging in defensive motivated cognition that helps them to believe that their successes are due to ability. This is the kind of inflated and defensive self-esteem that, as I argued above, is characteristic of arrogance.

These tendencies are deepened in those privileged individuals who are bullies. They act with impunity because they expect, in the sense of predict, that others will not call them to account for their actions. When these expectations become habitual, these individuals might behave as if they are unaccountable. They are thus likely to develop hubristic arrogance.

4 Concluding Remarks

This chapter has shown that the four intellectual character vices of arrogance, narcissism, timidity and servility can be caused by structural relations of oppression. On the one hand, exploitation, marginalisation, powerlessness and violence facilitate the formation of arrogance in the privileged whilst cultural imperialism fosters narcissistic tendencies in members of dominant groups. On the other hand, exploitation, powerlessness and cultural imperialism can lead oppressed people to suffer from intellectual servility, while marginalisation and violence promote timidity.

This chapter has focused on the causal role of structural relations of oppression in the formation of distinctive psychologies of privilege and of oppression.

But it has also offered some of the conceptual tools necessary to understand the causal roles of these psychological tendencies in the reproduction of oppressive structures. While I cannot fully substantiate the point here, I hope that one example will suffice to demonstrate the fruitfulness of the approach adopted here. I have argued above that relations of epistemic exploitation can generate in those who benefit from it normative expectations that they are fully deserving of preferential treatment. Those who arrogate exploitative entitlements, I have claimed, also engage in motivated cognition whose function is either to assert that one's privilege is earned or to deny that one has been privileged. These have been described respectively as the maintenance and the innocence motive (Phillips and Lowery 2018). Individuals who have acquired these, and are therefore arrogant, play a driving role in the perpetuation of the exploitative system. In particular, they are causally responsible, together with purely structural phenomena such as segregation, for the creation of a kind of herd invisibility that hides the true nature of exploitation. In this manner, other privileged individuals who are complacent about their privileges, but lack the motivations characteristic of arrogance can benefit from structural oppression whilst remaining unaware of its existence. The idea that some individuals play a pivotal role in the maintenance of oppression and that motivated cognition is a distinctive feature of their psychology has received some empirical support in the case of white privilege (Phillips and Lowery 2018). If these ideas are on the right track, there is a feedback loop between structural oppressive relations and individual vices.[38] While structures promote the development of vices of arrogance or servility in some individuals, the presence of people who have acquired the motivations characteristic of these vices is a distinctive driver in the reproduction of these structural relations.

These considerations also offer some suggestions for how to think about interventions aimed to reduce epistemic injustices. If, as I have proposed, injustices are reproduced by means of reinforcement loops between individual psychologies and structures, then we should aim to intervene at all stages in these loops. Some interventions will require collective action to impact on existing structures. These measures will seek to address misinformation and barriers to evidence and to change the incentives that guide individual decision making. These inter-

---

38 It might be remarked that there are intermediary entities between individuals and structures that also play significant causal roles in these processes. These include government, institutions, groups and civil associations, to name but a few. This is undoubtedly true and their roles in these processes deserves further exploration that I cannot hope to offer in this chapter. But see Medina (2022) for an obstructivist account of institutional and collective epistemic vices. Thanks to the editors for pressing this point.

ventions might be efficacious in changing the behaviour of those who are complicit with, or complacent about, oppressive structures without being actively motivated to preserve them. However, different interventions would also be required targeting individuals who to different degrees have acquired the motivations typical of epistemic vices.[39] These might include education, the promotion of reflection on the values that are defining of one's identity and exposure to admirable exemplars (Medina 2013; Tanesini 2016b; and Zagzebski 2015).

The arguments developed in this chapter go some way towards illustrating the explanatory power of a vice theoretical framework to understand the effects of epistemic oppression. They also show that this framework is not irredeemably individualistic. On the contrary, these arguments show that individual vices have impacts on structures and that structural relations have effects on individuals' characters. These considerations suggest, but do not establish, that if we want to address oppression, we must take measures that target structures but also adopt strategies to help individuals to improve their intellectual characters. These strategies are likely to involve structural changes but also the creation of collaborative and supportive epistemic communities of resistance.

Bibliography

Abramson, Kate (2014): "Turning up the Lights on Gaslighting". In: *Philosophical Perspectives* 28, pp. 1–30.

Anderson, Elizabeth (2012): "Epistemic Justice as a Virtue of Social Institutions". In: *Social Epistemology* 26. No. 2, pp. 163–174.

Applebaum, Barbara (2017): "Comforting Discomfort as Complicity: White Fragility and the Pursuit of Invulnerability". In: *Hypatia* 32. No. 4, pp. 862–875.

Berenstain, Nora (2016): "Epistemic Exploitation". In: *Ergo, an Open Access Journal of Philosophy* 3, pp. 569–590.

Cassam, Quassim (2019): *Vices of the Mind*. Oxford: Oxford University Press.

Césaire, Aime (1972): *Discourse on Colonialism*. Translated by Joan Pinkham. New York: Monthly Review Press.

Coady, David (2010): "Two Concepts of Epistemic Injustice". In: *Episteme* 7. No. 2, pp. 101–113.

Coady, David (2017): "Epistemic Injustice as Distributive Injustice". In: Ian J. Kidd, José Medina and Gaile Pohlhaus, Jr. (Eds.): *The Routledge Handbook of Epistemic Injustice*. London, New York: Routledge, pp. 61–68.

Collins, Patricia H. (2000): *Black Feminist Thought: Knowledge, Consciousness, and the Politics of Empowerment* (10th Anniversary ed.). New York, London: Routledge.

---

39 These kinds of interventions are best thought as complementary and as interacting with each other.

Cooper, Joel (2008): *Cognitive Dissonance: Fifty Years of a Classic Theory*. Los Angeles, London: SAGE Publications Limited.

Corcoran, Katja, Jan Crusius and Thomas Mussweiler (2011): "Social Comparison: Motives, Standards, and Mechanisms." In: Derek Chadee (Ed.): *Theories in Social Psychology*. Oxford: Wiley Blackwell, pp. 119–139.

Cudd, Ann E. (2006): *Analyzing Oppression*. New York: Oxford University Press.

Cull, Matthew J. (2019): "Dismissive Incomprehension: A Use of Purported Ignorance to Undermine Others". In: *Social Epistemology* 33. No. 3, pp. 262–271.

Dotson, Kristie (2011): "Tracking Epistemic Violence, Tracking Practices of Silencing". In: *Hypatia* 26. No. 2, pp. 236–257.

Dotson, Kristie (2014): "Conceptualizing Epistemic Oppression". In: *Social Epistemology*, 28. No. 2, pp. 115–138.

Du Bois, W. E. B. (1990): *The Souls of Black Folk*. New York: Vintage Books/Library of America.

Fanon, Frantz (1986): *Black Skin, White Masks*. Translated by Charles L. Markmann. London: Pluto Press.

Freeman, Lauren (2015): "Confronting Diminished Epistemic Privilege and Epistemic Injustice in Pregnancy by Challenging a 'Panoptics of the Womb'". In: *Journal of Medicine and Philosophy* 40. No. 1, pp. 44–68.

Fricker, Miranda (1998): "Epistemic Oppression and Epistemic Privilege". In: *Canadian Journal of Philosophy Supplementary Volume* 25, pp. 191–210.

Fricker, Miranda (2007): *Epistemic Injustice: Power and the Ethics of Knowing*. Oxford: Clarendon Press.

Fricker, Miranda (2017): "Evolving Concepts of Epistemic Injustice". In: Ian J. Kidd, José Medina and Gaile Pohlhaus, Jr. (Eds.): *The Routledge Handbook of Epistemic Injustice*. London, New York: Routledge, pp. 53–60.

Haraway, Donna (1991): "Situated knowledges: The Science Question in Feminism and the Privilege of Partial Perspective". In: *Simians, Cyborgs and Women: The Reinvention of Nature*. London: Free press, pp. 150–183.

Harding, Sandra G. (1993a): "Rethinking Standpoint Epistemology: 'What Is Strong Objectivity?'". In: Linda Alcoff and Elizabeth Potter (Eds.): *Feminist Epistemologies*. New York, London: Routledge, pp. 49–82.

Harding, Sandra G. (Ed.) (1993b): *The "Racial" Economy of Science: Toward a Democratic Future*. Bloomington: Indiana University Press.

Harding, Sandra G. (1998): *Is Science Multicultural?: Postcolonialisms, Feminisms, and Epistemologies*. Bloomington: Indiana University Press.

Hartsock, Nancy (1983): "The Feminist Standpoint: Developing the Ground for a Specifically Feminist Historical Materialism". In: Sandra G. Harding and Merril B. Hintikka (Eds.): *Discovering Reality*. Dordrecht: Reidel, pp. 283–310.

Hochschild, Arlie R. (2012): *The Managed Heart: Commercialization of Human Feeling*. Berkeley: University of California Press.

Intemann, Kristen (2010): "25 Years of Feminist Empiricism and Standpoint Theory: Where Are We Now?". In: *Hypatia* 25. No. 4, pp. 778–796.

Jones, Karen (2004): "Trust and Terror". In: Peggy DesAutels and Margaret U. Walker (Eds.): *Moral Psychology: Feminist Ethics and Social Theory*. Lanham: Rowman & Littlefield Publishers, pp. 3–18.

Kahan, Dan M. (2013): "Ideology, Motivated Reasoning, and Cognitive Reflection". In: *Judgment and Decision Making* 8. No. 4, pp. 407–424.

Kidd, Ian J. (2020): "Epistemic Corruption and Social Oppression". In: Ian J. Kidd, Quassim Cassam and Heather Battaly (Eds.): *Vice Epistemology*. London: Routledge, pp. 69–86.

Kidd, Ian J., and Havi Carel (2017): "Epistemic Injustice and Illness". In: *Journal of Applied Philosophy* 34. No. 2, pp. 172–190.

Kunda, Ziva (1990): "The Case for Motivated Reasoning". In: *Psychological Bulletin*, 108. No. 3, pp. 480–498.

Lackey, Jennifer (2020): "False Confessions and Testimonial Injustice". In: *The Journal of Criminal Law & Criminology* 110. No. 1, pp. 43–68.

London, Bonita et al. (2012): "Gender-based rejection sensitivity and academic self-silencing in women". In: *Journal of Personality and Social Psychology* 102. No. 5, pp. 961–979.

Ludwig, David (2018): "How Race Travels: Relating Local and Global Ontologies of Race". In: *Philosophical Studies* 176. No. 10, pp. 2729–2750.

Lupien, Shannon P., Mark D. Seery and Jessica L. Almonte (2010): "Discrepant and Congruent High Self-Esteem: Behavioral Self-Handicapping as a Preemptive Defensive Strategy". In: *Journal of Experimental Social Psychology* 46, pp. 1105–1108.

Lynch, Michael P. (2019): *Know-It-All Society: Truth and Arrogance in Political Culture*. New York, London: Liverlight Publishing Corporation.

Medina, José (2013): *The Epistemology of Resistance: Gender and Racial Oppression, Epistemic Injustice, and Resistant Imaginations*. Oxford, New York: Oxford University Press.

Medina, José. (2017): "Varieties of Hermeneutical Injustice". In: Ian J. Kidd, José Medina and Gaile Pohlhaus, Jr. (Eds.): *The Routledge Handbook of Epistemic Injustice*. London, New York: Routledge, pp. 41–52.

Medina, José (2022): "Political Epistemology". In: David Bordonaba-Plou, Victor Fernández-Castro and José R. Torices (Eds.): *The Political Turn in Analytic Philosophy: Reflections on Social Injustice and Oppression*. Berlin, Boston: De Gruyter Academic Publishing, pp. 55–76.

Mignolo, Walter (2011): "Decolonizing Western Epistemology/Building Decolonial Epistemologies". In: Ada M. Isasi-Diaz and Edoardo Mendieta (Eds.): *Decolonizing Epistemologies: Latina/o Theology and Philosophy*. New York: Fordham University Press, pp. 19–43.

Mills, Charles W. (2007): "White Ignorance". In: Shannon Sullivan and Nancy Tuana (Eds.): *Race and Epistemologies of Ignorance*. Albany: State University of New York Press, pp. 13–38.

Nussbaum, Martha C. (1995): "Objectification". In: *Philosophy & Public Affairs* 24. No. 4, pp. 249–291.

Olson, Philip R. (2016): "Epistemic Burdens and the Value of Ignorance". In: James H. Collier (Ed.): *The Future of Social Epistemology: A Collective Vision*. London, New York: Rowman & Littlefield, pp. 107–116.

Phillips, L. Taylor, and Brian S. Lowery, (2018): "Herd Invisibility: The Psychology of Racial Privilege". In: *Current Directions in Psychological Science* 27. No. 3, pp. 156–162.

Pohlhaus, Jr., Gaile (2011): "Relational Knowing and Epistemic Injustice: Toward a Theory of Willful Hermeneutical Ignorance". In: *Hypatia* 27. No. 4, pp. 715–735.

Schiebinger, Londa (2008): "West Indian Abortifacients and the Making of Ignorance". In: Robert N. Proctor and Londa Schiebinger (Eds.): *Agnotology: The Making and Unmaking of Ignorance*. Stanford: Stanford University Press, pp. 149–162.

Spelman, Elizabeth V. (1990): *Inessential Woman: Problems of Exclusion in Feminist Thought*. London: Women's Press.

Strawson, Peter F. (2008): *Freedom and Resentment and Other Essays*. London: Routledge.

Sullivan, Shannon (2006): *Revealing Whiteness: The Unconscious Habits of Racial Privilege*. Bloomington: Indiana University Press.

Tanesini, Alessandra (2016a): "I – 'Calm Down, Dear': Intellectual Arrogance, Silencing and Ignorance". In: *Aristotelian Society Supplementary Volume* 90. No. 1, pp. 71–92.

Tanesini, Alessandra (2016b): "Teaching Virtue: Changing Attitudes". In: *Logos & Episteme* 7. No. 4, pp. 503–527.

Tanesini, Alessandra (2018a): "Arrogance, Anger and Debate". In: *Symposion: Theoretical and Applied Inquiries in Philosophy and Social Sciences* 5 (Special issue on Skeptical Problems in Political Epistemology, edited by Scott Aikin and Tempest Henning), pp. 213–227.

Tanesini, Alessandra (2018b): "Intellectual Servility and Timidity". In: *Journal of Philosophical Research* 43, pp. 21–41.

Tanesini, Alessandra (2019): "Reducing Arrogance in Public Debate". In: James Arthur (Ed.): *Virtues in the Public Sphere*. London: Routledge, pp. 28–38.

Tanesini, Alessandra (2021): *The Mismeasure of the Self: A Study in Vice Epistemology*. Oxford: Oxford University Press.

Vaditya, Venkatesh (2018): "Social Domination and Epistemic Marginalisation: Towards Methodology of the Oppressed". In: *Social Epistemology* 32. No. 4, pp. 272–285.

Veselka, Livia et al. (2010): "Phenotypic and Genetic Relations Between the HEXACO Dimensions and Trait Emotional Intelligence". In: *Twin Research and Human Genetics* 13. No. 1, pp. 66–71.

Young, Iris M. (1990): *Justice and the Politics of Difference*. Princeton: Princeton University Press.

Young, Iris M. (1997): "Unruly Categories: A Critique of Nancy Fraser's Dual Systems Theory". In: *New Left Review* 1/222, pp. 147–160.

Zagzebski, Linda T. (2015): "I-Admiration and the Admirable". In: *Aristotelian Society Supplementary Volume* 89. No. 1, pp. 205–221.

Manuel de Pinedo and Neftalí Villanueva
Epistemic De-Platforming

Abstract: The goal of this paper is to argue for a particular epistemic policy, epistemic de-platforming, according to which nobody is a priori entitled to automatically turn their assertions into epistemically relevant alternatives, with the power to question what we know. We start with a presentation of our take on the political turn in analytic philosophy: what characterizes this trend is that philosophical concepts and theories are evaluated according to their power to help detecting hidden forms of injustice (generalizing over our pre-theoretical perceptions of injustice) and intervening to alleviate them. Building on our conception of the political turn, we highlight two epistemic paradoxes. One amounts to the idea that agents subjected to an excess of epistemic friction, for instance, by not receiving the credibility or authority that they deserve, may be said to simultaneously increase their epistemic standing and suffer an epistemic harm. The other is related to the sound attitude of disregarding any evidence that goes against what we know, leading to the allegedly dogmatic stance of rejecting beforehand future evidence and making our present knowledge an obstacle to the future acquisition of further knowledge. We contrast some epistemic policies in terms of their capacity to offer a way out from these paradoxes. Neither dogmatism nor libertarianism take into account which are, and which are not, the contextually relevant epistemic alternatives that should be considered. As an example of a pluralistic, context-dependent, epistemic policy, we explore epistemic de-platforming, a policy that, in line with our characterization of the political turn, entitles agents to ignore possibilities presented by groups whose aim is to further oppressive agendas, while recommends them to always take into account the ideas and opinions of those who have earned the right to be heard by them: their friends, allies, and loved ones.

1 Introduction

We cannot but feel perplexity on the face of the following contrast in contemporary epistemology. A lot of epistemologists have extracted the following lessons from contextualism:

1. You are driving through an isolated road, trees on each side. You take a look to the left and see a barn – you know there is a barn. But if you are crossing through Fake-Barn County, a remote countryside location where fake barns have mushroomed out, then your quick look to the side of the road is not

https://doi.org/10.1515/9783110612318-007

evidence enough to discard the possibility that what you are seeing is a fake barn, rather than a real barn. Thus, you do not know that there is a barn. Furthermore, if anyone even wonders whether it was Fake-Barn County that you were driving through and you cannot rule it out, you cannot claim to have knowledge that there is a barn in from of you either.

2. You go and see an eye doctor. The doctor observes through a machine while she places a lens in front of your left eye. 'What colour did you see?', she asks. 'Blue', you reply. 'Are you sure?' 'Positive'. 'Well, if it really is blue you saw, you must undertake immediate surgery, it will be a very painful procedure for which anaesthesia cannot be used . . . are you sure it was blue that you saw?' 'Please, let me check again, I am not completely sure, after all'. When you risk your rent money, you must fold'em.

3. Given what you believe about the solar system and the properties of the planets, the possibility that the Earth is flat is very, very low. Only a huge conspiracy would have been able to instil such a series of beliefs incompatible with the Earth being flat. It is only by chance that you received this kind of education, though, and it is not your epistemic merit to have enjoyed the educational system of a Western country in the twenty-first century. You are, nonetheless, perfectly justified in paying no attention whatsoever to flat-earthers, no matter how much they yell at you from their YouTube quarters.

In contrast, what follows is taken to be bad epistemology:

> The idea behind the old Kennedy maxim, "a rising tide lifts all boats", was embraced, and this became the dominant liberal response to any demand for a "black agenda" or directly "anti-racist" program. What passed for anti-racist programs – affirmative action, for instance – came under attack, not just from conservatives, but from liberals who believed that the true vector for attacking all that bedeviled African Americans was class, not race.

> Affinity to the "rising tide" theory was genuinely felt. It also offered certain advantages. Though often proffered by the self-styled "New Democrats," the theory connected to the ancient leftist dream of a broad coalition of working people. For those interested in electoral politics, the "rising tide" theory meant never having to confront white voters, still the mass of voters, with the weight of ancestral sins and all the privileges accrued from them. If race was declining in significance, then there really would be no need to talk about it. All one need do is urge the white working class, so often cruelly tricked into acting "against their interest," to see that they had cast their lot with their oppressors.

> The promise of a cost-free escape from history should have made me suspicious. But what ultimately made me question the "rising tide" idea was not the theory itself but all the attendant theories that so often went with it. There is a long tradition among liberal intellectuals, and even among black intellectuals, of insisting that some amount of the racial chasm is the fault of black people themselves. (Coates 2017, p. 154)

It seems bad epistemological practice to reject any view conflicting with our own simply because of the group that massively holds that view, or because of the adjacent beliefs that such a view is systematically accompanied by. However, why is the mere possibility that someone has planted fake-barn façades or that the stakes are high a perfectly rational trigger for doubting my belief that I saw a barn or that I saw blue, but the possibility of enlarging a group of neo-fascists is of no epistemological significance? And in the third example above, despite luck playing an essential part in my acquisition of the belief that the Earth is round, how can it be that I am entitled to have my belief to the point that it constitutes knowledge, while it is bad practice to epistemically rely on the luck of having been educated to despise fascist or racist worldviews? We want to argue that, far from being bad epistemology, rejecting or ignoring views pressed by groups with oppressive agendas or accompanied by discriminatory beliefs is both good epistemology and good politics, and it is no coincidence that it is both at the same time. A theory, such as the rising tide theory mentioned above, may be good on its own merits: if things improve globally for the worst off, everyone is in an equal position to benefit and, hence, they should all join in the common cause. However, we will recommend as an epistemic policy that you are entitled to ignore or oppose a theory if it is associated with people and with other theories that you have every reason to distrust. The political turn, as we outline it in section 2, advises us to embrace theories by their capacity to bring our focus towards hidden forms of injustice as well as for its potential to intervene on them. And, as in Coates's quote, the political turn also advises us to reject those theories that do the opposite.

What seems to go against our proposal is that following it as a policy can have a negative effect on our epistemic practices. We may miss information, arguments or evidence that could improve what we know. To counterbalance this possibility, we will also put forward a positive policy: in line with the idea that, in order for your opinions and arguments to deserve public attention, you need to earn the right to be heard, we will argue that your friends and allies and loved ones already possess that right by default. Trusting those that are closer to you can give you access to ideas that you would otherwise have all the reason to ignore if they only reach you through suspicious sources. We welcome a consequence of the criticism mentioned in the previous paragraph: neither of the two policies are universally conducive to improving our epistemic standing. As we shall see, this is also something implied by the Wittgensteinian understanding of rule-following that we believe sustains the political turn: policies are not virtuous or vicious considered in the abstract, but their worth has to be assessed contextually. And that also goes for the positive and negative policies that we discuss in this paper.

In section 3, we argue that our proposal is in a very good position to overcome some deep epistemological paradoxes. One of them is the following: on the one hand, as Almagro, Navarro, and Pinedo (2021) have argued, all forms of epistemic injustice affect the victim not only as a giver of knowledge but also as an enquirer and as a possessor of knowledge. There are many ways in which you know less if you suffer from epistemic injustice. Nevertheless, as José Medina (2012) has forcefully argued, (epistemic) oppression may result in the oppressed being epistemically better off than the oppressor. First, their own survival may depend on knowing how things look from the top and what the privileged expect from them. Second, and this is where the paradox may be felt more strongly, being constantly questioned, doubted or dismissed may provide an extra epistemic friction and put the subject in a better position to acquire knowledge and to improve their capacity to justify and offer reasons in favour of their beliefs. The paradox, in a nutshell, is that the victims of epistemic injustice receive less but also more epistemic friction, they are allowed to participate in fewer knowledge-related practices but, at the same time, they receive a more varied diet of perspectives.

One of the advantages of the position that we will present in the chapter is that it allows us to respond to this challenge without abandoning any of the apparently conflicting insights, as we well see in section 4. Firstly, we can defuse the tension by admitting, much in line with our general argument, that the epistemic advantage of the oppressed should not be overestimated: to overcome the multiple obstacles that a situation of economic, social and epistemic oppression brings about involves a great amount of work and, often, epistemic luck, for instance, being surrounded by a sympathetic (marginal) epistemic community where experiences can be shared. Secondly, we can show that the very same instance of epistemic injustice may have radically different effects on the subject with respect to the acquisition, maintenance, improvement and sharing of knowledge. This is the tragic nature of epistemic friction: being systematically disbelieved may lead the subject to doubt what she previously believed (and knew), but also to have access to a higher variety of sources of knowledge. She may lose knowledge and, at the same time, acquire better knowledge-conducing habits. In some cases, quite the opposite may happen, she may stubbornly retain her belief against unfair and prejudicial doubts but, also, end up developing an unhealthy uncertainty towards her own thinking.[1] Besides, as we will

---

1 It is no accident that Amia Srinivasan's examples in favour of externalism are politically loaded versions of neutral cases traditionally used in favour of internalism (Srinivasan 2020). In her examples, involving racist, classist and sexist features, people's knowledge or lack thereof regarding such features is not determined by their being able to justify their beliefs or by dogma-

see in section 3, the effects on a single one of the dimensions of knowledge might also be mixed. Being exposed to many different alternatives might make a better inquirer of you, one that is able to assess and discard a great number of possibilities, and therefore acquire better quality knowledge. But it can also make you more polarized. The systematic exposure to opposing views has been shown to be a source of polarization, a way through which subjects end up attributing way too high subjective probability to the beliefs that are part of the core of the political – social or otherwise – identity that is active in that particular context. A more polarized subject, in this sense, is one that becomes impervious to the other's reasons, and thus becomes a less effective enquirer.

2 The Political Turn in Analytic Philosophy

From the inception of the tradition, there has been no shortage of political engagement from analytic philosophers. Without dwelling on what differentiates analytic philosophy from other schools, whether the themes, methods, underlying metaphysical commitments or even bedside reading, it is clear that some of the most recognizable amongst the foundational figures (e. g., Russell and the members of the Vienna Circle) saw their philosophical work as an instrument at the service of transforming the social world. One could even argue that analytic philosophy suffered an *a*political turn after the Second World War when its centre of influence moved from Europe to America. Those philosophers who were forced to leave continental Europe, mostly analytic ones, had to pay the price of becoming less politically vocal in their land of adoption (Reisch 2005, pp. 370 – 388).[2] In a sense, what the CIA and the witch hunt took away, feminism and the civil rights movement have given back. As the introduction to this volume highlights, by political turn we mean something more specific than either the focus on political issues from a philosophical perspective (as we can find in Bernard Williams, John Rawls or Martha Nussbaum) or the emphasis on practical, political matters by philosophers of the analytic tradition, such as Michael Dummett's work on the rights of immigrants and refugees, Noam Chomsky's political writings or even Jason Stanley's books on propaganda and fascism. In

---

tism and stubbornness on their part. Being dogmatic or stubborn is compatible with having knowledge. In contrast, having done one's epistemic best is compatible with not knowing.
2 Although there are some extremely notorious examples, we do not want to suggest that philosophers from the varied traditions that fall under the label continental could remain in Central Europe because they shared sympathies with the totalitarian regimes that emerged in the 1930s. We are convinced that there are other more complex reasons to explain this.

these cases, the interest in and contribution to political issues can be perfectly understood independently of the nature of their academic work, despite some noble attempts to show the connection between one and the other. You can wholeheartedly commit to a Russellian theory of proper names, the unity of science, the existence of a universal innate grammar or semantic anti-realism and be politically indifferent or even profoundly reactionary (we are sure that each reader will be able to find examples that fit these patterns).

Because the guiding principle of the political turn is to always keep in mind the project of detecting and intervening on actual oppressive practices, one of the most common features of work that fits this pattern is the shift from a psychologistic and individualist attention to the mental states that underlie instances of oppression (say, prejudices, biases, racist or sexist beliefs and attitudes, and the like) to a more structural take that emphasizes social practices and structures. This, in turn, leads us directly to embrace the thesis that our normative assessments cannot be replaced by other forms of explanation. For instance, in contrast with mainstream philosophy of mind, where the identification of mental states in materialist or functionalist terms has dominated (something analogous could be claimed about the notion of literal meaning in the philosophy of language, or the search for the internal or external conditions for the attribution of knowledge in epistemology), the emphasis on the social and normative aspects of specific instances of injustice has come to the fore as a consequence of a radical change in epistemic goals: questions such as what is a belief?, what is meaning? or what is justification? are being replaced by a focus on detecting and intervening on hidden forms of injustice. The assessment of the relative value of our theoretical tools for analysis is thus a normative one, since our new epistemic goals are entangled with practical political interests. Throughout the paper, we will insist on several roles played by the normative with respect to our evaluation of theories in terms of their liberating potential. On the one hand, some distinction between evaluative and non-evaluative uses of language will be needed to unmask attempts at presenting as merely technical questions that involve values, norms and interests or, by contrast, presenting as matters of opinion issues that have already been firmly established by science. On the other hand, there is an inescapable normative dimension to our choice of theories if these must prove their worth with respect to their power to respond to our perceptions of injustice by generalizing upon them and offering tools to intervene. And, finally, this choice will not be made in terms of a pre-established rank of values or explicit norms, but it will always depend on contextually salient practical, moral and political demands: when we engage in evaluative debates, we only exceptionally discuss our standards and norms rather than whatever thing or situation we are judging. Furthermore, having different values is not an obstacle for evaluative agreement, nor does sharing values imply agreement.

As we mentioned above, what we take to be the defining novelty of the recent political turn in analytic philosophy is the centrality of the practical utility of our philosophical tools in bringing to the surface forms of injustice that may have remained invisible and, whenever possible and no matter how modestly, in contributing to the improvement of the social, political and epistemic standing of the oppressed. In this sense, we believe that our philosophical instruments are as good as the role that they can play in the identification of social and economic inequalities and in supporting the project of intervening to eradicate or alleviate them. The force of this political drive can be felt in many areas of philosophical reflection. In the rest of this section, we offer a brief presentation of the philosophical fields where we can localize strong currents within the political turn and which play an important role for our argumentation. We will mainly focus on recent ideas within epistemology (2.1), on the conception of the mind that we take to ground the turn (2.2), and on the approach to normativity that makes room for our proposal (2.3).

2.1 Epistemology

During the last fifteen years, epistemology, which will be the main focus of this chapter, has witnessed a dramatic departure from abstract, individualistic concerns towards social issues related to the ethics of knowing. As some authors keep reminding us, the idea that knowledge is essentially related to politics and morals has been close to the surface from the beginning of the European philosophical tradition (see, e. g., Moreno Pestaña 2019 and Broncano 2020). Miranda Fricker (2007) is a well-known example of the recent explicit concentration on this relationship: her commitment to a form of virtue ethics always seems parasitic on her project of identifying forms of epistemic oppression and injustice that were much less visible, if visible at all, before her work. Within the philosophy of language, the objective of bringing to light analogue discursive and linguistic forms of injustice is at the centre of recent work on speech act theory (see, e. g., Kukla 2014 and Ayala 2016). It matters not whether we can accommodate our diagnosis of injustice within the best theory in the market, be it Austin's or someone else's, but rather which are the most fruitful concepts for our practical political concerns. Even in the philosophy of mind, long suspected to be the least political area of philosophy, debates regarding the individualistic or externalist character of mental states are giving way to an approach that emphasises the structural conditions that determine our attributions of such things as implicit attitudes or prejudices (see Ayala and Vasilyeva 2016; Haslanger 2016; and McGeer 2015). Even in metaphysics, gender, race and their intersection are amongst the hottest topics in recent years (see Díaz León 2018 and Jorba and Rodó-Zárate 2019).

The arguments of this chapter are mainly relevant to our understanding of knowledge: in particular, we offer an overview of the type of contextual possibilities that should either be taken into account or ignored in assessing whether to consider something as knowledge. Our aim is to promote and facilitate forms of epistemic resistance that manage to increase the number of opinions and points of view that are opened to discussion precisely, contrary to many mainstream views, by recommending not paying attention to certain ideas. We take it that no reasonable defender of freedom of speech could be happy with the consequence that, given that opinions cannot ever be silenced, if someone has enough power, his opinion will be heard over and over and over again. Sometimes defending free speech involves shutting up for a while so that others may finally express themselves. But, before we set up this epistemology argument, we would like to insist on two issues, one about how to understand our use of mental vocabulary, at least in this context and, relatedly, one regarding the role of normativity in the political turn.

2.2 Philosophy of Mind

A broad variety of theoretical positions regarding they way that we talk about the mental and, more generally, our normative discourse, ranging from eliminativism to realism, assume that there must be some (mental, normative) entities, properties or facts that we refer to if what we say when we explain behaviour in mental terms – or when we evaluate something as correct or incorrect – is suitable to be true or false. An initial worry with this sort of ontological commitment is that, if we cannot know such mental or normative facts, we may need to suspend our judgement: our disagreements regarding how someone thinks, why they act the way they do or whether a given situation is fair or unjust, inasmuch as this is a function of intentions, will have to be put on hold until such facts are discovered. If the function of our mental vocabulary is understood as aiming to refer to mental facts, we may end up having to admit that we can never know what is going on mentally beyond our own skulls. This is not to say that facts are not relevant for our normative assessments. For instance, we may be more inclined to give moral consideration to creatures after discovering that they communicate (e.g., plants, see Baluška, Mancuso and Volkmann 2006 and Calvo 2016). But, in any case, facts of this kind have a normative impact only inasmuch as we antecedently take beings who communicate to be endowed with some (moral) dignity. There is no direct, immediate path from facts to evaluations.

In this sense, namely, thinking of the political turn in relation to the philosophy of mind, mental state attribution is a fundamentally normative rather than descriptive task. It has to do with what we expect people to do or with what it makes sense to think that they believe or want given how they act: normative aspects of the situations a person finds herself in are part and parcel of her mental life. This is one of the ineliminable roles of normativity when we appeal to psychological aspects in diagnosing cases of injustice: there is no mechanical explanation going from mental facts to unjust situations. Without a descriptivist commitment regarding the mind, the separation between psychological and structural approaches to injustice becomes blurry. Think of racism: debates regarding where we should put our emphasis, whether in changing people's minds, changing their explicit racist beliefs or other kinds of mental states, such as implicit biases or prejudices, or on the social conditions and norms that give rise to instances of discrimination, often assume that mental states are understandable independently of the social normative context within which subjects are placed. Even more ecumenical stands on this issue, which demand both individual/psychological and structural/normative interventions, often assume that they are two separate spheres. But once the idea that being minded is nothing over and above being recognized by others as capable of adopting commitments, and of taking responsibility for one's actions and thoughts, the possession of psychological states needs to be viewed as an aspect of being situated within a social normative network.

This shift, from an ontological to a normative conception of the mind, has been traced back to Kant's replacement of the mind/body dualism by the person/object dualism. For instance, Bjørn Ramberg (2000) attributes such a shift to Donald Davidson. The presumption that those whose behaviour we understand by means of mental and intentional vocabulary are, all in all, rational agents, means that we do not treat them as subject to prediction and control, but rather as having a dignity that demands understanding rather than mechanical explanation.[3] This is no ordinary, theoretical move, according to Ramberg.

3 The central role of normative considerations for our understanding of mind and cognition has been a hot topic in recent discussions of situated normativity and unreflective action, non-factualist and expressivist conceptions of the mind, and the regulative character of mental-state ascriptions. Some examples can be found in Rietveld (2008); McDowell (2007); Heras Escribano and Pinedo (2018); Frápolli and Villanueva (2012); McGeer (2007); and Fernández Castro and Heras Escribano (2019).

Much in line with the characterization of the political turn that we and the editors of the volume embrace, he claims:

> By contrast, if we describe this tool – construe this vocabulary – as of a piece with natural science, we might make ourselves more likely to employ such vocabularies for purposes where, so I think, the vocabulary of agency leaves us better off, better in the sense of "politically more free". I see Davidson as providing a tool, a marginal tool, to be sure, since he is a theorist, in a struggle against the steady spread of dehumanizing, homogenizing management of human existence that is the real threat of scientism. Scientism is not bad, I am sure Rorty would agree, because it gets the world wrong, or even because it is a rehash of Kantian and Platonic ontology, but because it renders us subject to certain forms of oppression. (Ramberg 2000, p. 367)

2.3 Rule following

Not any characterization of the normative will do, though. Davidson was, we believe, on the right track when he insisted that our interpretation of behaviour, linguistic or otherwise, other people's or our own, can at best depend on a provisional theory, a theory that, furthermore, incorporates the very behaviour that is being interpreted rather than being ready to be applied in advance (Davidson 1986). It is, however, in Wittgenstein's (1953) discussion of rule following and in Kripke's (1982) controversial but paradigm-shifting interpretation of it where we think that the politically more powerful presentation of the idea can be found. In §§201–202 of *Philosophical Investigations*, Wittgenstein presents an apparent paradox (if every course of action can be made to accord with a rule, there would be no possibility of either accord or of conflict) as one horn of a false dilemma the other horn of which is a commitment to the need of practice-transcendent norms to ground our normative judgements. This is how the dilemma is meant to look: without such universal, independently established rules, we are condemned to a never-ending regress of interpretations: on this take, normativity has more to do with negotiating the rules than with evaluating particular actions, situations or things. Both horns share the problematic premise that a fully determinate, explicit rule must be part of our assertions if we are to make sense of the distinction between right and wrong.

The consequences of thinking of normative uses of language in these terms are, we believe, politically disastrous (see Pinedo 2020). If every evaluative statement includes, one way or another, a reference to standards (whether the speaker's, the community's or whoever's), a twofold danger emerges: evaluations risk becoming descriptions (to say that according to my values such and such is correct seems very different to saying that such and such is correct) and people with different value systems or cultural backgrounds or, simply, with a different taste, could at most engage on explorations of each other's general principles, but never discuss whether an action is cruel, a movie funny or a rice pudding too dry (see Carmona and Villanueva Manuscript). Perhaps we can escape from the first danger: Saul Kripke's discussion shows that to elucidate what counts as correct or incorrect according to a rule is no factual matter, it is itself a normative question. However, it is hard to see how to avoid the second danger. We do agree and disagree regarding normative matters and, most of the time, we are not interested in general rules, but in particular actions, movies or dishes. And we do engage in productive normative arguments with people that are far from sharing a worldview with us – for instance, with children. This way of understanding the nature of evaluative practices as not depending on general principles or rules is, in our view, both the key to the relationship between the political turn that we welcome in this paper and the pride of place we give to the practical utility of philosophical instrument over their purely theoretical virtues. Furthermore, we will claim that the policies that we defend, epistemic de-platforming and the need to see our friends as companions in the search for knowledge and justice, are not commendable because they systematically lead to virtuous ways of thinking, but because they will show their value in a great variety of importantly political and epistemic contexts.

3 Two Paradoxes

We want to expand on the challenge that we presented at the beginning of this chapter by focusing on two different paradoxes that concern our concept of knowledge. They both hinge on the way in which the acquisition of new knowledge via the assessment of new evidence interacts with the knowledge that we already have. As we will see, socio-normative differences might introduce differences in the epistemological status of agents. These differences influence the way you tackle the acquisition of new knowledge, but also your ability to retain the knowledge that you already possess.

The first epistemic paradox that we want to pay attention to we dubbed the Medina-ANP paradox. Epistemic agents need to get their views questioned on a regular basis, yet this can be too much – in certain situations, lacking the appropriate respect for the things that you know might ultimately result in the loss of your knowledge. Not only does your confidence become diminished when you are constantly questioned, but in a more basic sense, you can only have knowledge when you are met with some degree of recognition (see Frápolli 2019, p. 561).

3.1 Medina-ANP Paradox

Disenfranchised agents are subjected to a higher degree of epistemic friction – for example, as a consequence of their receiving less credibility. This might provide them with some epistemic advantage, one that more privileged agents cannot access (Medina 2012). At the same time, testimonial injustice – systematic attribution of low credibility to members of an already discriminated group, for the kind of reasons that chase them in other spheres of activity – ultimately robs agents of their knowledge. There is no knowledge without acknowledgement (see Almagro, Navarro and Pinedo 2021).

In a socio-normative space, opportunities and mishaps are unevenly distributed. What can be done from every node in that space is severely limited by the socio-economic conditions associated with that position, but also by the norms that are in place. One of these unevenly distributed resources is friction. On the one hand, the amount of pushback that you receive from other nodes, the pressure to change your previously held beliefs, is inversely proportional to your privilege. This friction between different perspectives is essential for non-ideal forms of democracy (see Medina 2012, p. 11), and is also epistemically beneficial. This provides an epistemic advantage to the disenfranchised.

On the other hand, when your perspective is systematically put into question, friction might turn against your ability to maintain the knowledge that you have properly acquired. As Medina also recognizes, friction can be epistemically detrimental:

> the external epistemic resistances one encounters can be positive insofar as they offer beneficial epistemic friction, forcing one to be self-critical, to compare and contrast one's beliefs, to meet justificatory demands, to recognize cognitive gaps, and so on; but they can also be negative insofar as they produce detrimental epistemic friction, censoring, silencing, or inhibiting the formation of beliefs, the articulations of doubts, the formulation of questions and lines of inquiry, and so on. (Medina 2012, p. 50)

As Almagro, Navarro and Pinedo argue, knowledge requires acknowledgement. It is not possible to stand alone, an epistemic hero, on the hill of the things that you know, completely unrecognized by the rest of the world. Knowledge is a social product, and as such cannot be held in isolation. When friction becomes too rough, agents might lack the appropriate recognition to maintain the knowledge that they possessed, which is epistemically detrimental.

A possible way to accommodate this seemingly paradoxical result would be to distinguish between different dimensions of knowledge. Knowledge can be acquired, retained and transmitted, as Quassim Cassam points out (2019, p. 7), and it might as such be positively affected, in one sense, while it is negatively affected, in another sense. Given that knowledge a multi-dimensional concept, friction might turn out to be epistemically beneficial because it makes us better inquirers, but also epistemically detrimental because it poses a difficulty for the retention of knowledge. This provides a way out of the paradox because it shows that our intuitions can be accommodated in a way that makes our concept of knowledge internally consistent. And yet, it leaves us with nowhere to go on a more practical level. Once we know friction can affect knowers in two different opposite directions, what should we do? As we will see below, Medina defends the idea that the promotion of friction must be counterbalanced with the goal of keeping tabs between new evidence and the things that we know. Other authors, such as the epistemic libertarians (for some recent examples, see Caplan (2007) and Somin (2013)), will stress the benefits of open-mindedness, and our epistemic duty to take into consideration every possible perspective. Finally, we will defend the idea that epistemic agents need to give careful consideration to the contextual circumstances under which new evidence is presented to them, before being forced to accept the ideas pushed forward by others as epistemically relevant alternatives.

Let us now move to the second epistemic paradox that we want to consider in this chapter, the Kripke-Harman paradox on knowledge and dogmatism. When faced with new evidence, it is impossible to distance ourselves from the things that we already think that we know, and this becomes epistemically troublesome when the new evidence that we are set to ponder contradicts some of the things that we think we know.

3.2 The Kripke-Harman Paradox

> If I know that h is true, I know that any evidence against h is evidence against something that is true; so I know that such evidence is misleading. But I should disregard evidence that I know is misleading. So, once I know that h is true, I am in a position to disregard any future evidence that seems to tell against h. This is paradoxical, because I am never in a position simply to disregard any future evidence even though I do know a great many different things. (Harman 1973, p. 148)

This paradox presents an extreme way in which the things that we already know affect what we might know. The genealogy of our beliefs determines our ability to acquire and assess new information and therefore determines our epistemic habits. Kripke's result is not only paradoxical because it renders the counterintuitive conclusion that I am somehow in a position to disregard any future evidence that I might encounter that happens to contradict something that I think I know now, no matter how strong that evidence is. It is also problematic because it turns what we already know into an obstacle for the acquisition of new knowledge in general. The genealogy of my beliefs constitutes an essential part of the context that is going to determine whether a given possibility can be properly ignored,[4] and in that context, the subjective probability that might be rational to attribute to a new belief that seemingly contradicts something that we know might be extremely low. Under these circumstances, we do not just cherry-pick the evidence to assess and eliminate that new possibility, we simply ignore it altogether.

A possible way out of this apparently paradoxical result would be to enlarge the context with respect to which new evidence must be assessed. It is not only with respect to our previous beliefs and their history that a particular piece of new evidence has to be weighed in, but in relation to many other factors that might decisively determine the outcome. Engaging in discussions with those that hold views different from our own, maintaining an open-minded attitude towards new information, enriches the context, and therefore allows us to properly appreciate new evidence, despite our previous history. Open-minded people are, in this sense, in a much better epistemic position than closed-minded ones. Closed-minded people are 'disposed to freeze on a given conception, to be reluctant to consider new information, to be intolerant of opinions that contradict their own, and so on' (Cassam 2019, p. 33). Closed-mindedness is, in this sense, according to Cassam, a character trait that systematically gets in the way of knowledge, an epistemic vice.

Against this vice, open-mindedness can inspire an approach to knowledge, an epistemic policy, that might get us out of the trouble. This epistemic policy is epistemic libertarianism.

3.3 Epistemic Libertarianism

Any new information should be equally assessed if knowledge is going to be properly acquired. Picking and choosing or reinterpreting new evidence is irrational from an epistemological point of view. The result of such a process

---

4 According to David Lewis (1996), when determining whether a given subject actually knows a proposition p, it is necessary to make sure that the evidence at their disposal is enough to discard every possibility where not-p holds, except for those possibilities that can be properly ignored. Lewis provides a set of rules to help establishing whether, in a given context, possibilities can be safely ignored. As an anonymous referee points out, Lewis's contextualism might offer a solution to the paradox, one that is similar to the one spell out in the following paragraph.

would be considerably short of knowledge and, given that you already know that, it is irrational to call itknowledge.

In his *Vices of the Mind*, Cassam discusses whether open-mindedness is a liberal virtue, an epistemic virtue particularly linked to a given ideology. He compares (see Cassam 2019, 40 – 42) Mill's commitment to the idea that knowledge requires that individuals be willing to listen to any opinion, to anything that can be said against the views that they already hold, with Sayyid Qutb's idea that 'the enemy of knowledge and the best protection against being led astray in one's thinking is to be closed-minded about certain fundamental matters' (Cassam 2019, 40). When the arguments of both scholars are evaluated on their merits, independently of their ideological affiliations, our conclusion must be, according to Cassam, that closed-mindedness is an exemplary epistemic vice. Remaining open-minded could thus constitute a virtuous way out from the Kripke-Harman paradox. Against our inclination to stick to our previous beliefs, we must cultivate the virtuous character trait of open-mindedness. As stressed above, epistemic libertarianism is an epistemic policy, a general approach to the acquisition, maintenance and sharing of knowledge. It can be cashed out as a set of practical conditionals of the form "if the situation is so and so, do this". Epistemic libertarianism contains a set of instructions regarding how to behave epistemically, how to address other people's statements, and how to question what you think you know. In a nutshell: you only have knowledge when every point of view is taken into consideration, and you know it.

Epistemic libertarianism not only constitutes a possible response to the Kripke-Harman paradox, it also affects Medina-ANP. According to the Medina-ANP paradox, as we saw above, epistemic friction affects knowledge in two opposite directions: it improves knowledge acquisition by honing our skills as enquirers but, at the same time, it damages our ability to maintain and share the knowledge that we already have, not just by reducing self-confidence, but by the sheer fact that knowledge requires recognition. Epistemic libertarianism doubles down on the idea that knowledge requires friction. Taking into consideration every perspective is the only way to get real knowledge, and you should not settle for less. Constant exposure to the views of others will ultimately provide you with the only real way to obtain knowledge. Therefore, you should as much as possible engage in disagreement and evaluate the merits of those who do not think as you do.

However, as Thomas Kelly (2008, p. 612) reminds us, this is not how people actually behave: disagreement does not always promote a better epistemic position.[5] Sharing new neutral evidence in a discussion might lead the participants to be farther away from an agreement, rather than the opposite. Disagreement can make us more, rather than less, polarized. Being open to the opinions of other people might result in a more dogmatic attitude towards my previously held beliefs, rather than the opposite.

Opposing epistemic libertarianism requires resisting the idea that taking every perspective into consideration is epistemically beneficial.[6] This rings counterintuitive to most of us: it seems natural to think that getting to have a balanced opinion requires engaging in disagreements and discussions with those who hold different points of view, exposing our reasons to other people's scrutiny via an open debate. These commitments make Cassam, for example, praise open-mindedness as an epistemic virtue and closed-mindedness as an 'exemplary vice' (Cassam 2019, p. 41). This impulse should nevertheless be resisted. First, by taking into consideration a different kind of deep-rooted intuitions that we also have with respect to the impact of different points of view on our own thinking. Second, by paying attention to empirical studies on heterogeneity and group polarization. To introduce the kind of intuition that we want to bring to the fore, let us first mention Buccola's rendition of the young William Buckley:

> Buckley, "born to absolutes" and "nurtured on dogma," did not accept everything his parents taught him, but the basic outline of his worldview was set at a young age. This worldview consisted of an amalgamation of devout Catholicism, antidemocratic individualism, hostility to collectivism, and a strong devotion to hierarchy – including racial hierarchy – in the social sphere. . . . While at Yale, Buckley ended up pursuing a "divisional" major in the social sciences, he did not have any desire to utilize his coursework to rethink the religious and political commitments that had been inculcated in his household as long as he could remember. During his freshman year, one of his instructors encouraged him to take a course in metaphysics. "I have God and my father," Buckley replied. "That's all I need". (Buccola 2019, pp. 16 and 31)

Buckley had already been brought up a high-praised debater in high school, even before he attended Yale. He spent most of his professional life reading, writing about and discussing views in public that he did not share. Yet, according to

---

5 As an anonymous referee points out, Kelly's own position amounts to defending that conciliating in the face of disagreement is not the right thing to do is the agent is right from the beginning. As we will see more clearly below, the epistemic policy that we are advocating for in this paper differs from this form of anti-conciliationism in that we assume that it is correct for every epistemic agent, no matter what their starting point is, not to concede the a priori right to introduce epistemically relevant alternatives to every other agent.
6 And, with it, the myth that truth is the solution to a jigsaw made from pieces belonging to multiple points of view. We owe José Luis Moreno this elegant way of putting the point.

Buccola, we do not see a lot of changes in his views, views that remained heavily influenced by his early education in the South and the teachings of his father. This clearly shows in Buckley's reaction to the suggestions that he should perhaps take a metaphysics course while attending Yale: 'I have God and my father'. A whole life exposing his views to the scrutiny of others, taking in the views of his most respected opponents, such as he did with James Baldwin in Cambridge in 1965, did not make Buckley a more balanced, epistemically more nuanced individual. Rather, he kept pounding over the same core ideas for most of his life, developing new arguments and repeating old ones.

Disagreement did not have the expected effect on Buckley, one could argue, because of his particularly dogmatic upbringing, but perhaps apart from this rather extreme environment, things would have turned out differently, perhaps in more normal environments disagreement actually has an automatic positive effect on the epistemic status on those who engage in the exchange of ideas. See what happened to journalist Talia Lavin, a person with a less radical upbringing, when she decided to dive in for a year in the world of online white supremacism:

> The more I grew to know this movement, the less patience I had for it; and still less for those who tolerate it. Studying the far right taught me what it means to have an enemy to whom one must give no quarter, because any ground given allows them to accrue power; and any increment of power they receive they will use toward violent ends. Over the course of the research that I did for this book and the gonzo journalism-cum-activism it entailed, I became radicalized. (Lavin 2020, p. 18)

Disagreement, again, does not have to show the kind of positive effect that the epistemic libertarian poses. Lavin's examination of the views of white supremacists, her exposure to their behaviour for thousands of hours, only brought her anger and anxiety. Of course, not every attempt at 'reaching out to someone from another world' (Hochschild 2016, p. xi) leads to more anger or to lack of empathy. Arlie Hochschild spent several years in Tea Party territory exploring the role of feelings for the political beliefs of people very different to herself and, without actually getting closer to agreeing politically with them, she manages to offer a highly balanced and empathetic portrait of the persons themselves. We believe that the positive policy that we will endorse at the end of this section helps to make sense of Hochschild's respectful attention to ideas that, although alien to herself, are held by people she learned to appreciate.

The classic study questioning the positive impact of disagreement is Lord et al. (1979), where they show that 'the result of exposing contending factions in a social dispute to an identical body of relevant empirical evidence may be not a narrowing of disagreement but rather an increase in polarization' (Lord et al. 1979, p. 2098).[7] As Thomas Gilovich (1991, p. 54) points out, this is not the result of misconstruing or ignoring the evidence that seems to go against our previously held beliefs. We might have some room for that when the information is ambiguous, but whenever ambiguity is sufficiently controlled, our reaction is not to make this new information compatible with what we know by reinterpreting it. Rather, we scrutinize it closely, in order to discover any plausible flaw that we might encounter in it, and we develop argumentative strategies to reinforce our previous positions. All this explains, according to Bail (2018), why Twitter users openly affiliated with a political identity – republican or democrat – become more polarized, instead of less polarized, when they are forced to follow a Twitter bot that retweets information from the other camp. We do not necessarily ignore or reinterpret this new information, but our reaction to it de facto augments the size and density of the argument pool supporting our political identity's core beliefs. This resistance makes it really difficult to produce effective fact-checking, in spite of what the common folk allegedly believes (see Lewandowsky et al. 2012, p. 2017).

These studies regarding the reaction to information that seems to contradict our initial beliefs made some researchers postulate the existence of a backfire effect (Nyhan and Reifler 2010): every attempt to correct misperceptions and unsubstantiated beliefs is doomed to fail, it is bound to produce more polarization. The epistemic anti-libertarian could take this line of reasoning to support her views – more disagreement has no epistemic positive effect at all. Our duty is to progress from what we have, exposing yourself to the views of other people is not just an insufficient epistemic strategy. It is something to be avoided altogether. Nevertheless, this pessimistic approach should not be taken that far, as shown by Wood and Porter (2019), the aforementioned backfire effect is difficult to replicate widely, and there are reasons to consider that in many different contexts the effect of new information is actually epistemically very positive. The key might perhaps be here: even though getting more information by entering on debates makes our factual views more nuanced and more accurate, it might not

---

7 Even if we agree with Kevin Dorst's idea that polarizing in this context is rational (https://medium.com/science-and-philosophy/reasonably-polarized-why-politics-is-more-rational-than-you-think-178810fefdb7), our position in this paper is neutral with respect to this point. According to the view on affective polarization defended in Osorio and Villanueva (2019), becoming impervious to the reasons of others follows rationally from an abnormally high attribution of subjective probability to the beliefs that are part of the core of your active political identity in a given context. This process would be rational both for those that start from a true belief and for those whose starting premise is false. Thanks to an anonymous referee for pointing out the connection to Dorst's position.

change our political preferences (Porter et al. 2019). Political polarization might in turn make us less willing to engage seriously with the reasons of others, and this has substantially negative epistemic effects. It might thus be the case that even though there is nothing like a backfire effect supporting the epistemic anti-libertarian, the epistemic libertarian intuition could be still severely damaged by the connection between public disagreement and polarization.

The libertarian strategy thus faces a problem when confronted with how actual epistemic agents behave. Our tendency to become more, rather than less, polarized when we systematically subject our views to the scrutiny of others trumps an epistemic policy that is based on the benefits of friction. Therefore, the extreme endorsement of open-mindedness, epistemic libertarianism, does in the end have a negative effect on both dimensions of knowledge discussed above – acquisition and maintenance. Epistemic anti-libertarianism is not a sensible strategy either – more information might actually make our view of the world more accurate, even if we can still become more polarized. We will move now to a different strategy, one that does not necessarily constitute the other extreme of an imaginary spectrum, but the negation of the extreme position that epistemic libertarianism embodies. Instead of arguing for the idea that dogmatism should be turned into an epistemic policy, along the lines of epistemic anti-libertarianism, our aim is simply to question the assumption that everybody, all the time, no matter where you find yourself, has the a priori right to question what you know.

4 Epistemic De-Platforming

In this section, we examine a rival epistemic policy, one that is able to overcome some of the shortcomings mentioned in the previous section for epistemic libertarianism and to provide a practical way out of the Medina-ANP paradox and the Kripke-Harman paradox. Our defence of this epistemic policy is built upon its ability to allow us to detect and intervene in situations that we perceive as unfair. It thus constitutes an example of philosophical practice under the political turn, as presented in section 2.

When confronted with new evidence, it is not the case that everyone has an a priori right to turn their opinions into epistemic possibilities that cannot be properly ignored. Being able to take part in a meaningful epistemic discussion is a right that can be earned, and it can be lost as well. Trust is not only relevant to determine the ability of a subject to share testimony, but is also crucial to determine whether the information provided by a given knower is going to constitute a relevant epistemic possibility or it is going to be properly ignored.

When the preservation of knowledge is crucial for your community to survive and thrive, epistemic de-platforming becomes a form of resistance. How can knowledge be threatened at all? The sheer addition of new possibilities might move a belief from the realm of knowledge to the realm of doubt if you lack the evidence to discard those possibilities. This is not per se epistemologically detrimental. There is a sense in which you are closer to the truth if you doubt, if you are acting on the ratio evidence/possibilities at your disposal.

When the amount of information (true, false, and otherwise) is unlimited, attention becomes a priced commodity. There is a really high incentive to grasp your attention, and the general augmentation of knowledge is not the primary goal of those willing to monetize your attention. Again, straight lies are not necessary to achieve this effect:

1. You might be overflown by information tactically provided to make the balance evidence/epistemic possibilities shift, so that you might end up believing a proposition with too high a degree of confidence or reducing your confidence in your initial views as the new possibilities lead to doubt or to never-ending discussions.

   One of the earliest means of protest on the internet by dissidents had been theDDOS, or the denial-of-service attack in which thousands of people, bots, or scripts would repeatedly ping a website, overwhelming its capacity to respond, thus taking it down. For example, the website of a corporation that had just undertaken an unpopular action may go down as people around the world coordinate to drown it in requests. Nowadays, however, governments and the powerful, along with other authoritarians, have adopted parts of this playbook: by kicking up massive clouds of claims, accusations, misinformation and controversies they can overwhelm the capacity of the public and traditional media to respond to any one of them, thus causing a type of paralysis. It is as if the networked public sphere, and indeed traditional institutions of democracy, can be DDOSed via releasing large numbers of flares, each attracting and consuming attention, thus making focus and sustained censorship via blocking is not an end but a means to prevent political action; if hiding information is not feasible, confusion may work just fine. Indeed, confusion and misinformation, whether deliberate or a byproduct of information glut, have emerged as significant political problems for social movements. (Tufekci 2017, 273–274)

2. The goal of the informational strategy might not be to convince you of a proposition, but rather to affect the level of confidence that you had on something that you thought that you knew, so that you stop acting on the basis of those beliefs. As we mentioned below, this might be turned into a very effective electoral strategy.

In Michigan, a state that Trump won by 10,000 votes in 2016, 15 % of voters are black. But they represented 33 % of the special deterrence category in the secret database, meaning black voters were apparently disproportionately targeted by anti-Clinton ads.

In Wisconsin, where the Republicans won by 30,000 votes in 2016, 5.4 % of voters are black, but 17 % of the deterrence group. According to Channel 4, that amounted to more than a third of black voters in the state overall, all placed in the group to be sent anti-Clinton material on their Facebook feeds.[8]

Our attention is scarce, and it is literally monetized. The same media we use to get our information also make money out of keeping us on their sites, and are able to feed us with the kind of informational diet that guarantees grabbing our attention, while at the same time promoting the interest of those who pay for this free service. It is in this kind of context that epistemic de-platforming becomes a form of resistance.

Is de-platforming a form of closed-mindedness? Is it just another way of highlighting the epistemological significance of genealogy? We claimed above that epistemic de-platforming is not the result of turning dogmatism into an epistemic policy. Epistemic de-platforming does not contain a set of recommendations to cancel every interlocutor that happens to disagree with the things that you know; it is simply an invitation to ponder whether that person is worth your time and attention in an epistemic environment where both are severely limited.

Maintaining that epistemic de-platforming is either a vice or a virtue has a clearly counterintuitive ring to us. Even though the discussion of its merits might be linked to the debate about dogmatism, epistemic de-platforming does not look like an attitude, a character trait or a way of thinking. Epistemic de-platforming is a policy – a plan that can be specified in a series of conditional statements, containing particular situations in the antecedent, and practical recommendations in the consequent. Each of these conditionals tells you what to do in certain given circumstances. Epistemic de-platforming and epistemic libertarianism could only count as epistemic virtues or vices if epistemic stances, such as the policies that we are discussing here, are taken to be potential virtues or vices, as Cassam (2019, p. 83) does. This move seems odd to us. Epistemic de-platforming does not encourage you to become closed-minded, it simply reminds

---

8 https://www.theguardian.com/us-news/2020/sep/28/trump-2016-campaign-targeted-35m-black-americans-to-deter-them-from-voting accessed on 19 November 2021.

you that you are in an environment where new information might be posed to you for a variety of motives and that this needs to be taken into consideration to see whether a particular exchange is going to be epistemically beneficial or detrimental.

The epistemic policy that we defend in this chapter should be distinguished from the ecumenical position that both Medina and Lewis seemingly uphold when confronted with this tension between enquiry and maintenance of knowledge. An open attitude, they seem to claim, needs to be balanced with our ability to keep what we already have. Here is how Medina puts it:

> I propose two guiding principles for the epistemology of resistance, principles that will facilitate our analyses of cognitive forces and their interplay. These are regulative principles that will enable us to dissect and assess different forms of epistemic friction and to navigate our cognitive lives through them. They are the principle of acknowledgement and engagement, and the principle of epistemic equilibrium. The former dictates that all the cognitive forces we encounter must be acknowledged and, insofar as it becomes possible, they must be in some way engaged (even if in some cases only a negative mode of engagement is possible or epistemically beneficial). The latter principle lays out the desideratum of searching for equilibrium in the interplay of cognitive forces, without some forces overpowering others, without some cognitive influences becoming unchecked and unbalanced. In particular, as we shall see, it is important to aim at equilibrium between the internal and the external, or between the perspective of the subject and that of others. (Medina 2012, p. 50)

In Lewis's terms, what alternatives are appropriate to ignore depends on two rules – amongst many others – that seem to be pulling in opposite directions.

> Rule of attention. 'A possibility not ignored at all is ipso facto nor properly ignored'. (Lewis 1996, p. 559)

> Rule of conservatism. 'Suppose that those around us normally do ignore certain possibilities, and it is common knowledge that they do. (They do, they expect each other to, they expect each other to expect each other to, …) Then – again, very defeasibly! – these generally ignored possibilities may be properly ignored'. (Lewis 1996, p. 559)

Only when every alternative where p is false is eliminated by the evidence that you have, can you that you know that p, but there must be alternatives that can be properly ignored in a given context to make knowledge possible. You need to pay attention to every possibility that is put in front of you, or so the Rule of attention seems to require,[9] but it is also okay to ignore the possibilities that are

---

9 As an anonymous referee points out, a common intuition behind this rule is that for a possibility to be relevant is not enough that we pay attention to it, we need to have reasons to take it into consideration. We take these reasons for granted, since we assume that information strat-

commonly ignored by those around you. The commonsensical nature of Lewis's Rule of conservatism brings us to one of the important features of epistemic de-platforming: it is okay to ignore those possibilities that are systematically ignored by those close to you.

As we saw at the beginning of the chapter, when we introduced the quote by Ta-Nehisi Coates, the idea that when the tide rises all boats rise together, when presented in socio-epistemic terms, is problematic because it assumes that all nodes in a socio-economic space are sufficiently similar to each other, such that what benefits one of them necessarily benefits the others. We want to pay attention now to another feature of the paragraphs we quoted from Coates. Coates points out that what made him originally dubious of the idea was the kind of people that were defending it. More precisely, the kind of ideas that seem to be systematically connected to the rising tide intuition. Even if at first sight you cannot see what is wrong with the rising tide analogy, inasmuch as it is systematically associated with the kind of ideas that you strongly oppose, or with the kind of people that you know that want to further oppressive agendas, you need to be careful with it, no matter how appealing or hard to dismount it might initially look to you. In doing so, Coates not only provides an example of the negative stance that epistemic de-platforming promotes – it is okay to ignore those who hurt you – but also an implicit instance of the positive policy that might be derived from it. One of Coates's biggest influences, James Baldwin, had already rejected the rising-tide hypothesis during his 1965 interview with *The New York Times* (see Buccola 2019, p. 336), and he opened the second part of his 1972 *No Name in the Street* with the following paragraph:

> All of the Western nations have been caught in a lie, the lie of their pretended humanism; this means that their history has no moral justification, and that the West has no moral authority. Malcolm, yet more concretely than Frantz Fanon – since Malcolm operated in the Afro-American idiom, and referred to the Afro-American situation – made the nature of this lie, and its implications, relevant and articulate to the people whom he served. He made increasingly articulate the ways in which this lie, given the history and the power of the Western nations, had become a global problem, menacing the lives of millions. "Vile as I am," states one of the characters in Dostoevski's *The Idiot*, "I don't believe in the wagons that bring bread to humanity. For the wagons that bring bread to humanity, without any moral basis for conduct, may coldly exclude a considerable part of humanity from enjoying what is brought; so it has been already". Indeed. And so it is now. (Baldwin 1998, p. 404)

---

egies such as the ones mentioned in this paper already work through media that the agent has reasons to ponder, given that she has previously selected them.

Coates is thus not only rejecting the concerns of those who are far away from him, he is also in line with one of his closest allies. We find this positive policy as attractive and commendable as the negative one discussed above. You do not have an a priori obligation to give consideration to anything someone says, especially if they are at the service of dubious interests and against yours.[10] Your friends and loved ones have earned a special right to your attention: most times you can trust that they will call your attention to the interesting ideas that you have correctly ignored following the negative policy.

Note that the right to our attention granted by our positive policy does not depend on any narrowly construed epistemic credentials possessed by our close ones. While a certain level of irreflexive (although not blind: see Crary 2018 and McDowell 1993) trust seems to be behind the success of many of our testimonial practices, when it comes to our friends and companions, blind trust is indeed part of the essence of our very relationship with them. A lot of the recent discussions on the second person have centred on her role in our introduction into normative practices: morality, self-knowledge, joint actions, and belief (self-)attribution (see, e. g., Gomila 2002; Pinedo 2004; Ferrer 2014; Darwall 2006; Bilgrami 2006; and Velleman 2009). One dimension of this emphasis on the second-personal role of normative practices that we find particularly attractive is the idea that, even regarding my own mental life (what I believe, what I like, what my preferences are, what emotion or sensation I am feeling), someone who is very familiar with me and cares for me may sometimes be in a better position than myself to know what is going on and, crucially, learning to listen to what such a person has to say about myself may often be the best path to self-understanding.

Here is the briefest sketch of how this can work: knowledge, like any other normatively evaluable sphere, implies the possibility of being wrong, of thinking that one knows when one does not. Grasping such a possibility, that is, realizing that there may be a gap between seeming right to me and being right, is something we learn socially by means of being trained, corrected, punished or congratulated. Self-knowledge, if it is knowledge at all, also implies that sometimes one can be wrong about oneself. How do I gain the necessary distance from myself to countenance the possibility of being wrong? People that we have every reason to trust, either during our upbringing or at any point in our life, are in an ideal position to play the role of offering us alternative perspectives even

---

10 In fact, white supremacists are not acting irrationally if they ignore what the NYT or other liberal media have to say: they are not even under the epistemic obligation to check the fiability or the motives of the source in order to ignore it. The problem starts when they do not recognise any way for liberals to win the right to be heard.

on ourselves. Knowledge and belief, no matter whether directed to oneself, to others or to the world, demand the possibility of triangulation, of friction between different points of view (see Davidson 1991 and Bensusan and Pinedo 2007). If we discard the all too present perspective of the oppressor and place our attention instead on the perspectives of our friends and beloved, in all likelihood we will end up drawing far more original and subversive triangles. The idea, of course, is as old as philosophy itself: although we would like our proposal to ring a relativist bell that would have horrified Plato, he himself offers a beautiful version of our positive, erotic proposal: when we love somebody, we do not just discover a person, we discover a world and get as close as it is possible to understand what is, as Plato argues in *Symposium*, forever ungraspable by our mortal self: beauty, goodness and truth (Plato 2008).

This policy might strike some as a piece of epistemic irrationalism. There cannot be guilt by association when pondering the epistemic merits of a given position. Every idea, if it is logically independent of those surrounding it, might be scrutinized on its own merits, a grain of truth can always be found next to the most outrageous positions. Epistemic libertarianism compels you to keep this in mind and assess every bit of new information on its own. As the explicit denial of epistemic libertarianism, epistemic de-platforming questions this assumption, and thus needs to qualify in which sense it is not irrational to claim that the right to be taken seriously as an epistemic partner is something to be earned. We will finish this chapter with a few notes on this topic, on why it might be rational to ignore the alternatives pushed forward by those who pose a menace to you and your keen.

5 Conclusion

Epistemic de-platforming is a form of epistemic resistance that provides us with a practical way out both from the Medina-ANP paradox and the Kripke-Harman paradox. The positive effects of epistemic friction have to be balanced with our ability to preserve the knowledge that we already have. It is a practical policy that allows us to detect and intervene more effectively on situations that we perceive to be unfair. Protecting ourselves from manipulation, and from the rise of affective polarization, is compatible with being constantly open to consider new perspectives. Properly pondering the reasons provided by those who oppose my views does not require, though, the disposition to question anything I know at the same exact moment that I am presented with alleged counterevidence from any kind of source. If the metaphor of the socio-normative space is still going to be of use to model our epistemic practices, the parameters that characterize the nodes must not only include socio-economic status, normative status, speech affordances, technological affordances, etc., but also membership to partitions that reflect our identities and mega-identities. Relative distance between identities can be modelled by paying attention to the amount of overlap between the beliefs that constitute the core of the identity at a given time, but also by the attitudes that members of one group are willing to express with respect to members of a different group. The closer the nodes are, the easier it is to make the information provided by one epistemically relevant to the other.

Crucially, none of the above implies that the right to partake in an exchange cannot be earned. The distance between nodes does not depend on the beliefs that we actually hold, but the amount of overlap between the beliefs that constitute the core of the identity, and the affective attitudes that are exhibited from one node towards the other. The amount of overlap between identities actually changes over time. This means that two political identities, for example, that might look miles apart in a given context, might in a different one look actually closer. This should reflect, and be reflected by, the affective attitudes of a group towards the other. Closing the distance is actually possible, it just might take some work, sometimes hard work, if we are going to convince certain knowers that the information that we provide in a discussion is epistemically relevant. The mistake of those who decry epistemic de-platforming as an irrational policy is to suppose that socio-economic power automatically obliterates the effects of distance. No, no matter how huge your reach is, nobody is a priori forced to listen to whatever you have to say.

Why is then arguing for epistemic de-platforming, as we do in this chapter, different from defending dogmatism? Even though we presented Mill's quote as a target, note that it is only a very particular, highly radical, ahistorical interpretation of the paragraph that we are criticising. Dogmatism as a character trait that systematically stands in the way of knowledge is an epistemic vice (Cassam 2019, pp. 100–114), and we do not question that. If you lived in a world where only two books existed, and you refused to read one of them to protect what you think you know after reading the first one, you would be a dogmatic person because of your dogmatic attitudes. We do not live in such a world, though. The amount of accessible information and the presence of strategic attempts to tamper with our attention make the libertarian response to dogmatism naive at best and, very possibly, disingenuous. Every person is not, cannot be, automatically granted the right to get your attention. This is the gist of epistemic de-platforming. But this right can be acquired. A dogmatic person, in this context, is one who refuses to take into consideration the alternatives put forward by those who would have earned the right to be epistemically relevant, had it not been for an excessive attachment to her previous beliefs, or for biases against the new source. Epistemic de-platforming is not a blanket rationale to justify epistemic discrimination. When the epistemic alternatives brought to the fore by a person are ignored for the wrong reasons, reasons that might concern features of that person's identity that chase the source through other spheres of her life, then we are facing a case of testimonial epistemic injustice, one that should not be justified under any sensible epistemic policy, including epistemic de-platforming.

It should be possible to gain the right to be epistemically relevant almost from every point of departure, from every node in the socio-normative-identitarian space. When that right is systematically denied for the wrong reasons, one is a victim of a dogmatic attitude. Appealing to the wrong reasons might sound disappointing to some, but it should not. As we mentioned above, the role of normativity is ineliminable – thinking about knowledge is normative in nature. No epistemic policy can save us from the hassle of thinking case-by-case and assessing our behaviour in normative terms. Epistemic de-platforming is no exception. What epistemic de-platforming does is shifting the focus of your normative epistemic thinking. It questions the a priori right to be a provider of epistemically relevant alternatives, and in doing so it opens up a new chance to discuss what we take to be knowledge, in normative terms. In doing so, it also brings to the fore, justly, the epistemic role of those close to you.[11]

Bibliography

Almagro, Manuel, Llanos Navarro and Manuel de Pinedo (Forthcoming): "Is Testimonial Injustice Epistemic? Let Me Count the Ways". In: *Hypatia* 36. No 4, pp. 657-675.

Ayala, Saray (2016): "Speech Affordances: A Structural Take on How Much we Can Do with Our Words". In: *European Journal of Philosophy* 24. No 4, pp. 879–891.

Ayala, Saray, and Nadya Vasilyeva (2016): "Explaining Injustice in Speech: Individualistic vs. Structural Explanation". In: *Proceedings of the 37th Annual Conference of the Cognitive Science Society*.

---

11 This research is partially funded by the "Desacuerdos públicos, polarización afectiva e inmigración en Andalucía" project (B-HUM-459-UGR18, Consejería de Economía, Innovación y Ciencia. Junta de Andalucía. FEDER), and the "Desacuerdo en actitudes. Normatividad, desacuerdo y polarización afectiva" project (PID2019–109764RB-I00, Ministerio de Economía y Competitividad). We would like to thank María José Frápolli Sanz, Manuel Heras-Escribano, José Luis Moreno Pestaña, and the editors of this volume for their comments. We are especially grateful to an anonymous referee for their insightful criticisms and suggestions.

Bail, Christopher, et al. (2018): "Exposure to Opposing Views on Social Media Can Increase Political Polarization". In: *Proceedings of the National Academy of Sciences* 115. No 37, pp. 9216–9221.

Baldwin, James (1998): "No Name in the Street (1972)". In: *Collected Essays*. New York: Literary Classics of the United States. Library of America, pp. 349–477.

Baluška, Franktišek, Stefano Mancuso and Dieter Volkmann (Eds.) (2006): *Communication in Plants: Neuronal Aspects of Plant Life*. Berlin: Springer-Verlag.

Bensusan, Hilan, and Manuel de Pinedo (2007): "When My Own Beliefs are not Third-Personal Enough". In: *Theoria* 22. No 58, pp. 35–41.

Bilgrami, Akeel (2006): *Self-Knowledge and Resentment*. Cambridge: Harvard University Press.

Broncano, Fernando (2020): *Conocimiento Expropiado. Epistemología Política y Democracia Radical*. Madrid: Akal.

Buccola, Nicholas (2019): *The Fire is Upon Us. James Baldwin, William F. Buckley and the Debate over Race in America*. Princeton: Princeton University Press.

Calvo, Paco (2016): "The Philosophy of Plant Neurobiology: A Manifesto". In: *Synthese* 193, pp. 1323–1343.

Caplan, Bryan (2007); *The Myth of the Rational Voter*. Princeton: Princeton University Press.

Carmona, Carla, and Neftalí Villanueva (unpublished manuscript, September 24th, 2021): "Situated Judgments as a New Model for Intercultural Communication". Portable Document Format.

Cassam, Quassim (2019): *Vices of the Mind. From the Intellectual to the Political*. Oxford: Oxford University Press.

Coates, Ta-Nehisi (2018): *We Were Eight Years in Power: An American Tragedy*. London: Penguin Random House.

Crary, Alice (2018). "The methodological is political: What's the matter with "analytic feminism"?. In: Radical Philosophy 2.02.

Darwall, Stephen L. (2006): *The Second-Person Standpoint*. Cambridge: Harvard University Press.

Davidson, D. 1986, A nice derangement of epitaphs, in Truth, Language, and History, Oxford, Clarendon Press, 2005, pp. 89-107.

Davidson, Donald (2001): "Three Varieties of Knowledge (1991)". In: *Subjective, Intersubjective, Objective: Philosophical Essays*, vol. 3. Oxford: Oxford University Press, pp. 205–220.

Díaz-León, Esa (2018): "On Haslanger's Meta-Metaphysics: Social Structures and Metaphysical Deflationism". In: *Disputatio* 50, pp. 201–216.

Ferrer, José (2014): "El Papel de la Segunda Persona en la Constitución del Autoconocimiento". In: *Daimon* 62, pp. 71–86.

Fernández Castro, Víctor, and Manuel Heras Escribano (2019): "Social Cognition: A Normative Approach." In: *Acta Analytica* 35. No 1, pp. 75–100.

Frápolli, María J. (2019): "The Pragmatic Gettier". In: *Disputatio* 8. No 9, pp. 558–586.

Frápolli, María J., and Neftalí Villanueva, (2012): "Minimal Expressivism". In: *Dialectica* 66. No 4, pp. 471–487.

Fricker, Miranda (2007): *Epistemic Injustice. Power and the Ethics of Knowing*. Oxford: Oxford University Press.

Gilovich, Thomas (1991): *How We Know What Isn't So. The Fallibility of Human Reason in Everyday Life*. New York: The Free Press.

Gomila, Antoni (2002): "La Perspectiva de la Segunda Persona de la Atribución Mental". In: *Azafea* 4, pp. 123–138.

Harman, Gilbert (1973): *Thought*. Princeton: Princeton University Press.

Haslanger, Sally (2016): "What is a (Social) Structural Explanation?" In: *Philosophical Studies* 173. No. 1, pp. 113–130.

Heras Escribano, Manuel, and Manuel de Pinedo (2018): "Naturalism, Non-Factualism, and Normative Situated Behaviour". In: *South African Journal of Philosophy* 37. No. 1, pp. 80–98.

Hochschild, Arlie L. (2016): *Strangers in Their Own Land: Anger and Mourning on the American Right*. New York: The New Press.

Jorba, Marta, and Maria Rodó-Zárate (2019): "Beyond Mutual Constitution: The Properties Framework for Intersectionality Studies". In: *Signs: Journal of Women in Culture and Society* 4. No. 1, pp. 175–200.

Kelly, Thomas (2008): "Disagreement, Dogmatism, and Belief Polarization". In: *The Journal of Philosophy* 105. No. 10, pp. 611–633.

Kripke, Saul (1982): *Wittgenstein on Rules and Private Language*. Oxford: Blackwell.

Kukla, Quill (2014): "Performative Force, Convention, and Discursive Injustice". In: *Hypatia* 29. No. 2, pp. 440–457.

Lavin, Talia (2020): *Culture Warlords: My Journey into the Dark Web of White Supremacy*. New York: Hachette Books.

Lewandowsky, Stephan, et al. (2012): "Misinformation and its Correction: Continued Influence and Successful Debiasing". In: *Psychological Science in the Public Interest* 13. No. 3, pp. 106–131.

Lewis, David (1999): "Elusive Knowledge" (1996). In: *Papers in Metaphysics and Epistemology*. Oxford: Oxford University Press, pp. 418–45.

Lord, Charles G., Lee Ross and Mark R. Lepper (1979): "Biased Assimilation and Attitude Polarization: The Effects of Prior Theories on Subsequently Considered Evidence". In: *Journal of Personality and Social Psychology* 37. No. 11, pp. 2098–2109.

McDowell, John (2007): "What Myth?" In: *Inquiry* 4, pp. 338–351.

McDowell, John (1993). "Knowledge by hearsay". In: Meaning, Knowledge and Reality, Cambridge, Mass., Harvard University Press, 1998, pp. 414-443.

McGeer, Victoria (2007): "The Regulative Dimension of Folk Psychology". In: Daniel D. Hutto and Matthew Ratcliffe (Eds.): *Folk Psychology Re-Assessed*. Dordrecht: Springer, pp. 137–156.

McGeer, Victoria (2015): "Mind-Making practices: The Social Infrastructure of Self-Knowing Agency and Responsibility". *Philosophical Explorations* 18. No. 2, pp. 259–281.

Medina, José (2012): *The Epistemology of Resistance: Gender and Racial Oppression, Epistemic Injustice, and Resistant Imaginations*. Oxford: Oxford University Press.

Moreno Pestaña, José L. (2019): *Retorno a Atenas. La Democracia como Principio Antioligárquico*. Madrid: Siglo XXI.

Nyhan, Brendan, and Jason Reifler (2010): "When Corrections Fail: The Persistence of Political Misperceptions". In: *Political Behavior* 32, pp. 303–330.

Osorio, Javier, and Neftalí Villanueva (2019): "Expressivism and Crossed Disagreement". In: *Royal Institute of Philosophy Supplements, Vol. 86: Expressivisms, Knowledge and Truth*, pp. 111–132.

Pinedo, Manuel de (2004): "De la Interpretación Radical a la Fusión de Horizontes: La Perspectiva de la Segunda Persona". In: Juan J. Acero et al. (Eds.): *El Legado de Gadamer*. Granada: Universidad de Granada, pp. 225–236.

Pinedo, Manuel de (2020): "Ecological Psychology and Enactivism: A Normative Way out from Ontological Dilemmas". In: *Frontiers in Psychology*. https://doi.org/10.3389/fpsyg.2020.01637, accessed September 24th 2021.

Plato (2008): *Symposium*. Oxford: Oxford University Press. Translated by Robin Waterfield.

Porter, Ethan, Thomas J. Wood and Babak Bahador (2019): "Can Presidential Misinformation on Climate Change be Corrected? Evidence from Internet and Phone Experiments". In: *Research & Politics*. https://doi.org/10.1177/2053168019864784, accessed September 24th 2021.

Ramberg, Bjørn (2000): "Post-ontological Philosophy of Mind: Rorty Versus Davidson". In: Robert Brandom (Ed.): *Rorty and his Critics*. Malden: Blackwell Publishers, pp. 351–369.

Reisch, George A. (2005): *How the Cold War Transformed Philosophy of Science. To the Icy Slopes of Logic*. Cambridge: Cambridge University Press.

Rietveld, Erik (2008): "Situated Normativity". In: *Mind* 117. No. 468, pp. 973–1001.

Somin, Ilya (2013): *Democracy and Political Ignorance*. Stanford: Stanford University Press.

Srinivasan, Amia (2020): "Radical Externalism". In: *The Philosophical Review* 129. No. 3, pp. 395–431.

Tüfekçi, Zeynep (2017): *Twitter and Tear Gas: The Power and Fragility of Networked Protest*. New Haven: Yale University Press.

Velleman, J. David (2009): *How We Get Along*. Cambridge: Cambridge University Press.

Wittgenstein, Ludwig (1953): *Philosophical Investigations*. Oxford: Blackwell.

Wood, Thomas, and Ethan Porter (2019): "The Elusive Backfire Effect: Mass Attitudes' Steadfast Factual Adherence". In: *Political Behavior* 41, pp. 135–163.

Cristina Borgoni
Philosophy of Mind after Implicit Biases

Abstract: This chapter aims at characterizing the impact of the discussion on implicit biases—a politically loaded phenomenon—for some fundamental questions in philosophy of mind and at exploring new lines of investigation in epistemology. In the first part of the chapter, I focus on the still lively debate about the type of mental state that implicit biases are. In contrast to the current views that analyze implicit bias against a traditional view of beliefs, I propose that a more promising position is one that accommodates implicit biases within the belief category, while proposing a new paradigm for what beliefs are. The second part explores some epistemological issues centered on the question about what is biased in implicit biases. Implicit biases seem to be epistemically and ethically mistaken. Implicit biases are prejudice-based mental states. Also, very often, implicit biases diverge from what people explicitly hold to be the case. However, it is unclear whether implicit biases are constitutively inaccurate or unjustified. Moreover, it is not easy to understand what the source of such failures are, whether they have their origin in the individual's cognitive system or in the individual's social environment. I aim to formulate these epistemological problems and explore some possible routes to their answer.

1 Introduction

'Implicit bias' is a term of art in philosophy that refers to prejudiced implicit attitudes towards members of socially stigmatized groups (see, e.g. Brownstein 2019). Implicit biases thus form a subset of implicit attitudes that operate in a variety of human behaviors that were discovered in the 1990s by empirical research in social psychology (see Greenwald and Banaji 1995).[1] Since then, interest in the topic among philosophers has increased and developed into challenging questions in the philosophy of mind, epistemology, and ethics. Implicit attitudes are relevant for understanding cases where a person's behavior and judgments contrast with their defended egalitarian ideas. Furthermore, understanding the functioning of implicit bias can help people to design concrete

---

[1] According to Michael Brownstein (2019), the other important root of the work on implicit social cognition is the earlier work on the distinction between controlled and automatic information processing made by cognitive psychologists (e. g., Shiffrin and Schneider 1977).

https://doi.org/10.1515/9783110612318-008

strategies for promoting egalitarian policies and procedures in various spheres of social life, such as hiring committees and other evaluation processes.

In this paper, I focus on the question of the nature of implicit bias and argue that answering it goes beyond identifying the type of mental state that implicit attitudes are. In order to give a full account of implicit bias, we also need to understand the *bias*—i.e., in which way they are wrong or deviate from a given norm—in implicit bias. For implicit attitudes do not need to be biased (see Greenwald and Banaji 1995). The interesting, more general result of this paper is that questions in the philosophy of mind that are in principle politically neutral begin to take on an intrinsic political aspect. Understanding the constitutive aspects of implicit bias, and thus its nature, requires understanding its political role in perpetuating injustice. With this said, it is not my aim in this paper to provide a full account of implicit bias. Rather, I aim to offer methodological and programmatic remarks regarding the direction of studies of this phenomenon in the philosophy of mind. I suggest that the social and political aspects of implicit bias help us to understand why implicit biases are *biased.*

2 The Structure of Implicit Biases

The traditional methodology for investigating human behavior in the philosophy of mind starts with the identification of the type of mental state that explains a particular individual's behavior. In addition to the paradigmatic pair of mental states—belief and desire—the philosophy of mind investigates a whole menu of different types of mental states that help to explain people's actions and judgments: intention, perception, emotions, feelings, etc. When implicit biases started to gain attention in the philosophy of mind, the discussion was largely guided by this methodology. If the leading issue has been the nature of implicit biases, the leading methodology has been an attempt to understand the type of mental state they are.

Research on the structure of implicit attitudes, as allegedly revealed by empirical experiments, has thus largely guided the discussion on the nature of implicit bias. The standard understanding of implicit attitudes holds that they have an associative structure. According to certain authors (e.g., Mandelbaum 2016), however, this assumption has been incorrectly taken for granted, since other studies show that implicit attitudes reveal an underlying propositional structure. In fact, there is an ongoing dispute on the nature of implicit bias guided by research on its structure. Identifying the structure of implicit biases involves looking at how their content is stored, activated, and modified. The philosophical analysis thus attempts to identify the mental state that best captures this structure and related patterns and traits.

In broad strokes, the majority of views on the nature of implicit bias can be divided into three groups, all of them identified in relation to beliefs: (i) implicit biases are not beliefs; (ii) implicit biases are only partly like beliefs; (iii) implicit attitudes are beliefs (Bendaña 2021). The first view is supported by a shared assumption about implicit biases as having an associative structure. Given that the distinguishing features of associations are that they do not encode inferential information and are sensitive to evaluative conditioning (Nosek et al. 2007), they stand in sharp contrast to the accepted view on the structure of beliefs. One key supporter of this view is Alex Madva (2016). Another proponent of the view is Tamar Gendler (2008),[2] who has canonically proposed the notion of 'alief.' According to her initial proposal, alief is a mental state the representational content of which has an associative structure, is automatically activated, and is a-rational. In addition to representational content, Gendler proposes that aliefs have two other components: an emotional component and an action-prompt component.

The second group holds that implicit biases are very similar to beliefs in some respects but do not have enough of their constitutive features to count as such. Neil Levy (2015), with his notion of patchy endorsements, is a representative of the second group. According to Levy (2015), implicit biases have content and can thus feature in inferential transitions, but they also have an affective component that explains their resistance to change and revision.

The third group of views holds that implicit biases are full-blown beliefs (De Houwer 2014; Frankish 2016; Mandelbaum 2016; and Borgoni 2018). It stands in sharp contrast to the first of positions and thus opposes the view that implicit biases have an associative structure. This group of positions is supported by recent data from social psychology, according to which implicit attitudes can feature in inferential transitions and exhibit sensitivity to semantic content. It is also supported by the available data on the malleability of implicit attitudes (see Bendaña 2021).

In sum, the underlying rationale for the connection between empirical research on the structure of implicit biases and the corresponding philosophical account is the following: if the empirical evidence about implicit biases strongly supports the hypothesis that they have an associative structure, positions like Gendler's (2008) are favored. In this case, implicit biases would differ from explicit beliefs in key respects. If, however, empirical evidence reveals that implicit

---

2 Pace Bendaña (2021), who classifies Gendler as belonging to the second group.

biases have a structure that is relevantly similar to beliefs, then positions such as Eric Mandelbaum's (2016) are favored. Both groups of positions have *beliefs* as the main contrasting mental states when it comes to understanding implicit biases. If there is one class of prejudiced mental states with which we are familiar, it is that of *explicit* prejudiced attitudes. These are commonly treated as beliefs.

In previous work (Borgoni 2015a, 2015b, 2016, and 2018), I have extensively defended the view that implicit attitudes are better understood as beliefs. As such, they make sense of people's actions, judgments, and feelings and allow us to hold people responsible for them, even when such implicit attitudes contrast with people's explicit views. Additionally, the particular biases that we can identify in our society correspond to existing kinds of prejudice that often take the form of explicit attitudes. This correlation supports the idea that implicit biases are likely beliefs even when we do not have full awareness or control of them.

The main objective of this chapter is nevertheless orthogonal to the dispute about which mental state better corresponds to implicit biases. Rather, its objective is to show that if our philosophical aim is that of identifying the nature of implicit biases, we will fall short if we only discuss the *type* of mental state they are. If implicit bias is to be understood as an independent phenomenon due to its relevance to daily life, we need to incorporate a discussion about what makes an implicit bias a "bias," so as to identify its nature. Independently of one's favored account of implicit bias (in terms of what kind of mental state it is), looking at its structure is insufficient to explain its nature (if implicit bias is to be accounted for as a proper class of phenomenon). We also need to look at its normative aspects. As such, the political aspects of implicit bias are not additional, independent factors, but rather central to the nature of the phenomenon.

Here is a brief argument in support of the critical part of my view. Biased and non-biased implicit attitudes are tested by the same indirect measurement tests. If such tests reveal the internal structure of implicit attitudes, they reveal the structure of both biased and non-biased implicit attitudes. Thus, the cognitive structure of such attitudes (be it associative or not) does not identify what makes an implicit attitude biased. If we fix the way the term is used in philosophy and why it matters, namely, that implicit biases are types of prejudice and thus involve ethical deviation (e. g., they perpetuate injustice; see Haslanger 2015), and perhaps also epistemic deviations, determining the cognitive structure of implicit attitudes falls short of identifying the nature of the phenomenon.

Here is the extended argument. Implicit measurement procedures measure a wide variety of implicit attitudes across a variety of topics (Nosek, Hawkins, and

Frazier 2011).[3] Implicit attitudes operate in a wide range of social behavior (Greenwald and Banaji 1995). Implicit bias is a subset of implicit attitudes that operate in prejudicial social behavior (definition). An implicit attitude is biased when it involves ethical and perhaps epistemological deviations from the relevant norms: it perpetuates injustice and may lack epistemic warrant (definition). Implicit measurement procedures demonstrate the presence of implicit attitudes in the individual's psychology and their cognitive structure (philosophical assumption).An implicit attitude can be non-biased. If implicit measurement procedures demonstrate the psychological structure of implicit attitudes, then they reveal the structure of both biased and non-biased implicit attitudes. Thus, the psychological structure of an implicit attitude does not tell us anything about why it is biased. Understanding the psychological structure of an implicit attitude does not explain the ethical (and possibly epistemological) deviation that can be ascribed to an individual who has implicit biases and who thereby engages in prejudicial behavior that operates in an implicit way.

The critical point of this chapter relies on the idea that the identification of the nature of a given phenomenon is partly a function of the role that phenomenon plays in our theories. If the interest in implicit bias were merely an interest in discovering parts of our unconscious mind and how they work, then perhaps identifying its structure and its corresponding mental state would suffice. However, interest in implicit bias is primarily rooted in our interest in learning about the subtle ways in which people think, act, and judge in prejudicial ways. Thus, understanding the phenomenon on its own requires dealing with its social and political aspects. This is why it is so important to provide an account of the bias in implicit bias.

If the cognitive structure of implicit biases does not explain why they are biased, one might think that the locus of the deviation lies in the cognitive processes involved in holding and activating implicit biases. On such an account, certain implicit attitudes are biased due to mistakes in the cognitive processing of information. One possible way to substantiate this account is to consult the literature on cognitive biases. I will analyze this possibility in what follows.

---

3 Here are some examples of implicit measurement: the Implicit Association test (Greenwald et al. 1998); the Affect Misattribution Procedure (Payne et al. 2005); the Weapon Identification Task (Payne 2001 and Correll et al. 2002). One criticism that has been raised against such tests is that, although implicit measurements are internally reliable, they do not correlate highly with each other (Fazio and Olson 2003).

3 Cognitive Biases

In this section, I examine an alternative way to identify the biased nature of implicit biases by looking at the cognitive processes that may be involved in their storage and activation. In the previous sections, I argued that if implicit bias deserves to be called a phenomenon of its own (given its theoretical and practical relevance), part of the theoretical task of identifying its nature lies in determining the sense in which implicit biases are *biased*. I have already argued that the mere psychological structure of implicit biases, which is an individualistic and apolitical factor, is not able to provide us with such an account. The structure does not identify implicit biases exclusively. In this section, I investigate whether research on cognitive biases, which are centered on processes and heuristics, can be of any help.

First, however, allow me to provide an explanation on the terminology I will be using. 'Cognitive bias' and 'implicit bias' pick out different phenomena. Implicit biases form a subset of implicit attitudes that involve prejudices toward members of socially stigmatized groups. By contrast, the term cognitive bias is applied to a number of tendencies (heuristics) that lead to systematic deviations from a given pattern of reasoning (see, e. g., Tversky and Kanneman 1974 and 1983). These tendencies are better understood as processes than as attitudes. Thus, although both phenomena involve implicit layers of cognition, they have different metaphysics. Additionally, whereas all cognitive biases seem to deviate in some way from accepted patterns of reasoning, it is still an open question whether all such deviations consist in epistemic mistakes. In fact, current research on heuristics and ecological rationality has tried to vindicate their rationality (see Gigerenzer and Goldstein 1996; Samuels et al. 2002; and Chater et al. 2003). By contrast, implicit biases seem necessarily to involve a deviation. They certainly involve a moral deviation to the extent that they are a species of prejudice. It is an open question whether they involve epistemic deviation as part of their nature. In what follows, I will examine possible ways of accounting for implicit biases by looking at cognitive heuristics that seem to generate epistemic deviations.

One cognitive bias that involves an epistemic deviation is the so-called base-rate fallacy. This heuristic is characterized by the "failure to take into consideration background information about the relative distribution of properties [or, the base rate]" (Gendler 2011, p. 34) when individuals calculate probabilities under uncertainty. Some philosophers (e. g., Gendler 2011) have suggested that looking at this heuristic is illuminating when it comes to understanding the epistemic aspects of implicit biases. One paradigmatic form by which implicit biases are activated in action and judgment is via the generalization of some negative trait to all members of a stigmatized group, who as a consequence are unjustly mistreated. One might therefore suggest that the storage and activation of implicit biases is the output of particular cognitive biases that have to do with a biased calculus of probability. The base-rate fallacy heuristic is a potential candidate for accounting for this.

Gendler (2011) refers to this heuristic in her vehement denial that it can be involved in the generation, storage, and activation of implicit biases. Against this possibility, she holds that implicit forms of prejudice (her most discussed example being racial prejudice against black people) are in place because our cognitive system encodes existing patterns. According to her argumentation, implicit biases do not arise in the base-rate fallacy and instead actually represent epistemic success in encoding existent patterns. This is the basis of her suggestion that trying to avoid encoding such patterns is *epistemically* costly.

Gendler's line of reasoning is quite debatable, if not altogether troubling.[4] The central problem with her argument, however, is that the base-rate fallacy is not the only heuristic involved in the probability calculus that might involve some sort of epistemic deviation. Thus, from the fact that implicit biases do not arise in the base-rate fallacy, it does not follow that they represent cognitive success. This is so because there are other ways in which implicit biases might count as *biased*, in epistemic terms.

---

4 Here is one example of a troubling aspect of Gendler's argument. Gendler explicitly argues that an implicitly racist person (in the US) whose actions are guided by the stereotype of black people's being violent and criminal does not commit the base-rate fallacy. Rather, the racist person has correctly encoded the pattern generated from the numbers according to which, at least in the US, there are more black people in prisons than white people. What Gendler does not explain, however, is why one undertakes (or should undertake) this particular probability calculus (i.e., whether a black person encountered in the street is criminal) rather than any other probability calculus. A less problematic way of accounting for Gendler's intuition is the following: an implicit bias can be stored and activated via fully adequate processes of information acquisition and processing. Someone who lives in a society where it is believed that green veggies are good and healthy sources of vitamins and nutrients does not commit a mistake in relating the concepts 'green vegies' and 'good health' more quickly than the concepts 'fast food' and 'good health.' Not only is such a connection reasonable, but the person's cognitive system is successful in maintaining the learned relation. A similar pattern seems in principle to occur with implicit biases. Someone living in an ageist society, who connects the concepts 'elderly' and 'not fully competent' more quickly than others, does not seem to make a mistake when activating the learned relation at the individual level (cf. Gendler 2011). In response one might insist, however, that not committing this particular mistake is not enough to render the person in question cognitively successful.

Bordalo et. al. (2016) propose a model of stereotypes based on Tversky and Kahneman's representativeness heuristics (1983). According to Tversky and Kahneman, "an attribute is representative of a class if it is very diagnostic; that is the relative frequency of this attribute is much higher in that class than in the relevant reference class" (Tversky and Kahneman 1983, p. 296). The person overweighs its representative types when assessing a target group via the application of the representativeness heuristics. "The most representative types come to mind first and so are overweighed in judgments" (Bordalo et. al. 2016, p. 1755), which result in distorted distributions or stereotypes. Here is one illuminating example from (Bordalo et. al. 2006, p. 1756):

Consider the stereotype "Florida residents are elderly." The proportion of elderly people in Florida and in the overall United States population is shown here:

Age	0 – 19	20 – 44	45 – 64	65+
Florida	24.0 %	31.7 %	27.0 %	17.4 %
US	26.9 %	33.6 %	26.4 %	13.1 %

The table shows that the age distributions in Florida and in the rest of the United States are very similar. Yet someone over 65 is highly representative of a Florida resident, because this age bracket maximizes the likelihood ratio Pr (t/Florida)/Pr (t/US).[5] When thinking about the age of Floridians, then, the 65+ type immediately comes to mind because in this age bracket Florida is most different from the rest of the United States, in the precise sense of representativeness. Representativeness-based recall induces an observer to overweight the 65+ type in their assessment of the average age of Floridians.

Bordalo et. al (2006) argue for the idea that there is something true at the basis of all stereotypes, including those involving prejudiced belief, whether explicit or implicit. In their own words, "Stereotypes formed this way contain a 'kernel of truth': they are rooted in true differences between groups" (Bordalo et. al 2006, p. 1753). This seems to bring their analysis close to Gendler's (2011). In sharp contrast with Gendler (2011), however, Bordalo et. al. (2016) highlight that a stereotype may be wrong due to a heuristic that amplifies the difference between the representative types of different groups. The person encodes a

---

5 Bordalo et. al. (2006, p. 1755) follow the terminology used by Gennaioli and Shleifer (2010) in assuming "that a type t is representative for group G relative to a comparison group −G if it scores high on the likelihood ratio: Pr(t/G)/Pr(t/ −G)."

given pattern of reality using a heuristic that can itself be taken to be epistemically faulty. In reference to the above example, they write the following:

> This example also illustrates how stereotypes can be inaccurate. Indeed, and perhaps surprisingly, only about 17% of Florida residents are elderly. The largest share of Florida residents, nearly as many as in the overall U.S. population, are in the age bracket 19–44, which maximizes Pr(t/Florida). Being elderly is not the most likely age bracket for Florida residents, but rather the age bracket that occurs with the highest relative frequency. A stereotype-based prediction that a Florida resident is elderly has very little validity. (Bordalo et. al. 2006, p. 1756)

One consequence is that, if implicit biases involve implicit and prejudiced stereotyping, and the model proposed by Bordalo et. al. correctly characterizes how social stereotypes are formed, they arise in a particular form of epistemic deviation, which is the output of the representativeness heuristic. Nevertheless, their approach still falls short of our objective of characterizing the nature of implicit biases, in particular, since the representativeness heuristic may result in both biased and non-biased stereotypes. The inaccurate judgment that Florida residents are elderly is not biased in the sense that implicit biases are to the extent that they are not prejudicial.

The conclusion of this section is that focusing on the cognitive process that may be involved in implicit biases does not help us to identify either the ways in which they might succeed or the ways in which they fail. The mere fact that they are neither specific nor exclusive to implicit biases shows that a deviation at the level of the cognitive process does not explain when and why implicit attitudes count as biased. In the next section, we will turn to social perspectives on implicit biases in an attempt to identify what is biased about implicit biases.

4 Social Accounts of Bias

The two previous sections suggest that a purely individualistic account of implicit biases cannot explain the nature of the phenomenon. Individualistic aspects of implicit biases, such as their cognitive structure and any related implicit cognitive mechanisms, cannot provide an account of what is biased about implicit biases. As previously noted, however, if implicit bias is to be treated as a phenomenon of its own, given its theoretical and practical importance, a full account of its nature needs to explain what is biased about implicit biases. In a nutshell, saying that these are implicit, automatic attitudes that may be the output of some heuristic falls short of explaining what implicit biases are.

Let us turn to a different perspective, based on non-individualistic factors. Sally Haslanger (2015) is a central author who proposes that there is an intrinsic connection between implicit bias and social factors. She provides a picture of the nature of implicit bias based on social factors by defending the idea that implicit biases are social schemas: "Roughly, schemas are clusters of culturally shared (public) concepts, propositions, and norms that enable us, collectively, to interpret and organize information and coordinate action, thought, and affect" (Haslanger 2015, p. 4. In this sense, Haslanger's account is able to explain both the frequent dissonance) between individuals' explicit and endorsed ideas (e. g., about 'egalitarianism') and their implicit biases. It also provides a hint regarding the source of these biases. As social schemas, implicit biases have a different nature when compared with explicit beliefs: they do not really belong within the individual's psychology. By moving from the individual to the social sphere, as Haslanger's picture does, we can perhaps find a way to explain what is biased in implicit biases by understanding how they feed and perpetuate unjust practices.

Haslanger's view is illuminating when it comes to understanding the social aspect of implicit biases. Nevertheless, if we treat her view—which denies that implicit biases are proper beliefs held by individuals and are instead social clusters of concepts—in the way that the current discussion on implicit bias within the philosophy of mind does, it remains trapped within the traditional discussion of the type of mental state that implicit biases are. As argued above, deciding on the type of mental state that implicit biases are will not provide us with an explanation of what is *biased* about them. Although I do not share the metaphysics of implicit bias proposed by Haslanger, this is not a proper criticism of her account. The dispute about the metaphysics of implicit biases, which involves the question of what kinds of mental states they are, must be decided elsewhere. What I am claiming is that shifting the locus of implicit bias from the individual's psychology to the social realm does not seem to explain why it constitutes a deviation unless we properly address that question. If there are concepts that belong to a society's repository, it is very unlikely that this fact will be exclusive to implicit biases. A society's conceptual repository will also likely have accurate shared concepts. Thus, even if we avoid resorting to a purely individualist discussion of implicit bias, this approach will not lead us to a precise answer to the question about the deviation at the heart of implicit biases.

5 The Political Turn in the Philosophy of Mind after Implicit Biases

Research on the nature of prejudice is not new to philosophy, or to psychology (see Allport 1954), and some of the recent disputes on the nature of prejudice incorporate clear questions from the philosophy of mind. Jorge Garcia (1996), for example, holds that prejudice is an affective mental state directed against the welfare of those who belong to the stigmatized group; it is centrally an affective state of hatred. According to Garcia, racism, as a central case of prejudice, is always morally wrong. In opposition to Garcia's view, Tommie Shelby (2002) argues that prejudices are corrupted beliefs. According to Shelby, racism is fundamentally a type of ideology and, as such, a set of "widely accepted illusory systems of belief that function to establish or reinforce structures of social oppression" (Shelby 2002, p. 415). Racism, in particular, is organized around the idea that the essential nature of some groups (as races) explains why it is appropriate to exploit or disadvantage them. If cases of prejudice involve the affective mental state that is central to Garcia's view, this is because prejudiced beliefs lie at their basis.

A third view on prejudices can be derived from Kate Manne's account of misogyny (2017), according to which misogyny, that is, hostility towards women and girls because of their gendered social role, serves the function of enforcing the patriarchal order. According to Manne, misogyny does not need to arise from any particular psychological attitude, such as belief (as in Shelby's view) or hatred (as in Garcia's view). What is central to understanding misogyny and related prejudiced phenomena such as sexism is their role in the maintenance and enforcement of a certain order, such as patriarchal norms with gendered content.

My suggestion in briefly citing these three views on prejudice is that the current discussion about implicit attitudes in the philosophy of mind would make progress if it were informed by the question of the political role of prejudices. Because implicit biases involve a particular form of prejudice, and prejudices concern corrupted ways of treating certain social groups, they are political phenomena. Having them and acting on them promote certain social structures that benefit certain groups at the expense of others. Precisely because the primary question in the philosophy of mind concerns the constitutive aspects of certain psychological phenomena, it is crucial to understand implicit biases as phenomena that play this particular political role. I suggest that the question of the deviation of implicit biases connects to this aspect. Apolitical discussions of implicit bias, which seem to populate the current discussion on the topic, risk distorting the phenomenon as a result of this methodological failure.

The above-cited work by Haslanger is a clear exception to the standard apolitical treatment of implicit biases. Because my view of the metaphysics of the phenomenon contrasts with hers, however, allow me to attempt to delineate an alternative way to account for implicit biases, following the program proposed in this paper (according to which discussion of the nature of implicit bias should include discussion about what makes it biased).

As previously mentioned, my view is that implicit biases are beliefs. In contrast to Haslanger's view, I find it difficult to deny that implicit bias is partly a psychological phenomenon that has its locus within an individual's psychology. The fact that the individual's endorsed attitudes contrast with her implicit biases is not a reason against its belonging to the individual's psychology. The fact that the subject ignores and lacks full control over her implicit biases indicates that we might ignore a lot about our psychology, including our held attitudes. I take this to be a fact about our psychology. Furthermore, conceiving of implicit biases as a type of attitude that belongs within an individual's psychology is compatible with their playing a special social and political role in perpetuating injustice.

Based on my favored framework on implicit biases, let me *sketch* a view of what is biased about implicit biases, based on Shelby's view of prejudices (2002). Due to social practices in which certain social groups take advantage of the injustices promoted by implicit biases, their ordinary function of being true and warranted is corrupted. As belief states, one of the functions of implicit biases is to be true. Beliefs fail when they are false. Because the functions of implicit biases are corrupted by social and political agendas, however, they do not per se aim at the truth when disseminated in society. It is as if, when encoding a certain implicit bias and acting on it, one were encoding a malfunctioning belief. The malfunction of such a belief is due to the illusory system of beliefs that promotes social injustice. If one implicit bias happens to have a true content, its being true seems to be a matter of coincidence, or the result of doxastic luck, so to speak.[6]

This is just a sketch of what a full account of implicit biases could look like—one that takes into account their role in the social sphere in which they operate. The aim of this paper is programmatic to the extent that it calls attention to the current state of the research on implicit biases. Traditional questions in the philosophy of mind concerning the nature of our mental states are legitimate. Nev-

---

6 One way in which an implicit or explicit biased attitude can be true is when the generalization that it embeds corresponds to reality. However, I am claiming that the main function of implicit biases seems to be that of supporting a socially corrupted view about certain social groups rather than accurately describing them. This is why I am suggesting that if a given implicit bias turns out to be true, it is a matter of doxastic luck.

ertheless, implicit biases are not merely mental states, if by that we mean a neutral, apolitical phenomenon.[7] Implicit biases are centrally political: they have to do with how social groups are conceived of in ways that promote injustice. Implicit biases are particularly interesting because they involve implicit layers of cognition. They are particularly challenging because the process of fully updating and revising these implicit layers is still a mystery. When it comes to implicit biases, however, questions about the control and revision of implicit layers are not merely theoretically important. They are important because implicit biases play a key role in perpetuating injustice. As I have argued, understanding the bias in implicit biases is thus part of the task of understanding implicit biases.

The general lesson one can draw from these observations is that once discussion of implicit biases entered the philosophy of mind, previous apolitical and neutral psychological phenomena began to be reconsidered. If in the past much of the discussion on belief and belief content involved neutral propositions such as 'The grass is green' and 'The snow is white', nowadays, interestingly, philosophers have turned to the constitutive aspects of prejudicial propositions that seem to be part of the many implicit biases at work in society. The result is that there is no going back to engaging in the philosophy of mind in a way that does not take into account the context, function, and political aspects of such states.

Bibliography

Allport, Gordon W. (1954): *The Nature of Prejudice*. Cambridge: Perseus Books.
Bendaña, Joseph (2021): "Implicit attitudes are (probably) beliefs." In: Cristina Borgoni, Dirk Kindermann, and Andrea Onofri (Eds.): *The Fragmented Mind*. Oxford: Oxford University Press.
Bordalo, Pedro, Katherine Coffman, Nicola Gennaioli, and Andrei Shleifer (2016): "Stereotypes." In: *The Quarterly Journal of Economics* 131. No. 4, pp. 1753–1794.
Borgoni, Cristina (2015a): "Dissonance and Moorean Propositions." In: *Dialectica* 69. No. 1, pp. 107–127.
Borgoni, Cristina (2015b): "Dissonance and Doxastic Resistance." In: *Erkenntnis* 80. No. 5, pp. 957–974.

---

7 To give a comparison class, delusion (see Bortolotti 2009) is a further phenomenon the nature of which goes beyond the identification of its corresponding mental states. Analogously to how I argue about implicit biases, it looks like one falls short of explaining what delusions are if one limits oneself to the debate on what mental states they are. Key to understanding delusions is understanding their specific clinical aspects, and perhaps the source of the relevant psychological inaccuracies.

Borgoni, Cristina (2016): "Dissonance and Irrationality: A Criticism of the In-Between Account of Dissonance Cases." In: *Pacific Philosophical Quarterly* 97. No. 1, pp. 107–127.

Borgoni, Cristina (2018): "Unendorsed Beliefs." In: *Dialectica* 72. No. 1, pp. 49–68.

Bortolotti, Lisa (2009): *Delusions and Other Beliefs*. Oxford: Oxford University Press.

Brownstein, Michael (2019): "Implicit Bias." In: *The Stanford Encyclopedia of Philosophy* (Fall 2019 Edition), Edward N. Zalta (Ed.), URL = <https://plato.stanford.edu/archives/fall2019/entries/implicit-bias/>

Chater, Nick, Mike Oaksford, Ramin Nakisa, and Martin Redignton (2003): "Fast, frugal, and rational: How rational norms explain behavior." In: *Organizational Behavior and Human Decision Processes* 90, pp. 63–86.

Correll, Joshua, Bernadette Park, Charles M. Judd, and Bernd Wittenbrink (2002): "The police officer's dilemma: Using ethnicity to disambiguate potentially threatening individuals." In: *Journal Pers Soc Psychol.* 83. No. 6, pp. 1314–1329.

De Houwer, Jan (2014): "A propositional model of implicit evaluations." In: *Social and Personality Compass* 8. No. 7, pp. 342–353.

Fazio, Russell, and Michael Olson (2003): "Implicit Measures in Social Cognition Research: Their Meaning and Use." In: *Annual Review of Psychology* 54. No. 1, pp. 297–327.

Frankish, Keith (2016): "Playing Double: Implicit Bias, Dual Levels, and Self-Control." In: Michael Brownstein and Jennifer Saul (Eds.): *Implicit Bias and Philosophy Volume 1: Metaphysics and Epistemology*. Oxford: Oxford University Press, pp. 23–46.

Garcia, Jorge L. A. (1996): "The Heart of Racism." In: *Journal of Social Philosophy* 27. No. 1, pp. 5–45.

Gendler, Tamar (2008): "Alief and belief." In: *The Journal of Philosophy* 105. No. 10, pp. 634–663.

Gendler, Tamar (2011): "On the epistemic costs of implicit bias." In *Philosophical Studies* 156, pp. 33–63.

Gennaioli, Nicola, and Andrei Shleifer (2010): "What Comes to Mind." In: *Quarterly Journal of Economics* 125, pp. 1399–1433.

Gigerenzer, Gerd, and Daniel Goldstein (1996): "Reasoning the Fast and Frugal Way: Models of Bounded Rationality." In: *Psychological Review* 103. No. 4, pp. 650–669.

Greenwald, Anthony, and Mahzarin Banaji (1995): "Implicit Social Cognition: Attitudes, Self-esteem, and Stereotypes." In: *Psychological review* 102. No.1, pp. 4–27

Greenwald, Anthony, Debbie E. McGhee, and Jordan L. K. Schwartz, (1998): "Measuring Individual Differences in Implicit Cognition: The Implicit Association Test." In: *Journal of Personality and Social Psychology* 74. No. 6, pp. 1464–1480.

Haslanger, Sally (2015): "Social Structure, Narrative, and Explanation." In: *Canadian Journal of Philosophy* 45. No. 1, pp. 1–15.

Levy, Neil (2015): "Neither Fish nor Fowl: Implicit Attitudes as Patchy Endorsements." In: *Noûs* 49. No. 4, pp. 800–823.

Madva, Alex (2016): "Why implicit attitudes are (probably) not beliefs." In: *Synthese* 193. No. 8, pp. 2659–2684.

Mandelbaum, Eric (2016): "Attitude, Association, and Inference: On the Propositional Structure of Implicit Bias." In: *Noûs* 50. No. 3, pp. 629–658.

Manne, Kate (2017): *Down Girl: The Logic of Misogyny*. Oxford: Oxford University Press.

Nosek, Brian, Carlee Hawkins, and Rebecca Frazier (2011): "Implicit social cognition: From measures to mechanisms." In: *Trends Cognitive Science* 15. No. 4, pp. 152–159.

Nosek, Brian, Anthony Greenwald, and M Mahzarin Banaji (2007): "The Implicit Association Test at Age 7: A Methodological and Conceptual Review." In John Bargh (Ed.): *Automatic Processes in Social Thinking and Behaviour.* New York: Psychology Press, pp. 265–292.

Payne, B. K. (2001): "Prejudice and perception: The role of automatic and controlled processes in misperceiving a weapon." *Journal Pers Soc Psychol.* 81. No. 2, pp. 181–192.

Payne, B. Keith, Clara M. Cheng, Olesya Govorun, and Brandon D. Stewart (2005): "An Inkblot for Attitudes: Affect Misattribution as Implicit Measurement." In: *Journal of Personality and Social Psychology* 89, pp. 277–293.

Samuels, Richard, Stephen Stich, and Michael Bishop (2002): "Ending the Rationality Wars: How to Make Disputes About Human Rationality Disappear." In: Renee Elio (Ed.): *Common Sense, Reasoning and Rationality.* New York: Oxford University Press, pp. 236–268.

Shelby, Tommie (2002): "Is Racism in the 'Heart'?" In: *Journal of Social Philosophy* 33. No. 3, pp. 411–420.

Shiffrin, Richard, and Walter Schneider (1977): "Controlled and Automatic Human Information Processing: Perceptual Learning, Automatic Attending, and a General Theory." In: *Psychological Review* 84, pp. 127–190.

Tversky, Amos, and Daniel Kahneman (1974): "Judgement under Uncertainty: Heuristics and Biases." In: *Science* 185, pp .1124–1131.

Tversky, Amos, and Daniel Kahneman (1983): "Extensional vs. Intuitive reasoning: The Conjunction Fallacy in Probability Judgement." In: *Psychological Review* 90, pp. 293–315.

Emily C. McWilliams
Ameliorative Inquiry in Epistemology

Abstract: Recently, some work in feminist epistemology has received more up-take from mainstream western analytic epistemology than it had in the past. There has been recognition of the importance of topics like epistemic injustice, standpoint epistemology, and epistemologies of ignorance, for instance. But these discussions are still often seen as orthogonal to core epistemic theorizing. Generally, they have not received uptake as fundamental contestations of the ways we understand epistemic value, or core normative epistemic concepts. I suggest that one reason for this is the perception that insofar as feminist theorizing is responsive to moral and political concerns, it is not doing epistemology because it is not theorizing about epistemic value. This perception assumes a specific kind of epistemic value monism – a view whose popularity, I argue, derives at least partly from features of the methodologies that are popular in mainstream epistemology. I show that by using a different type of methodology—ameliorative inquiry—we see that there is principled reason to doubt epistemic value monism. We can thus understand feminist theorizing in ways that are richer, more accurate, and that contribute to our understanding of the ways in which our practical, moral, and epistemic agency are intertwined.

> to suppose that what we value epistemically is what we ought to value epistemically is to leave the normative part of normative epistemology undone. (Haslanger 1999, p. 466)

1 Introduction

Recently, some work in feminist epistemology has gotten more uptake from mainstream western analytic epistemology than it has in the past.[1] There has been more widespread acceptance of the relevance and importance of topics like epistemic injustice and epistemic oppression, standpoint epistemology, and epistemologies of ignorance, for instance. But even as the importance of these discussions is increasingly granted, they are still often seen as orthogonal to core epistemic theorizing—theorizing about concepts like rationality, justification, and knowledge. As Lisa Miracchi has recently explained, "It is common

---

1 Hereafter, I will use *mainstream epistemology* as a shorthand for contemporary western analytic epistemology.

https://doi.org/10.1515/9783110612318-009

these days to accept the value and importance of this work, but to think that the normative issues these theorists are concerned with are really ethical and political, rather than epistemic (Mirachhi 2020, p. 4)." Miracchi argues that this is the wrong way to see things, and I agree. She focuses on arguing against the idea that feminists' responsiveness to moral, political, and practical concerns undermines the objectivity of their inquiry.[2] The present chapter turns our attention to a related concern, one that I take to be even more fundamental: the idea that insofar as their theorizing is responsive to these practical and moral concerns, feminists are simply not doing epistemology, because they are not theorizing about epistemic value. I will respond to this claim that feminist epistemology fails to be epistemology.[3]

The claim relies on a specific picture of epistemic value: a monist picture on which there is just one fundamental epistemic value, which is autonomous of the practical and moral values that feminist work is responsive to. Though this monist picture of epistemic value is popular in many parts of contemporary analytic epistemology, I will argue that we have principled reason, independent of its dialectical role vis-à-vis feminist epistemology, to doubt it. Moreover, I suggest that monism's popularity is at least partly explained by the methodologies that we often use in mainstream epistemology. These methodologies are conservative in their approach to theorizing epistemic value, so that entrenched assumptions about it tend to stick. I will show that by using a different type of methodology— ameliorative inquiry—we make space to critically examine these assumptions and ultimately show that we have principled reasons to doubt epistemic value monism. If so, then the fact that feminist theorizing responds to non-monist values does not entail that feminists are not theorizing about epistemic value, and not doing epistemology.[4]

---

2 As the above quote demonstrates, Miracchi initially frames the concern as one about epistemic vs. moral and practical values. But she then focuses on the claim that responding to these values undermines objectivity. The worry about objectivity is not reducible to the worry about epistemology. I say more about how her focus differs from mine in section 3.

3 Thanks to the anonymous referees for pointing out that the way I am using the term 'feminist epistemology' is not solely related to gender issues. Indeed, I take it to include any epistemic theorizing that is done from a feminist standpoint. Roughly, a feminist standpoint is one that uses an intersectional lens to attend to how social power and social categories operate to cause and perpetuate oppression and injustice. Feminist epistemology applies this lens to the epistemic realm.

4 The claim that in responding to non-monist values, theorists fail to do epistemic theorizing need not be exclusively applied to feminist epistemology. It would, for the same reasons, apply to work on pragmatic and moral encroachment in epistemology (see, e.g., Moss 2018 and Stanley 2005. Nonetheless, the effects of making this claim against feminist epistemology may differ from

2 Veritistic Value Monism

The idea that either truth itself, or true/accurate belief is the sole fundamental epistemic value has been popular in mainstream epistemology.[5] Indeed, a number of epistemologists either argue or stipulate that the value of having true beliefs and avoiding false ones is the sole fundamental epistemic value, in that it explains why other things have epistemic value.[6] Their underlying assumption is that there must be some fundamental, distinctly epistemic value, to which all other epistemic values are reducible. Some refer to the idea that the fundamental distinctly epistemic value is truth or true/accurate belief as *veritism*, following Alvin Goldman (1999), while others take veritism to include an additional claim that truth is the norm and the aim of belief.[7] To avoid this ambiguity, I will borrow the term *veritistic value monism* (hereafter, *monism*, for short) from Kristoffer Ahlstrom-Vij (2013).[8] For our purposes, I take the monist's thesis to be that all epistemic value derives from and is reducible to the final value of either truth or true/accurate belief.

From the monist's perspective, feminists' inclusion of practical and moral interests in their theorizing means that they are not doing epistemic theorizing because they are not theorizing about epistemic value. Which feminist projects might the monist include here, and in what ways are their projects responsive to values that monism counts as non-epistemic? Miracchi points out that "we can think of the work of such thinkers as Dotson (2011), Fricker (2007), Jenkins (2016), McKinnon (2016), Medina (2012), and Peet (2017) as developing concepts that help us understand the ways in which our epistemic practices are unjust, and which help us reason individually, interpersonally, and publicly in ways that resist such injustice" (Miracchi 2020, p. 4). Insofar as justice is not reducible

those of making it against the encroachment literatures, given the ways in which feminism is politicized both within philosophy and beyond. I will say more about these effects in section 4.
5 There is some debate about whether truth or true belief is the fundamental epistemic value. I follow Duncan Pritchard (2014) in referring to it as truth, as he coined the term *truth monism*. But this will not make a difference for my purposes.
6 See, e. g., Alston 1985; Ahlstrom-Vij 2013; Bonjour 198; Leplin 2009; Sosa 2007; and Stapleford 2016.
7 See Tamer Nawar (Forthcoming) for more on these two distinct theses, and how the term veritism has been applied to them.
8 Duncan Pritchard (2014) coined the term *truth monism* to refer to this position. I find the term *veritistic value monism* useful as it highlights the connection to veritism, which has been a subject of robust debate in contemporary analytic epistemology.

to the final value of truth, monism says these thinkers are not doing epistemic theorizing, as the concepts they develop are not concerned with epistemic value.

How might we understand the situation differently, and what might it contribute to our understanding of these thinkers' concepts, if we do not assume, with the monist, that epistemic value is autonomous of moral and pragmatic values? I will take Miranda Fricker's (2007) foundational concept of epistemic injustice as an illustrative example, as most of the other work that Miracchi mentions responds to Fricker in meaningful ways. Fricker says that epistemic injustice is distinctly epistemic in kind because it consists "most fundamentally in a wrong done to someone specifically in their capacity as a knower" (Fricker 2007, p. 1). But if the monism is correct, then this is not sufficient, since an epistemic problem only arises if suffering an epistemic injustice prevents one from gaining some true belief(s). Otherwise, epistemic injustice is simply a moral wrong that happens to befall one in the course of seeking or sharing knowledge.

Though it is not clear whether Fricker herself would reject the foregoing analysis,[9] I think there is a deeper sense in which the concept of epistemic injustice names a hybrid moral/epistemic problem, and thus illustrates an important way in which the monist picture of epistemic value is inadequate. Although a full defense of the following claim is beyond the scope of this chapter, I submit that part of what underlies the significance of epistemic injustice, and shows that it is not simply a moral problem that happens to occur in the course of epistemic activity, is that final epistemic value inheres in being able to exercise our epistemic agency, both independently and in cooperation with other people, given the kinds of creatures that we are.[10] Since this final value is not reducible to the value of the true beliefs that we gain by doing so, it can only count for the monist as practical or moral value. I think this fails to capture an important part of what is significant and interesting about epistemic injustice. I will return to the idea that final epistemic value inheres in exercising our epistemic agency in section 5.6. For now, the more general point is that the possibility of understanding these feminist concepts in ways that may be richer, more accurate, and contribute more to our understanding of the ways in which our practical, moral, and epistemic agency are intertwined requires moving beyond a monist picture of epistemic value.

---

9 In the beginning of her chapter on the genealogy of testimonial justice, Fricker says that "testimonial justice is a genuine hybrid—both ethical and intellectual—for it aims at once at truth and justice" (Fricker 2007, p. 109). This is certainly consistent with conceiving of these as separate realms of value, in the way monism does.
10 This goes beyond what Fricker herself says about why epistemic injustice is a distinctly epistemic problem. It is my own view.

Having established that I will take veritistic value monists to be my interlocutors, I am now in a position to say more precisely how this chapter's argument and upshot differs from that of Miracchi's (2020) recent and convincing argument that social justice concerns have a valid place in our epistemic theorizing. While Miracchi responds to a concern that feminist epistemology's responsiveness to broader values like social justice undermines its objectivity, I respond to a concern that it undermines its status as epistemology. Miracchi's opponent thinks that to be objective, epistemic inquiry must be prior to and independent of ethical and social justice concerns. Miracchi responds by explaining that the same kind of responsiveness to broader values routinely and helpfully occurs in scientific inquiry, where the kinds and methodology are nonetheless agreed to be objective.[11] In fact, responding to such values in scientific inquiry positively contributes to its objectivity, by helping us reveal the structures we aim to understand (for example, since medical categories like health and disease are ethically loaded, ethical considerations are crucial in helping us reveal them). The monist's concern is different. They can agree with Miracchi that broader values like social justice help us to theorize in ways that better capture the structure of the objective, scientific world, particularly as concepts like health are ethically loaded. This does not bear on their view that such values have no place in helping us theorize the epistemic world. To think otherwise, they insist, is to make a category mistake.

3 Ameliorative Inquiry and Methodology in Mainstream Epistemology

Sally Haslanger's (2005 and 2012) notion of *ameliorative inquiry* describes one way that philosophers can go about developing concepts and theories of kinds. The methodology's distinguishing feature is that it is normative and revisionary. Ameliorative inquiry starts by asking: What legitimate reasons do we have for wanting a concept or theory of X in the first place? What work should we want it to do for us? Then, it aims to develop a theory that is up to the task. Haslanger explains that by using this two-step process, ameliorative inquiry can enhance our conceptual and theoretical resources to serve our (critically examined) purposes. The methodology has gained significant uptake amongst philosophers working on social kinds and social ontology. Such projects have a natural

---

11 See Elizabeth Anderson (1995) for a related argument that science, when done correctly, is value-laden.

answer to the question of what work we should want a theory of X to do, as their ultimate aim is to make progress toward social justice by improving our concepts and theories of social kinds. Much of feminist epistemology also aims to make progress toward social justice by attending to the social and political aspects of our epistemic practices, and many projects in feminist epistemology have drawn either implicitly or explicitly on ameliorative methodology. Ameliorative inquiry has not been used as much in mainstream epistemology.

Haslanger distinguishes ameliorative inquiry from two other methodologies that philosophers use in developing concepts and theories of kinds. I will outline them. Then, I will point out how their use in epistemology may partly explain the ongoing popularity of the monist picture of epistemic value.

The first methodology is *conceptual inquiry*, which aims to achieve an articulation or a clarification of the concept we apply when we use a corresponding term in ordinary language. Conceptual inquiry proceeds by reflection on examples of how concepts are used, with the goal of determining whether there are principles that unite them. This *a priori* methodology has often been used in mainstream epistemology to help us to articulate precisely what we mean by terms like justification and knowledge.

The second methodology is *descriptive inquiry*. It seeks to determine whether there is any objective kind out there in the world that our use of a concept or kind term tracks. It starts by identifying paradigm cases that function to fix the referent of the term, and then it draws on empirical research to determine whether there is an objective kind to which the paradigms belong, and where its boundaries are. The goal is to develop concepts and theories of kinds that are more accurate than the ones we currently operate with. This methodology has been used by projects in naturalized epistemology, which draw on empirical research from fields like cognitive science, to see, for instance, whether there is an objective type that paradigm cases of knowledge belong to.

Both conceptual and descriptive inquiry have often been used by mainstream epistemologists in theorizing concepts like justification, rationality, and knowledge. Both methods are useful and important. But they are also inherently conservative, in that they start by fixing the subject matter with respect to the descriptive question of what our extant concepts of epistemic value are, rather than the normative question of what they should be. This can preclude us from seeing feminist epistemology as theorizing about epistemic value, insofar as its understandings of what that consists in are revisionary.

To explain how these methods are conservative, I will first examine conceptual inquiry, and then explain the relevant similarity to descriptive inquiry. Conceptual inquiry looks for an articulation or an analysis of the concept we take ourselves to be applying when we use the corresponding term.[12] There are two ways one could go about this: either by using reflective equilibrium, or not. When we are not using reflective equilibrium, we look for an analysis that both captures our ordinary notion exactly, in that it gives us necessary and sufficient conditions for something's being in the extension of the concept, and that is exhaustive, in that it does not admit of counterexamples. This is a very conservative methodology, since it leaves no room for revision of the target concepts.

When we are using reflective equilibrium,[13] we look to our ordinary use of the term or concept not as a final court of appeals, but rather to fix the subject matter. In epistemology, we look to the values embedded in our ordinary normative epistemic concepts to figure out what it is that we value in epistemic matters, and we use this to fix ideas about what is in fact valuable epistemically. We can see this, for example, in Lawrence Bonjour's (1985) widely cited argument for a monist view of epistemic justification. He argues that we can distinguish epistemic justification from non-epistemic species of justification by reflecting on the implicit rationale of the concept of knowledge itself. He observes that the core idea embedded in this concept is that our main cognitive goal is to have true belief (Bonjour 1985, p. 7).[14] He uses this to argue that the basic role of epistemic justification is that of a means to truth. In this way, conceptual inquiry with reflective equilibrium also uses the epistemic values embedded in our ordinary concepts to fix ideas about what is in fact valuable epistemically.

Descriptive inquiry in epistemology also uses the epistemic values embedded in our extant concepts—either ordinary or theoretical ones—to fix the subject matter. For instance, projects in naturalized epistemology do not start with empirical information and use it to construct normative theories from the ground up. They start by using the epistemic values embedded in extant normative epistemic concepts to fix the subject matter. The key difference between conceptual and descriptive inquiries is only that conceptual inquiry uses *a priori* methods to refine the concepts and theories, while descriptive inquiry uses empirical information to do so. But both are doing such refinement only after having used the descriptive question of what epistemic values are embedded in our established concepts to fix ideas about what is in fact epistemically valuable.

---

12 This is what Haslanger (1995) calls the *manifest* concept as opposed to the *operative* one.
13 Distinctions between the many different versions of reflective equilibrium will not matter for my purposes. We can understand it simply as the process by which one aims to balance judgments about particular cases with theoretical commitments.
14 Bonjour does not explicitly identify himself as doing conceptual inquiry/analysis with reflective equilibrium. His book predates Haslanger's distinction amongst conceptual, descriptive, and ameliorative inquiry.

Haslanger points out that there is something strange about this:

> If we ... take the primary task of epistemology to be a normative investigation into knowledge—one investigating how we *ought to* reason, on what basis we *ought to* form beliefs, and more generally what is epistemically *valuable*—then there is something peculiar about pursuing an 'immanent' strategy in epistemology that undertakes simply to describe 'our' concept, or to discover the (natural?) kind we ordinarily refer to ... to suppose that what we value epistemically is what we ought to value epistemically is to leave the normative part of normative epistemology undone. (Haslanger 1999, p. 466)

Conceptual and descriptive inquiries are inherently conservative in that their primary approach to the question of what epistemic value consists in is a descriptive one that uses the values embedded in established normative epistemic concepts to fix the subject matter. So, insofar as work in feminist epistemology is responsive to additional values and interests, using these methodologies will tend to support the monist's claim that feminists are not doing epistemic theorizing. Haslanger's point pushes back on this. She points out that insofar as the primary task of epistemology is normative, it is strange to put so much weight on the descriptive question of what values are embedded in our extant concepts.

4 Why Does it Matter?

Given the ways in which they are conservative, the exclusive use of conceptual and descriptive inquiries makes it difficult or impossible to contest the dominant normative frameworks in epistemology. They reinforce the monist idea that insofar as a project challenges dominant normative frameworks, it is not doing epistemology.

This point has important resonances with insights that have come out of feminists' critical engagements in other areas of philosophy, and with critical theory more generally. To take just one example, Fiona Jenkins's (2019) work on gendered innovation argues that the operative model of innovation in many parts of social science makes a certain theoretical framework obligatory, namely, a certain positivist account of improving knowledge, which she argues has long benefitted those who study it.[15] Jenkins makes a convincing case

---

15 Jenkins argues that the accepted model is particularly pernicious because it embeds a worldview that has benefitted men and the dominantly situated, while at the same time being presumed neutral with respect to situated interests. That is, it benefits the dominant under a presumption of neutrality. Feminist epistemologists have argued that the same thing has happened

that, as a result, the fields' dominant discourse (a) fails to recognize the historical contingency in how its key terms have been defined; and, (b) disallows a fundamental contestation of those terms, since innovations that challenge the basics of mainstream theory cannot readily be mainstreamed. This leads to fragmentation of the discipline into disparate sub-fields with autonomous networks and reward systems, further preventing engagement with critical standpoints, and making it again more difficult to challenge prevailing ideas.

I worry that these same patterns of fragmentation are playing out in epistemology. Even as some feminist work has begun to receive more attention within dominant circles, it has not received uptake *as* a contestation of dominant normative frameworks, or relatedly, as part of the project of core epistemic theorizing, for reasons that Haslanger and Jenkins identify. This is bad for epistemology as a field because it restricts what could be a forum for a very rich set of discussions to focus instead on a narrow range of traditional concerns, such as what the correct monist analyses of knowledge and justification are. Meanwhile, it sidelines conversations that bear on the broader roles and importance of epistemic processes and activities in human life and society.

It is worth noting that the sidelined conversations are by no means limited to those that specifically understand themselves as feminist. And the argument that we have principled reason to doubt monism has much broader relevance. Nonetheless, one might worry more acutely about the effects of sidelining of feminist work because it manifests or compounds the ways that politicizing feminism can contribute to gatekeeping within the field. The complaint that feminist epistemology is not really epistemology is voiced primarily in offhand remarks, the general form of which is that while of course feminist work is interesting and important, at the end of the day it is concerned more with the sociology of knowledge and inquiry than with core epistemic theorizing. And while certainly there is nothing inherently wrong with work on the sociology of knowledge and inquiry, in practice these classifications are used to divide serious —that which is worthy of consideration by serious philosophers—from everything else.

---

with respect to dominant frameworks for understanding fundamental concepts like knowledge in epistemology.

5 Using Ameliorative Inquiry to Argue against Veritistic Value Monism

This section argues that using ameliorative inquiry in epistemology gives us the resources to argue that we have principled reason to doubt veritistic value monism. If so, then the fact that feminists and others are responsive to non-monist values does not entail that they are not theorizing about epistemic value, and thus not doing epistemology.

I mentioned that ameliorative inquiry starts by asking what legitimate reasons we have for wanting a concept or theory of X in the first place. Here, we are interested in discovering the proper subject matter of epistemology, or the nature of epistemic value. So, we begin with a suitably broad set of ameliorative questions: What should the purpose of our normative epistemic concepts and theories be? What do we want them to do for us? I will consider three different answers to this set of ameliorative questions, each of which serves to motivate a picture of epistemic value that is broader than truth monism—a picture on which epistemic value is not autonomous of moral and pragmatic values. For reasons that I will describe, I do not think the first two types of arguments will convince the truth monist. But I think the third one ought to.

5.1 Answer 1: Epistemic Flourishing

One thing we might want our repertoire of normative epistemic concepts to do for us is promote our flourishing qua epistemic agents. Different versions of this idea have been defended. For instance, Berit Brogaard (2014) argues that the fundamental epistemic norm demands that we do not hinder intellectual flourishing. And Miracchi (2020) suggests that we should conceive of epistemology as the study of the nature of individual and collective epistemic health. On a much broader level, one might view the field of virtue epistemology as fundamentally aimed at understanding how we can cultivate intellectual flourishing in human epistemic agents.

Given this answer to the ameliorative questions, one could argue that, insofar as feminist theorizing aims to promote epistemic flourishing, it is epistemic theorizing. And indeed, feminist ideas find a natural home in a framework of concepts and explanatory resources whose aim is to promote our flourishing qua epistemic agents. Many projects in feminist epistemology aim to help us understand how social and political factors might impact epistemic flourishing. For instance, theories of epistemic injustice and epistemologies of ignorance help us uncover the ways that our social and political situations can detract from epistemic flourishing, or health–either of certain individuals or of the epistemic community as a whole. And this puts us in a better position to do something about it.

But truth monists will object that this does not amount to a principled argument that all theorizing that aims at promoting our flourishing is genuinely epistemic theorizing, that is, theorizing about specifically epistemic value. Feminist theorizing is concerned with promoting equity in epistemic flourishing. And that, for the monist, is a moral and political concern, rather than an epistemic one.

5.2 Answer 2: Epistemic Value and The Good Life

A second answer to the ameliorative questions says that fundamentally, what we want our normative epistemic theorizing to do is understand how creatures like us can do things like reason, inquire, and represent the world well. Initially, this answer sounds uncontroversial, perhaps even platitudinous. But 'well' is ambiguous. This answer can serve as the basis for an argument against truth monism if we allow practical and social justice considerations to determine part of what it means to reason, inquire, and represent the world well. The idea is that, for creatures like us, part of what it means for things like reasoning and inquiry to go well is for them to proceed in ways that serve or partly constitute the wider values of a good life for such creatures. Therefore, there is no reason to understand epistemic value as fundamentally autonomous of these wider values.

Linda Zagzebski is a prominent example of an epistemologist who thinks that the kind of value that makes things like reasoning and inquiry in creatures like us fitting objects of philosophical inquiry is not independent of the wider value of a good life for such creatures. She argues that "the common view that epistemic good is independent of moral good is largely an illusion" (Zagzebski 2003, p. 12), and, "It is very unlikely that epistemic value in any interesting sense is autonomous" (Zagzebski 2003, p. 26). Zagzebski focuses on moral value, but a similar point can be made about pragmatic value by pointing out that some of the characteristic roles that reasoning and inquiry play in a good human life are practical.[16] From these perspectives, there is no reason to think that feminists' responsiveness to practical and social justice considerations is extrinsic to epistemic theorizing.

---

16 For an exploration of such a view, see Foley (2001), who argues that the epistemic goal of truth is instrumentally valuable as a means to other goals.

Again, truth monists will object. From their perspective, we do not just want epistemology to give a theory that understands how creatures like us can reason, inquire, and represent the world well. Rather, we want a theory that understands this from a perspective that is concerned with (and only with) their distinctly epistemic value. To avoid begging the question against monism, then, we would need to give some independent, principled reason to think that any value that attaches to paradigmatically epistemic states, processes, and activities should thereby be deemed epistemic value, rather than autonomous moral or pragmatic value. Bonjour makes this point in his discussion of the difference between epistemic justification and other species, like moral and pragmatic justification: "Now it might be thought that . . . epistemic justification is that species of justification which is appropriate to beliefs or judgments, rather than to actions, decisions, and so on . . . But there are other species of justification which also can apply to beliefs, so that mere applicability to beliefs cannot be the sole distinguishing characteristic of epistemic justification" (Bonjour 1985, p. 6). From this perspective, for instance, Zagzebski's point about the importance of knowledge might be orthogonal to understanding its distinctly epistemic value.

5.3 Why Do the First Two Answers Fail?

The way in which the first two answers to the ameliorative questions fail to provide a basis for a convincing argument against monism points to a broader worry about ameliorative inquiry. The objections I articulated on behalf of the monist in sections 5.1 and 5.2 are instances of a broader objection to ameliorative inquiry in general. The objection is that often, when ameliorative inquiry takes itself to be doing semantic amelioration, what it is actually doing is simply changing the subject—in this case, from epistemic value to some broader category of value. A stronger version of this objection says that semantic amelioration is not possible because concepts have their content essentially.[17] But the objector (in this case, the monist) might accept that semantic amelioration is possible while thinking that, in this specific instance, we have changed the subject by departing too far from the original meaning of epistemic value. This objection demands that the proponent of using ameliorative inquiry in epistemology (henceforth, the ameliorativist) be able to give a principled reason to think that whatever revisio-

---

17 For more on this objection and a response to it, see Haslanger (2020a).

nary concept of (in this case) epistemic value they propose has not simply changed the subject to a different kind of value.

This is where ameliorative inquiry gets tricky. On the one hand, its revisionary potential stems from the fact that it begins by asking normative questions. This is what might allow the ameliorativist to avoid the assumption that, as Haslanger puts it, "what we value epistemically is what we ought to value epistemically", and thus to avoid leaving "the normative part of epistemology undone" (Haslanger 1999, p. 466). On the other hand, this same fact means that ameliorativists must find a principled way to ensure that they are doing conceptual revision, rather than simply changing the subject.

It is beyond the scope of present discussion to attempt a principled account of what it means to change the subject in general, or to say whether one is possible.[18] But in what follows, I suggest a positive strategy for answering the ameliorative questions in a way that both (1) gives us a principled account of why the ameliorativist's revisionary contributions do not simply amount to changing the subject, and (2) leaves room to do the normative part of normative epistemology (in our case, by arguing that epistemic value need not be understood as autonomous of moral and pragmatic value).

5.4 The Bearers of Epistemic Value

Theorizing about what some kind of value consists in is bound up with theorizing about the kind(s) of things that bear that value. In our case, theorizing about what epistemic value consists in is bound up with theorizing about the epistemic states, processes, and activities (hereafter, *SPA*) that bear that value. The question of what epistemic value consists in is partly a question of what distinctive kind of value the epistemic SPA have, and the question of what makes something a distinctively epistemic SPA is partly a question of what kinds of things can be bearers of epistemic value.

Given this, my positive suggestion is that the ameliorativist satisfy the demand for a principled account of why revisionary theories of epistemic value do necessarily not amount to changing the subject by *first* fixing ideas about what makes something a distinctively epistemic SPA, rather than by fixing ideas about what counts as epistemic value directly. In other words, I suggest that the ameliorativist anchor their understanding of the subject matter in extant conceptions of what makes something an epistemic SPA–that is, what makes

---

18 For more on this, see Haslanger (2020a).

something a *bearer* of epistemic value, rather than in established conceptions of epistemic value directly. Section 5.7 will argue that this strategy allows us to grant the monist their preferred conception of what makes something an epistemic SPA, and still make a principled argument against the monist conception of epistemic value.

5.5 Answer 3: Evaluating Epistemic SPA qua Epistemic SPA

The answer I gave to the ameliorative questions in section 5.2 turned out to be objectionable because we cannot assume that any evaluation of a paradigmatically epistemic SPA is thereby necessarily an evaluation of its epistemic value. Given this, a better answer to the ameliorative questions says that we want epistemology to give a theory that understands how we should evaluate epistemic SPA qua epistemic SPA. That is, we want it to evaluate them *under the aspect of being* epistemic SPA; or, *with respect to what makes them* epistemic SPA. What I mean by this is that we want epistemology to give a theory that understands how we should evaluate epistemic SPA with respect to the distinct value(s) of the properties that *make* them epistemic, considered apart from any non-epistemic or extra-epistemic properties or features they may have. This satisfies the demand for a principled account of why we have not changed the subject because if we can show that some value attaches to epistemic SPA qua epistemic SPA in this sense, then this value has a principled claim to either being epistemic value or not being autonomous of[19] epistemic value.

The strategy of beginning with the question of what makes something a distinctively epistemic SPA also preserves space to do the normative work of normative epistemology. Having an account of what *makes* certain SPA distinctively epistemic ones, we can ask more specific ameliorative questions: What legitimate values do *these kinds* of SPA have for us qua *these kinds* of SPA, in virtue of the properties that make them epistemic?[20] What work do we want *these kinds* of SPA to do for us, qua *these kinds* of SPA? When we start by fixing ideas about the bearers of epistemic value rather than about epistemic value itself, we fix less of the conversation about epistemic value in advance, and leave room for

---

19 In what follows, I sometimes use *being epistemic value* as a less clunky shorthand for referring to this disjunction. In an important sense, the difference between 'being' and 'not being independent of' is verbal.

20 I ask the question this way, rather than asking what legitimate reasons we have to want an evaluative theory of *these kinds* of SPA qua *these kinds* of SPA, since we might have legitimate non-epistemic reasons to want to evaluate epistemic SPA qua epistemic SPA.

this normative theorizing. I discuss the kinds of answers that we can give to these questions—ones that I think the monist ought to accept—in section 5.7, after addressing an initial objection to this strategy in section 5.6.

5.6 Objection: Artificial Separation?

One might object that the appearance of separating inquiry about the bearers of epistemic value from inquiry about epistemic value itself is artificial. After all, I began section 5.4 by pointing out that the question of what counts as an epistemic SPA and the question of what counts as epistemic value are deeply interconnected.

Ultimately, this interconnectedness suggests that we should revise our answers to the two questions in light of one another until they reach something like a reflective equilibrium. We should not purport to give a final answer to either question before having considered the other. But it does not mean that we cannot or should not start with one question rather than the other—indeed, we have to start somewhere.

Of course, one might worry that, in the end, separating the two questions is meaningless because the same dialectic will play out no matter where we start. The idea is that the monist's theory of epistemic SPA will be informed by their idea about what epistemic value is, and will in turn support that idea; so, it does not matter which question we start with. Monists will define *epistemic* SPA as *truth-directed* SPA because truth-directed SPA are the most natural candidates for bearers of the type of value that is reducible to that of truth, or of acquiring true beliefs and avoiding false ones. Others may find this too restrictive, as there are difficult questions, for instance, about whether certain SPA like *understanding* are truth-directed.[21] They might prefer to define *epistemic* SPA as SPA that aim for a cognitive grasp of reality, or perhaps simply as *cognitive* SPA. These definitions leave more room for conceptions of epistemic value that are not entirely reducible to the inherent value of truth. Nonetheless, the next section argues that we can grant the monist their preferred definition of epistemic SPA, and still use ameliorative inquiry to argue against the monist conception of epistemic value.

---

21 For a detailed discussion of this issue, see Elgin (2017).

5.7 The Value of Truth-Directed SPA

Granting the monist their preferred definition of epistemic SPA does not imply a monist definition of epistemic value. To see this, it will be helpful to start by articulating why one might think otherwise. A natural thought is that whatever value accrues to a truth-directed SPA qua truth-directed SPA must be reducible to the inherent value of truth. After all, the truth aim is built into the very definition of the kind. The opening line of Duncan Pritchard's paper defending monism reads: "That truth is the fundamental epistemic good used to be orthodoxy within epistemology. Indeed, isn't it simply characteristic of the epistemic that it is directed at truth?" (Pritchard 2014, p. 112). Here, Pritchard gives voice to this natural thought, by moving directly from the idea that it is characteristic of the epistemic that it is directed at truth to a monist picture of epistemic value.

But this natural thought is mistaken. The claim that being a truth-directed SPA is both necessary and sufficient for being an epistemic SPA is a descriptive claim about what it means to be an epistemic SPA. To be an epistemic SPA is to aim at truth. The mistake that Pritchard and others make is to view it as smuggling in a strong normative claim, namely, that it also gives us a full explanation of the ontology of value that attaches to epistemic SPA qua epistemic SPA. Again, the assumption is natural, since the truth aim is built into the definition of the kind. But, while this definition does strongly suggest that epistemic SPA are made inherently better qua epistemic SPA when they get at the truth, it does not entail monism about epistemic value. It does not entail that the value of truth-directed SPA qua truth-directed SPA derives from and is reducible to the inherent value of truth. More generally, evaluating something with respect to the kind of thing it is *might* mean evaluating the extent to which it achieves characteristic aims that are built into the definition of that kind of thing. But it might also mean evaluating it with respect to the value that kind of thing has for us as such, and not necessarily in virtue of achieving the ends it aims for.

So, what sorts of value might truth-directed SPA have qua truth-directed SPA? First, they have monist value; that is, value that derives from *inherent* value of truth. Second, moving beyond monist value, they might also have value that derives from the *instrumental* values of truth. For instance, having a grasp of the truth helps us achieve things of practical and moral importance.[22] Such value is not reducible to the final value of truth. Third, truth-directed SPA might have value that derives from the value of *seeking* the truth—that is,

---

22 See Zagzebski (2003 and 2009) on the idea that true beliefs are valuable because we need them in order to achieve the things we morally ought to care about.

from the *directedness* part of truth-directedness. Such value can be final or instrumental. Indeed, I suggested in section 3 that final value inheres in the exercise of our intellectual, or epistemic agency in seeking the truth. Exercising such agency in seeking truth is partly (or perhaps on the monist's view, wholly) constitutive of our rational agency, which, according to a long tradition in philosophy, is valuable in itself. And finally, there is instrumental (and plausibly also final) value in the development of the capacities involved in exercising our rational agency to seek the truth.

So, not all sources of the value that truth-directed SPA have qua truth-directed SPA are reducible to the inherent value of truth. And many of these sources of value are, as Zagzebski puts it, not independent of the wider values of a good life. Exercising our agency in seeking the truth, for instance, is partly constitutive of a good human life. So, if we accept that the value that attaches to epistemic SPA qua epistemic SPA has a principled claim to being epistemic value, then we have principled reason to think that epistemic value need not be understood as autonomous of these wider values. For Zagzebski and many others, this is neither surprising nor radical. It simply acknowledges that the valuable roles and functions that truth-directed SPA have in human life—the reasons that we have to care about them qua truth-directed SPA are not reducible to the inherent value of truth.

Feminist work has a natural place in discussions that use this richer framework in theorizing epistemic value. For instance, I suggested that there is final epistemic value in exercising our intellectual agency in seeking the truth, as well as instrumental (and perhaps final) value in the development of the capacities involved in doing so. Feminist work on epistemic injustice and epistemic oppression aims fundamentally to understand what happens when the intellectual agency that we exercise in seeking the truth is systematically undermined, and to understand why and how it is bad. Understanding the value of our ability to develop and exercise our agency in seeking the truth is important. If we foreclose the possibility that discussion of our ability to develop and exercise our agency in seeking the truth concerns epistemic value, then we make it impossible to understand these as distinctly epistemic problems, rather than simply as problems that happen to occur in the course of epistemic activity.

5.8 Objection: Sources of Value vs. Kinds of Value

The monist may object that I have confused *sources* of value with *kinds* of value. I suggested that the sources of the value that truth-directed SPA have qua truth-directed SPA have a principled claim to either being or not being autonomous of epistemic value. The objector says that a condition that stipulates only the source of value is not strong enough; we must also restrict the *type* of value. They suggest instead that only the *final* values of truth-directed SPA have principled claim to either being or not being autonomous of epistemic value. Anything else counts as epistemically valuable only insofar as it is reducible to such values. So, any instrumental value of the truth-directedness of truth-directed SPA counts as value *of* the epistemic, not as epistemic value. This, I imagine, is what Pritchard and other monists would say in response to me.

The objection does not undermine my argument. I argued that the exercise of our epistemic agency in seeking the truth has final value, and pointed out that this should not be understood as autonomous of the wider values of a good life. Even if the objector is right, this part of the argument stands. So in the end my argument shows that even if we grant the monist their definition of the epistemic as the truth-directed, *and* allow them to stipulate that epistemic value must be reducible to the inherent value of truth-directed SPA as such,[23] there is *still* room to do the normative part of normative epistemic theorizing, and make a principled case that epistemic value it not autonomous of moral and pragmatic values. Some of the inherent value(s) of truth-directed SPA have are not independent of the wider value of a good life.

6 Further Benefits of Ameliorative Inquiry in Epistemology

I have emphasized that doing ameliorative inquiry in epistemology helps us respond to what I take to be a false and harmful claim: that much of feminist epistemology is not really doing epistemic theorizing. I do not mean to imply that this is the only benefit it provides to epistemology. I will close this chapter with three brief suggestions about other significant benefits that I think it might have. There is not space here to do full justice to any of them, but they are potential areas for future research.

First, the ameliorative argument against monism might help us intervene on contemporary debates in mainstream epistemology, over core normative epistemic concepts like justification, rationality, and knowledge. To give one example,

---

23 The only way for the monist to make the condition of what counts as having a principled claim to being epistemic value stricter would be to beg the question against the non-monist by stipulating that only value that is reducible to the inherent value of truth has such claim. But that is definition by fiat.

it might help us settle debate about whether diachonic agency is relevant to epistemic evaluation (contra some notions of evidentialism).²⁴ Insofar as we have reason to doubt monism, we have room to argue that epistemic value (as opposed to only autonomous practical or moral value) attaches to the way(s) that we use our agential capacities to gather evidence.

Second, ameliorative inquiry may help us re-frame long standing debates in 'mainstream' epistemology, specifically by helping us diagnose and deal with instances where we are talking past one another. Again, Haslanger helps us see why. She says, "Given the different projects of analysis and different subject matters for analysis, it is not surprising that philosophers who may appear to be asking the same question are in fact talking past each other" (Haslanger 2005, p. 20). For the reasons she describes, talking past one another is common in many areas of philosophy, epistemology included. Consider, for instance, the long-standing debates between internalists and externalists over the correct notion of core epistemic concepts like justification. Although there is not space to properly argue for it here, I suspect that participants in this debate often talk past one another because they are implicitly assuming different answers to ameliorative questions. That is, they focus on different reasons for which we might want an evaluative theory of the truth-directed SPA of human beings. Very roughly, internalism emphasizes the value of the agency that we have in relation to our belief-forming and other epistemic processes, perhaps because of its connection with valuable forms of human agency more generally. Externalism, meanwhile, emphasizes the value of reliability, which in some instances is motivated by a monist picture of epistemic value. Ameliorative inquiry helps us see that there may be different legitimate answers to ameliorative questions, so we may want a variety of normative epistemic concepts to help us understand these different aspects of what it means for things to go well epistemically. If so, then ameliorative inquiry may help explain that these different notions of justification are not always in conflict with one another.

Finally, ameliorative inquiry can help us add novel and useful notions to our repertoire of normative epistemic concepts and other explanatory resources. Many of the evaluative concepts that feminist epistemology has introduced may fall into this category. I will close the chapter by gesturing toward the possibility of adding one more.

---

24 For more on the idea that facts about one's diachronic agency are not relevant to epistemic evaluation, see Kelly (2003). See also the discussion of whether epistemic evaluation should include sensitivity to a subject's diachronic agency in Miracchi (2020).

There is a question that arises, in relation to discussions of epistemologies of ignorance, about whether there is anything epistemically (as opposed to just practically or morally) wrong with one's lacking knowledge about things that do not immediately bear on one's own experience. To borrow an example from Shannon Sullivan (2007), consider that most white Americans probably do not know exactly what Puerto Rico's relationship to the United States is. Suppose that it is not that they have unjustified beliefs about the relationship of Puerto Rico to the U.S., or that they have justified beliefs that fail to be knowledge. Suppose instead that they simply do not have any beliefs that bear on this issue.[25] We can then ask: If there anything *epistemically* (as opposed to just practically or morally) wrong with this? If the only normative epistemic concepts that we have available to us, in terms of which we can evaluate this scenario epistemically, are traditional notions like justification and knowledge, then it seems that the answer is no. My own intuitive judgment is that this answer seems wrong. The ameliorative argument against monism about epistemic value may put us in a position to understand why this is, and to add to our repertoire of normative epistemic concepts in order to explain it.

If we have principled reason to think epistemic value is not autonomous of moral and pragmatic values, and that some of the reasons that we want evaluative theories of the truth-directed SPA of humans have to do with the roles of those SPA in moral and practical life, then my intuitive judgment about this case may be correct. And we can add to our repertoire of normative epistemic concepts in order to capture this. I would suggest a notion that evaluates our truth-directed SPA's as either *apt* or *inapt* (which might come in degrees), where the evaluation targets the question of whether our truth-directed SPA bear a relationship of fit to the things that morally ought to be important to us.[26] And I would argue that, as white Americans, our nation's relationship to Puerto Rico ought to be important to us, given the colonial oppression that inheres in it.

---

25 This may or may not be true for a majority of people, but it only needs to be possible in order to illustrate this point. And it is certainly likely that it is actual in a least some cases.
26 Contrast this with Leplin, a monist, who says that "It can certainly be better not to believe . . . better to be out of the loop and so unaccountable (Leplin 2009, p. 19–20). This is of course not so according to a notion like the one I am describing here.

Bibliography

Ahlstrom-Vij, Kristoffer (2013): "In Defense of Veritistic Value Monism." In: *Pacific Philosophical Quarterly* 94. No. 1, pp. 19 – 40.

Alston, William (1985): "Concepts of Epistemic Justification." In: *The Monist* 68, pp. 57 – 89.

Anderson, Elizabeth. (1995): "Knowledge, Human Interests, and Objectivity in Feminist Epistemology." In: *Philosophical Topics 23*, pp. 27 – 58.

Bonjour, Laurence (1985): *The Structure of Empirical Knowledge.* Cambridge: Harvard University Press.

Brogaard, Berit (2014): "Intellectual Flourishing as the Fundamental Epistemic Norm." In: Clayto Littlejohn and John Turri (Eds.): *Epistemic Norms: New Essays on Action, Belief, and Assertion.* Oxford: Oxford University Press, pp. 11 – 32.

Dotson, Kate (2011): "Tracking Epistemic Violence, Tracking Practices of Silencing." In: *Hypatia* 26. No. 2, pp. 236 – 257.

Elgin, Catherine (2017): *True Enough.* Cambridge: MIT Press.

Foley, Richard (2001): "The Foundational Role of Epistemology in a General Theory of Rationality." In: Abrol Fairweather and Linda Zagzebski (Eds.): *Virtue Epistemology: Essays on Epistemic Virtue and Responsibility.* Oxford: Oxford University Press, pp. 214 – 230.

Fricker, Miranda (2007): *Epistemic Injustice: Power and The Ethics of Knowing.* Oxford: Oxford University Press.

Goldman, Alvin (1999): *Knowledge in a Social World.* Oxford: Oxford University Press.

Haslanger, Sally (1995): "Ontology and Social Construction." In: *Philosophical Topics 23.* No.2, pp. 95 – 125.

Haslanger, Sally (1999): "What Knowledge Is and What It Ought to Be: Feminist Values and Normative Epistemology." In: *Philosophical Perspectives 13*, pp. 459 – 480.

Haslanger, Sally (2005): "What Are we Talking About? The Semantics and Politics of Social Kinds." In: *Hypatia* 20. No. 4, pp. 10 – 26.

Haslanger, Sally (2020a): "How Not to Change the Subject." In: Teresa Marques and Åsa Wikforss (Eds.): *Shifting Concepts: The Philosophy and Psychology of Conceptual Variation.* Oxford: Oxford University Press, pp. 235 – 259.

Haslanger, Sally (2020b): "Going On, not in the Same Way." In: Alexis Burgess, Herman Cappelen, and David Plunkett (Eds.): *Conceptual Engineering and Conceptual Ethics.* Oxford: Oxford University Press, pp. 230 – 260.

Jenkins, Fiona (2019): "Gendered Innovation in the Social Sciences." In: Marian Sawer and Karryn Baker (Eds.): *Gender Innovation in Political Science: New Norms, New Knowledge.* Cham: Palgrave Macmillan, pp. 41 – 59.

Jenkins, Katharine (2016): "Amelioration and Inclusion: Gender Identity and the Concept of Woman." In: *Ethics* 126. No.2, pp. 394 – 421.

Kelly, Thomas (2003): "Epistemic Rationality as Instrumental Rationality: A Critique." In: *Philosophy and Phenomenological Research* LXVI. No. 2, pp. 612 – 640.

Leplin, Jarret (2009): *A Theory of Epistemic Justification.* Dordrecht: Springer.

McKinnon, Rachel (2016): "Epistemic Injustice." In: *Philosophy Compass* 11. No. 8, pp. 437 – 466.

Medina, José (2012): *The Epistemology of Resistance.* Oxford: Oxford University Press.

Miracchi, Lisa (2020): "A Case for Integrative Epistemology." In: *Synthese* 198. No. 12, pp. 12021–12039.

Moss, Sarah (2018): "Moral Encroachment." In: *Proceedings of the Aristotelian Society CXVIII*. No. 2, pp. 177–205.

Nawar, Tamer (2019): "Veritism Refuted? Understanding, Idealization, and the Facts." In: *Synthese* 198. No 5, pp. 4295–4313.

Peet, Andrew (2017): "Epistemic Injustice in Utterance Interpretation." In: *Synthese* 194. No. 9, pp. 3421–3443.

Pritchard, Duncan (2014): "Truth as the Fundamental Epistemic Good." In: Jonathan Matheson and Rico Vitz (Eds.): *The Ethics of Belief*. Oxford: Oxford University Press, pp. 112–128.

Sosa, Ernest (2007): *A Virtue Epistemology: Apt Belief and Reflective Knowledge*, vol. 1. Oxford: Clarendon Press; New York: Oxford University Press.

Stanley, Jason (2005): *Knowledge and Practical Interests*. New York: Oxford University Press.

Stapleford, Scott (2016): "Epistemic Value Monism and The Swamping Problem." In: *Ratio 29*. No. 3, 283–287.

Sullivan, Shannon (2007): "White Ignorance and Colonial Oppression: Or, Why I Know So Little about Puerto Rico." In: Shannon Sullivan and Nancy Tuana (Eds.): *Race and Epistemologies of Ignorance*. State University of New York Press. pp.153–172.

Zagzebski, Linda (2003): "The Search for the Source of Epistemic Good." In: *Metaphilosophy* 34. No. 1, pp. 12–28.

Zagzebski, Linda (2009): *On Epistemology*. Belmont: Wadsworth.

Part III: **Meaning, Politics, and Identity**

Deborah Mühlebach

Tackling Verbal Derogation: Linguistic Meaning, Social Meaning and Constructive Contestation

Abstract: Our everyday practices are meaningful in several ways. In addition to the linguistic meanings of our terms and sentences, we attach social meanings to actions and statuses. Philosophy of language and public debates often focus on contesting morally and politically pernicious linguistic exchanges. My aim is to show that this is too little: even if we are only interested in morally and politically problematic terms, we must counteract a pernicious linguistic practice on many levels, especially on the level of its underlying social meanings. Otherwise, the critique of specific words as the most salient fruits of this practice will be futile. I trace out two paths through which pernicious social meanings feed into linguistic meanings and make the case for constructive contestations of social meanings as an alternative to criticising the use of a few highly pernicious terms in which these social meanings are manifest. My investigation into how social structures shape both social and linguistic meanings sheds further light on the ways in which social meanings enter linguistic exchanges. Moreover, it reveals that what is said in specific situations is more closely connected to our non-verbal actions than the current literature on semantics and the social sciences allows.

1 Introduction

Most of our everyday practices are meaningful in several ways: we attach meanings to the actions themselves, to the social situations in which they take place, to interacting people as individuals and to these people as members of specific groups. For example, for a couple to have matching rings means to be married. Giving a small tip in a restaurant expresses dissatisfaction with the service in some countries, but not in others. And we ascribe numerous stereotypical meanings to ways of talking, walking, dressing or generally behaving. For instance, we take wearing a pink dress to signal girlhood. Call these different types of meanings social meanings.

Some initial characteristics of social meanings, borrowed from Sally Haslanger, are the following: Firstly, social meanings are usually part of a web of interdependent meanings. The larger and more coherent these webs of interde-

https://doi.org/10.1515/9783110612318-010

pendent meanings are, the more stable they are and the harder it is to change individual social meanings. Secondly, social meanings are often contested: 'Social meanings are malleable and contested, but when hegemonic, they function a bit like the local geography, i.e., as a "given" around which we structure our lives' (Haslanger 2019, p. 8). Thirdly, they depend on the way our world is, but they equally help to shape this world. As Haslanger notes:

> interpreting bunnies as pets may be apt, but wolves, not so much. As a result, the looping of social meanings and resources can function as a source of correction. But because we also shape the world to "fit" social meanings, the "correction" may not always be what we need to reimagine and reshape the world to be more just. For example, a disabled child may not learn to read in school and so face reduced employment options and need government support. But the conclusion should not be that such children are uneducable and a drain on society. Too often the social meaning of disability is disabling. (Haslanger 2019, p. 10)

Finally, if (webs of interdependent) social meanings guide us to engage in politically and morally problematic practices, they might be ideological.

A lot of philosophy of language has focused on contesting pernicious linguistic exchanges. My aim in this paper is to show that this is too little: even if we are only interested in morally and politically problematic terms, we must counteract a pernicious linguistic practice on many levels, especially on the level of its underlying social meanings. Otherwise, the critique of specific words as the most salient fruits of this practice will be futile. Very much in line with the literature on how ideologically misguided social practices enter the pragmatics of our speech situations (Kukla 2018) and broader communicative situations (Tirrell 1999a and 1999b), and how discourse may engender action (Tirrell 2012 and 2017), I take it that modelling language as a social practice provides us with the necessary tools to account for various ways in which our broader social practices enter our linguistic exchanges.

Proceeding from this body of literature that draws on resources of the inferentialist framework as it has been developed by Robert Brandom (1994), I focus on how morally and politically problematic social meanings, specifically, shape what is said in a given speech situation. The ways in which pernicious linguistic and social meanings are connected remain underexplored in the literature on philosophy of language and the social sciences. I trace out two paths through which pernicious social meanings feed into linguistic meanings, I show how power relations between different discursive (sub-)communities affect this relationship, and make the case for constructive contestations of social meanings as an alternative to criticising the use of a few highly pernicious terms in which these social meanings are manifest.

I shall proceed as follows. Firstly, I specify the notion of social meanings by critically engaging with a definition given by legal scholar Larry Lessig. I then outline how I take social meanings to relate to forms of life and communities of practice. Elsewhere (see Mühlebach 2021), I argue that in order to understand semantic contestations of politically significant terms, we best model the meaning (re)producing communities as (sub-)communities of practice with different, that is, contesting, conceptual norms. These diverging conceptual norms are manifest in different inferentially structured sets of commitments and entitlements that come along with the use of specific terms. For the purposes of this paper, the Brandomian inferentialist framework (Brandom 1994) also builds the background of my understanding of both linguistic and social meaning.[1] As a theory of conceptual content in terms of discursive practice it allows us to explain the connections between linguistic and social meaning and to see the striking parallels between the ways in which they are (re-)produced in communities of practice. Against this background, I shall develop the idea that social meanings are linguistically relevant in at least two different ways: on the one hand, what is said in a given context crucially depends on background commitments that, in turn, depend on what social meanings are available. I make use of Robert Stalnaker's notion of common ground and show how a practical common ground as a modified version of it illuminates the connection between linguistic and social meaning within a broad inferentialist framework. On the other hand, social meanings may turn into linguistic meanings if the practices in which they are invoked make their ascription uniform in specific ways. In both cases, so I shall argue, political contestations, especially constructive contestations, cut at the social roots of verbal derogation. They are thus key to both preventing some pernicious social meanings from entering our speech situations and promoting non-pernicious social and linguistic meanings.

2 What Are Social Meanings?

A promising definition of social meanings is given by Larry Lessig. According to Lessig, (i) social meanings are 'the semiotic content attached to various actions, or inactions, or statuses, within a particular context' (Lessig 1995, p. 951). (ii) They are not mere idiosyncratic associations, but widely shared among individ-

---

1 I do not aim to present an inferentialist account of pernicious linguistic meaning here, but rather draw on already existing sources such as Brandom's work on inferentialist semantics (1994 and 2000) and an inferentialist view of communicative situations involving derogation (Mühlebach Manuscript and Tirrell 1999b, 2012 and 2017).

uals. As such, they are contingent on a particular society or group within a society (see Lessig 1995, p. 952). (iii) Social meanings have to be widely accepted among members of a group in order to have their force within that group:

> They must be *taken for granted* by those within the group at issue, or put another way, they must be relatively uncontested in that context. It is not enough that individuals understand that a particular idea along with a given action may yield a given meaning. For it to function as a "social meaning," the individuals in this context must also accept it. For an action to convey a social meaning in the sense I want to use the term here, it must do so without appearing contingent or contested; it must do so in a way that feels natural. (Lessig 1995, p. 958)

Accepting (i) seems reasonable, perhaps because (i) is hardly informative. Social meanings are context- or situation-dependent. In order to avoid confusion in later parts of this paper, I shall use the term "situation" for the constitution of a specific social meaning ascription, and I shall restrict the term "context" to phenomena of linguistic exchange. With regard to their situation-dependence, social meanings differ from many linguistic meanings of the terms that we need to describe the actions which bear a social meaning. It is, however, likely that some social meanings are only effective in one type of situation, whereas others affect several type of situations. As I will show, repeated and uniform citation of the same social meaning across different situations may render a social meaning more and more situation-independent, thus changing the inferentially structured commitments and entitlements which come along with the ascription of the social meaning, and eventually turn formerly broader social norms into narrower conceptual norms.

With regards to (ii), taking social meanings to be shared associations does not seem to be controversial either. However, if we consider the fact that there is a continuum between loose and rather idiosyncratic associations and fixed and widely shared associations, different social meanings will most probably occupy different positions on this continuum. Moreover, as I shall argue in connection with my interpretation of (i), these positions may shift over time.

By contrast, we cannot take (iii) for granted. It is too strong and fails to accommodate cases in which a social meaning is operative even if none of the participants of a social interaction adheres to it. We would at least need to explain how those groups and specific contexts relate to each other.

Take the autobiographical example from Elizabeth Anderson, in which she recounts her night-time interaction with a young Black man at a gas station:

> One late night in 2007 I was driving in Detroit when my oil light came on. I pulled into the nearest gas station to investigate the problem when a young black man approached me to

offer help. "Don't worry, I'm not here to rob you," he said, holding up his hands, palms flat at face level, gesturing his innocence. "Do you need some help with your car?" I thanked him for his offer and told him I wasn't sure how much oil I needed. He read the dip stick, told me my car needed two quarts, and offered to do the job for free. From the look on his face when I paid him anyway, it was clear that he needed the cash. (Anderson 2013, p. 53)

As Anderson suggests, a plausible reading of this situation is that – pace Lessig's view – neither the young Black man nor Anderson herself agree with the social meaning of young Black men being criminal. Nevertheless, it affected the concrete social situation in a complicated way, so that both interlocutors had to shape their interaction in relation to this social meaning. This example requires us to give a more sophisticated account of how social meanings are operative in our social interactions. We need to explain how social meanings can be effective even if, in a given context, nobody accepts them.

Furthermore, Anderson's example highlights the importance of paying close attention to how we specify a situation in cases where politically significant meanings are operative. If Lessig rightly considers social meanings to be situation-dependent, Anderson's gas station case makes salient that it is not trivial to specify the relevant aspects that make up the situation of the interaction. It is likely that this social meaning would not be effective in the same way (or perhaps at all) if Anderson were not the middle-aged white woman, but a Black teenage boy. Note, however, that Anderson herself thinks that the social meaning of young Black men being criminal is so pervasive that it has default status, that is, it is common knowledge that it 'is taken for granted as a common premise of public discussion and interpersonal interaction, such that people must send countervailing signals to one another to establish a different common premise' (Anderson 2013, p. 54).

Regardless of whether this social meaning really has this default status, there are many other social meanings that do not. Rather than having a dispute over which interpretation of this specific case is correct, we should explain why some social meanings do have a default status and how such a status comes about. We should resist the idea that we can define a specific communicative situation regardless of the social positions of the interacting parties. And since social positions are usually shaped by complicated power relations, specifying a situation requires us to take those power relations into consideration.

Presumably, this holds equally for linguistic speech contexts. Pierre Bourdieu's *Ce que parler veut dire* (1982) is an example of theorising speech acts in a way that does justice to the political constitution of speech contexts. Instead of viewing linguistic exchanges as intellectual operations that mainly consist in the encoding and decoding of grammatically well-formed messages (see

Thompson 1991, p. 8), he frames them as exchanges of goods on a linguistic market which are determined by specific relations of power and authority. Bourdieu is mainly interested in the questions of who gets to say what in which speech contexts, and who or what decides which mode of expression is accepted in such contexts. According to him, answers to these questions must pay attention to the different forms of capital – economic, cultural, social and ultimately linguistic – that are associated with different social positions. It must also include the effects that these different forms of capital have on a specific speech situation.

I take from Bourdieu the thought that power relations have a determining influence on what can be said by whom in a given context and that this requires us to theorise these power relations and their effects on communicative situations in order to understand cases such as Anderson's gas station example. I share Bourdieu's interest in how the social positions of the interlocutors affect their power to use expressions with specific meanings in specific contexts, and I use his perspective to focus on the specific ways in which power relations already inform the production, negotiation, contestation and change of such social and linguistic meanings.

3 Communities of Practice, Forms of Life and the (Re-)Production of Social Meanings

Social meanings, as opposed to idiosyncratic associations, guide people's perception, thoughts and actions not only as individuals but as members of a community who hold each other accountable for their verbal and non-verbal actions. It is possible that there are few social meanings that are shared by a whole linguistic community, for example, by all British English speakers. It is also possible that some social meanings are shared among all people that fall within a specific social category, for example, among all transgender people. However, communities of people who mutually engage in meaningful practices do not always (and perhaps only rarely) form along the clear boundaries of linguistic, national or other social categories. Some meaningful practices are only effective locally, while some extend to a whole range of people across linguistic boarders.

In this section, I will introduce the idea of communities of practice – in the sense in which it has been discussed in sociolinguistics and has recently entered philosophical work on politically significant terms (see Anderson 2018 and Mühlebach 2021). As I shall argue, both social and linguistic meanings are created and effective in communities of practice and the power relations between different communities of practice will be crucial to the political contestation of social and linguistic meanings.

Communities of practice are groups of people who share beliefs, values, ways of acting and talking (cf. McConnell-Ginet 2011, Wenger 1998 and Anderson 2018). According to McConnell-Ginet (2011), the membership of such a community is not defined by specific characteristics common to all members such as geographical location, ethnicity or gender, but rather by participation in common practices with shared social meanings: 'A community of practice is different as a social construct from the traditional notion of community, primarily because it is defined simultaneously by its membership and by the practice in which that membership engages' (McConnell-Ginet 2011, p. 100). The same set of people may be members of two different communities of practice. If, for instance, all members of a Harley Davidson club were also stamp collectors, their practices with regards to riding Harley Davidson motorbikes would differ significantly from their stamp collecting practices.[2] These practices would then be the practices of two different communities even though their membership is the same. Moreover, each individual participates in a broad range of communities of practice, such as in an office work group, a feminist activist group, a movement against police violence, a nuclear family, a sex workers union or a further education class. As these examples suggest, communities of practice can be small or large, and they can be aggregates of several communities of practice.

This characterisation of communities of practice reveals that they are closely connected to the Wittgensteinian notion of forms of life. Forms of life are the frame in which meaningful action takes place. They set norms according to which actions can be justified or criticised. A dominant conservative reading of Wittgenstein treats forms of life as pre-reflexively given and immune to critique. According to this view, social practices can only be criticised with reasons internal to a form of life, such as if two social practices within the same form of life contradict each other. However, this conservative reading of Wittgenstein's remarks has been challenged by several scholars.[3] Without going into the details

---

2 I owe this example to Nico Müller.
3 José Medina's account of resistance as a contending with (as opposed to a contending against) dominant practices (2013) and his notion of speaking from elsewhere (2006) presuppose criticisability from within. Naomi Scheman (1997) uses Wittgensteinian resources in order to argue for queering the centre of mainstream practices. Jäggi (2014 and 2015) points to immanent – as opposed to internal or external – criticism of social practices: a form of life is criticisable if it does not live up to its own function. Robin Celikates (2015) argues for the claim that the subjects who engage in social practices are essentially able to take a critical stance towards their own practices. He thereby emphasises the essentially heterogeneous, reflexive and conflictual character

of attempts to model forms of life as in principle criticisable, I take it that these endeavours are promising. I will assume that there are reason-guided contesting practices which challenge forms of life from within.

My aim in connecting the notion of forms of life with that of communities of practice with regards to social meanings, lies in fleshing out what contesting practices from within consists in. I do this by paying close attention to both the social positions of the interacting parties and the power relations through which these positions are determined.

According to Rahel Jäggi, a form of life is:

> a culturally informed "order of human co-existence" that encompasses an "ensemble of practices and orientations" as well as their institutional manifestations and materializations. Differences between forms of life are not only expressed by different beliefs, values and attitudes; they are also manifest and materialized in fashion, architecture, legal systems and family organization. (Jäggi 2015, p. 16)

Like communities of practice, forms of life can be more or less comprehensive, and several forms of life can be part of a bigger form of life. Since forms of life are a set of interdependent practices and orientations in which common beliefs, attitudes, and values as well as ways of behaving are manifest, a community of practice shares a form of life. Consequently, communities of practice produce and change meanings by engaging in different practices, that is, by changing their form of life.

Members of communities of practice engage in patterns of behaviour and co-ordinate and evaluate their actions according to the social norms that are effective in their community of practice. Social meanings play a crucial role in structuring these processes of coordination and evaluation because members of a community of practice hold each other accountable for their performances with reference to (interdependent sets) of social meanings (see Haslanger 2019). Such meanings shape our social interactions in that they serve as shared reasons for certain actions and foreclose certain others. In our coordination, we have to rely on social meanings whose contents determine which moves in an interaction make sense, or, if brought into propositional form, which material inferences are valid. For example, it makes sense to buy a cat in order to take care of it, but not so much to buy a cat in order to eat it because we attach the social

---

of forms of life. And with regards to specifically conceptual practices, Queloz and Cueni (2020) argue for a non-foundationalist Left Wittgensteinianism which draws on the idea of both a horizontal, i.e., internal, and a vertical or genealogical critique of conceptual practices: if we can explain why and how we came to have specific concepts, these explanations may provide reasons against engaging in the relevant conceptual practices.

meaning of being a pet, rather than food, to cats. Social meanings thus help determine what performances are intelligible, permissible and appropriate.[4]

In turn, the content of these social meanings is determined by the commitments and entitlements that come along with specific moves according to the social norms of a given community of practice. If the social norms of a community of practice are such that they license the practical move from buying a cat to taking care of it, and they do not allow for the move from buying a cat to killing it so that it serves as food, they impose a set of commitments and entitlements to further practical moves on the person who buys the cat. This set of commitments and entitlements is what it is for an animal to have the social meaning of a pet rather than a working animal or food.

The expectations, commitments, and entitlements with regard to specific moves within a social practice of a given community constitute what I call the practical common ground, which will later become relevant as the background of linguistic exchanges.[5] I take a practical common ground to be the practical knowledge of how beliefs, values, attitudes and actions within a community of practice hang together. Because of the practical common ground of a community of practice, people practically know which social meanings are available and applicable, which performances are possible or appropriate and which commitments follow from specific performances. A crucial feature of the practical common ground is that it is manifest in the practical expectations, commitments and entitlements relevant in a social situation, rather than the consciously held beliefs or self-ascribed commitments of community members.

A further characteristic of social meanings and their functioning within communities of practice is their signalling function: for example, dressing a small child in pink signals that she is a girl. Strangers then address the child – who is otherwise not recognisably gendered – as a girl and interact with her accordingly. The subsequent interactions, in turn, shape and reinforce what girls are meant to be in many communities of practice. Their signalling function allows us to make subversive use of social meanings. By performing an action with a social meaning in a social situation different from the usual one, we disrupt

---

4 For a richer example of how social meanings crucially shape our social practices, see Robert Gooding-Williams's (Forthcoming) work on how the circulation of anti-Black stereotypes and practices of policing in the US are connected in a way such that anti-Black domination is not only an unwanted by-product, but rather integral to these practices.

5 I take a practical common ground to be what Charles Mills (2007, p. 27) calls a social mind-set. I prefer "practical common ground" for two reasons. First, it names the shared basis of coordinated and meaningful action. Second, it highlights the practical aspect of this coordination, rather than insinuating an explanatory primacy of the (conscious) mind.

the expectations of the participants in the practice and thus expose the social meaning as such. If a small boy wears a pink sweater, it is likely that strangers will misidentify his gender. The relationship between social meanings and social norms of a given community of practice suggest that only if more and more boys wear pink clothes and, hence, it becomes a counter practice to dress pink regardless of one's gender identity, the community of practice in question ceases to treat wearing pink clothes as a signal for girlhood. Or, depending on the specific constitution of the new set of commitments which comes along with dressing pink, this counter practice might as well lead to establishing a new social meaning.

Subversive use of social meanings does not automatically lead to engagement in a whole counter practice because the conditions for creating new social norms are often not given. If, however, members of a community of practice manage to turn their individual subversive use into a coordinated counter practice, they thereby create a new community of practice with new expectations, commitments and entitlements that come along with certain moves in their practices. Hence, social meanings may differ across various communities of practice and, in fact, they often do. However, not all of these differences generate political conflict. There are at least two ways in which social meanings can be different across different groups: a parallel and a contested way.

Let us turn to the parallel way first. Two or more communities of practice may ascribe different social meanings to the same action or one community may ascribe a specific social meaning to an action whereas another community does not. I call both parallel ascriptions of social meanings. They are parallel in the sense that none of the two communities of practice aspires to convince the other community of practice that their social meaning is the right one or one which everybody should apply. Take, for example, Jennifer Saul's discussion of George W. Bush using a dogwhistle in his 2003 State of the Union speech. In this speech, which was seemingly unrelated to religious issues, he invoked the idea of wonder-working power to signal to fundamentalist Christians that he spoke their language, that is, that he was one of them. For them, using "wonder-working power" in a speech means alluding to the power of Christ, for people who are not fundamentalist Christians, however, it is an empty phrase, or as Saul notes, 'an ordinary piece of fluffy political boilerplate, which passes without notice' (Saul 2018, p. 362). Due to their different practices, the fundamentalist Christian community of practice ascribes a meaning to invoking the idea of wonder-working power whereas other communities of practice do not. In fact, many dogwhistles, though perhaps not all, draw on different ascriptions of social meanings between two or more communities of practice.

Among two parallel social meanings, one may be dominant and the other subordinate. They do not have to be equally widespread in order to count as parallel social meanings. For instance, different religious communities of practice attach different social meanings to specific days in the year. The social meaning of December 25 as the birthday of Jesus Christ is a subordinate social meaning of a Christian community of practice in a predominantly Muslim country. The subordinate status of it is manifest, for example, in that the official calendar of holidays, according to which work, events and everyday activities are structured, does include Muslim holidays, but not those of Christianity. They are dominant and subordinate in the sense in which the community of practice with the subordinate social meaning has to take the social meaning of the dominant community of practice into account in their actions.

The second way of ascribing social meanings, by contrast, is to contest and challenge the appropriateness of a given competing ascription. There are at least two different types of contested social meanings: First, contestations can be about whether a specific social meaning applies to an action or a status. Stereotypical ascriptions are usually such that one community of practice treats a social meaning as appropriate whereas another community of practice does not. None of them would grant to the other community that their (non-)ascription is appropriate. Some communities of practice, for example, ascribe the social meaning of callous mother to women who work while raising children. Non-sexist communities of practice actively challenge this ascription. The second type of contested social meanings is such that the contestation is about whether one rather than another social meaning applies. Take the social meaning of wearing a hijab. There are some communities of practice in which wearing a hijab is viewed as an act of submission. In other communities of practice, such as Muslim feminist movements, this social meaning is not only denied, but rather replaced by the positive social meaning of an act of self-determination.

With regards to contested meanings, the distinction between dominant and subordinate social meanings becomes more politically charged. In most societies, for example, sexist and racist stereotypical ascriptions dominate our interactions. The social meanings provided by feminist and anti-racist communities of practice are not as pervasive as sexist and racist social meanings because the feminist and anti-racist communities of practice are themselves less powerful than the dominating sexist and racist communities. Domination and subordination of social meanings feeds upon the materially grounded structure of domination and subordination of groups of people. Note, however, that domination and subordination of social meanings are situation dependent. In the situation of a workshop about sex work led by sex workers, the social meaning of sex work *as work* dominates the setting whereas the forces in our larger society are distributed in a way such that stigmatising social meanings of sex work clearly dominate.

Let me introduce two complications to the general picture of meaning ascriptions along the two axes of parallel vs. contested and dominant vs. subordinate social meanings: First, there are what Haslanger calls manifest vs. operative meanings,[6] which must not be confounded with contested meanings. The distinction between manifest and operative meanings captures the gap between what people think they are doing when acting and what their practices reveal they are actually doing, or between their consciously held beliefs and their practical commitments. Some cases to which the distinction between manifest and operative meanings applies prima facie look like cases of contested meanings.

Take, for example, Marilyn Frye (1983) on door-opening practices. A woman might disagree about the social meaning of gallant gestures with a man who opens the door for her. He might take it as a helpful gesture in order to remove a barrier from the woman's path. By contrast, she may take it to be an act of oppression since it treats able-bodied women as incapable of accomplishing mundane tasks whereas no or little assistance is offered in real (and often shared) tasks, such as housekeeping, care work and situations of assault. As Frye notes, the meaning of the gesture is only understandable as part of a pattern of unneeded helpful acts that are carried out regardless of whether they make any practical sense in a particular situation (for a detailed analysis see Frey 1983, p. 6). What the man takes himself to be doing is in tension with the pattern of behaviour he is engaging in according to a sociological point of view. The effects of politically significant meanings do not so much depend on what people think they are, but rather on the function they have according to historical and sociological analyses.

The second complication of spelling out how social meanings operate in our social practices is the possibility of tension between an individual's different communities of practice. Think of a radical feminist activist who comes from a nuclear family with conservative family values. Interacting both with her home and with her feminist activist peers requires her to act according to rather different and sometimes contradictory social meanings. Given both that such tensions abound, and that people are usually very good at navigating them, I suggest that individuals often engage in an activity that I call practical common ground switching.

If people switch from one practical common ground to another, they change their behaviour in a way that takes new social meanings and different social

---

6 See Haslanger (2005, p. 14).

rules into consideration. In the case of the feminist activist from the highly conservative family, her practical common ground switching consists in the fact that she will take different social meanings for granted depending on whether she interacts with members of her home community of practice or her feminist community of practice. If she does not endorse many of the prevalent social meanings of her home anymore, she still has to take them into account in her interactions with her family, that is, she might start avoiding certain topics, consciously challenge certain social meanings and back her challenge up with reasons, etc. If she were not able to switch practical common grounds, the interactions she had would sometimes, or perhaps often, lead to misunderstandings and confusion. Framing the ability to interact with different people whose background assumptions strongly differ as the ability to switch practical common grounds allows us to make sense of why people sometimes seem to be committed to different and perhaps contradictory social meanings depending on whom they are interacting with. Furthermore, it lets us find regularities in the seemingly contradictory ways in which individuals usually act.

4 Preventing Pernicious Social Meanings from Entering the Practical Common Ground

Social meanings not only shape our non-verbal meaningful interactions, they are also linguistically relevant. In this and the next section, I discuss two ways in which this is the case. Both depend on the idea that background information and assumptions influence both *what can be said* in a specific situation and *what social meanings can be ascribed* to an action or a status. Firstly, I shall show in what ways the ascriptions of social meanings not only depend on background assumptions, but function as background assumptions in specific speech situations too. Ascriptions of social meanings are both situation-dependent and part of the linguistic context. This means that they depend on the social norms that are effective in a specific situation. At the same time, they are part of the set of social norms that constitute the linguistic context of specific utterances. I shall make use of Stalnaker's influential notion of common ground, discuss Rae Langton's further developments of this notion so to allow for specific contesting moves in a linguistic exchange, and ultimately argue that yet another modification is needed in order to account for the power dynamics that shape the communicative situation. Spelt out in an inferentialist vocabulary, the notion of practical common ground as already introduced above will capture the ways in which social meanings enter and shape linguistic exchanges. Moreover, it allows for suggestions of critical interventions which appropriately deal with the political constitution of the given context. Secondly, in the last section, I will explore the connection between situation-independence of social meanings through increasingly uniform ascriptions and the linguistic meaning of terms which describe the socially meaningful actions and statuses.

Let us turn to the first case. Social meanings are linguistically relevant in that they enter speech contexts as part of the practical common ground. Stalnaker's original notion of a common ground is usually taken to be a 'term for the presumed background information shared by participants in a conversation' (Stalnaker 2002, p. 701). According to this understanding, speakers presuppose certain background information on the basis of which meaningful utterances are then made.

Anderson's gas station example, however, requires us to consider social meanings that are effective even if they are not shared by any of the interlocutors. Stalnaker accounts for such situations with his notion of the common ground by not only including the presupposition of common beliefs, but also mistaken presuppositions that something is common belief:

> the logic of presupposition obeys positive introspection, but not negative introspection: if Alice presupposes that φ, then she presupposes that she presupposes that φ, but the fact that she does not presuppose that φ does not imply that she presupposes that she does not presuppose it. For even if Alice does not herself believe that φ is common belief, she may believe that Bob mistakenly believes that φ is common belief. (Stalnaker 2002, p. 708)

Thus, in Stalnaker's framework, Anderson's case would look as follows: even though the young Black man at the gas station himself does not presuppose that Black men like him are likely to have criminal intentions, he is familiar with the generalised presupposition. He acts according to his beliefs that Anderson mistakenly believes both that he is likely to be criminal and that this is a common belief between them. Unfortunately, Stalnaker's notion does not tell us anything about the mechanisms that lead both Anderson and her interlocutor to grasp the social meaning and act upon it without endorsing it. Thereby Stalnaker also remains silent about the ways in which social meanings as part of the common ground can be challenged effectively.

Modelling social meanings as part of the common ground of speech situations is more promising in light of Rae Langton's discussion of common ground and accommodating attitudes. Her notion not only includes beliefs, but also feelings and desires. Specific beliefs, feelings and desires may be altered by certain speech acts because every move in a linguistic exchange changes the common ground by adding or removing some presuppositions, that is, presupposed beliefs, feelings and desires (see Langton 2002). Social meanings do not only oper-

ate on the cognitive level, they also sometimes draw heavily on attitudes and desires and they call for attitudes and desires as responses to certain performances.
Take Langton's example of stigmatising stereotypes of Jewish people. In Nazi
Germany, the pernicious social meanings of being Jewish were wide-spread
and could easily enter speech situations as part of the common ground:

> A children's story might presuppose that "Jews often kidnap children"; that "It is appropri-
> ate to hate Jews"; that "Good Germans hate Jews"; that "Good Germans avoid Jews". True,
> the story is presented as fiction, but as fiction that says something about the world, and
> says it by presupposing it. The abstract score incorporates the fact-claiming proposition
> that Jews often kidnap children, the normative proposition that it is appropriate to hate
> Jews, and the proposition (factual and normative) that good Germans avoid Jews. Then,
> if the conversation is a successful one, the attitudes of hearers change, just as the attitudes
> of bystanders change in response to the score of the baseball game. (Langton 2002, p. 87)

If nobody intervenes, the presupposed social meanings are accommodated and
shape what further moves in a linguistic exchange can be made.

Blocking pernicious social meanings in specific speech situations thus
seems to be a promising way to counteract pernicious stereotypical ascriptions.
Langton shows how challenging presuppositions – for example by asking 'Wait,
do you mean good Germans avoid Jews?' (2018, p. 144) – is only one form of
blocking the course of the speech situation. Blocking as counter speech can
take on several forms: questioning the authority of a speaker, challenging presuppositions, and retroactively undoing a speech act. Elaborating on these
forms of counter speech is helpful for seeing which moves are possible in a specific speech context and what responsibility an individual participant can thereby assume.

However, Langton's framework of counter speech requires two disagreeing
parties to talk to each other. Unfortunately, our social interactions are often
such that people who disagree about social meanings tend not to engage in a
mutual conversation very often. Moreover, Langton's framework is too individualistic to account for the ways in which any speech context in which social meanings operate is informed by power relations that already partially determine
which moves are possible and which are not. Blocking social meanings as
well as challenging the authority of a speaker may not be successful at all if
the blocking is not backed up with favourable power relations between the interlocutors' social positions. Finally, even though Langton's notion of common
ground is not restricted to beliefs, but also includes feelings and desires, it
does not allow for social meanings that are operative, but not manifest in a
given speech context.

For the purposes of this paper, I propose to think of the common ground as a practical common ground in order to account for all the three issues mentioned here. As I have suggested above, the notion of a practical common ground is primarily determined by the practically shared commitments of the parties involved in a given speech situation. That is, in a specific speech situation the practical commitments that speakers undertake individually are in agreement with those of other speakers. These commitments are determined by the moves that are made within the practice in question that, in turn, are governed by the social norms underlying the practice. In some cases, these commitments exactly amount to the beliefs, desires and feelings proposed by Langton. However, there are cases in which in a given practice, agents have a poor understanding of their own moves in this practice or even of the practice as a whole. In such cases, some or even all parties involved in the communicative situation take themselves to be doing something quite different from what their sequences of moves are indicating if we try to rationalise their practice. It is a central idea of Brandom's *Making It Explicit* (1994) that discursive commitments are often only implicit in a given discursive practice, and that by making them explicit, we bring them into the realm of reasoning.

It is not an easy task to find out whether the agents in a given communicative situation really misunderstand their own practice. As a result, it is often difficult to assess whether different takes on a specific social meaning are merely apparent or real contestations. Moreover, this assessment is often itself subject to contestations and it might well be the case that the difference between apparent and real contestations is a matter of degree, rather than of kind. Nevertheless, if there is a difference between real contestations such as my hijab example from the previous section, and apparent contestations based on the manifest/operative distinction as I have suggested was the case in the door opening example, the notions of a practical common ground and of practical commitments allow us to capture it.

Take as an illustration the dogwhistle term "inner city" as it was used by Republican Congressman Paul Ryan in 2014. It is an example of a speech situation that accommodates the three issues mentioned above: in this example, the disagreeing parties do not directly talk to each other, it rests on power asymmetries that influence the consequences of the utterance in question, and it involves an operative social meaning that crucially structures the speech situation but is only acknowledged by some parties. In an interview, Ryan said:

> We have got this tailspin of culture, in our inner cities in particular, of men not working and just generations of men not even thinking about working or learning the value and the cul-

ture of work, and so there is a real culture problem here that has to be dealt with. (Blow 2014, par. 2)

By 2014, every politically conscious person of colour, and some white people, in the US knew that "inner city" is a racially charged term. Hence, it does not come as a surprise that Ryan's remarks have caused heated debates. However, these debates were rather local and did not prevent the dogwhistle from being successful.

Here, different communities of practice and the power relations between them enter the explanation. Even though Ryan's dogwhistle was criticised by some, its overall success was rather untouched. The community of practice to which the racialised social meaning of invoking "inner city" in a speech is clear – i. e. a community of practice predominantly consisting of people of colour – does not dominate public discourse. The dominant community of practice with regard to the social meaning of invoking inner cities may be largely ignorant of its racist social meaning or members of this community might as well pretend not to know the social meaning. The social meaning may not be manifest, but as many situations in which people talk about inner cities shows, it has increasingly become operative over the last few years. An indicator of this development is that in the community of practice whose members know about the social meaning, they tend to refrain from using the term "inner city".

Thus, our political contestations concerning social meanings should not only involve blocking them from further operation in a specific speech context, but also – and perhaps more importantly – preventing them from entering the practical common ground altogether. The latter requires us to engage in constructive contestations. It calls for participation in whole counter practices, in contrast to individual instances of counter speech. Communities of practice provide the resources for engaging in such counter practices because their members hold each other accountable for their actions with reference to shared social meanings. The more coherent and the larger these communities of practice with alternative social meanings become, the more these social meanings are able to enter, and help shape, various speech contexts involving members of the community of practice. Constructive contestations thus operate on the level of the pool of possible practical common grounds. They are attempts to establish one specific practical common ground rather than another.

As Anderson's gas station case suggests, however, these constructive contestations cannot only consist in building up a coherent net of social meanings to which many people adhere. As long as the relations between a racist community of practice and those of anti-racist counter practices are such that members of the former sit in positions of economic, cultural, educational, or even linguistic power, then their social meanings will enter contexts such as the gas station encounter where the participants do not know which community of practice their interlocutors belongs to. The participants need to take the social meaning into account even if they do not endorse it. In contrast to parallel ascriptions of social meanings, constructive contestations should thus not focus on the symbolic dimension of social (counter) practices alone. As Haslanger notes, 'broad social change requires change on multiple levels: change to agents, change to culture, and change to structures, policies, and laws' (Haslanger 2017, p. 15). If the task is to engage in a struggle over which social meaning dominates public discourse, communities of practice are well-advised to practically challenge material and symbolic structures at the same time.

5 Counteracting the Linguistic Conventionalisation of Pernicious Social Meanings

There is a second way in which social meanings may become linguistically relevant. Remember that they fall somewhere on the continuum between loose, and perhaps rather idiosyncratic, associations and fixed, widely shared associations. Since social meanings are the semiotic content that we attach to actions and statuses, then if they are more uniformly and commonly ascribed to the same action across different situations and communities of practice than others, we can say that one is more stably institutionalised than the other. Examples of social meanings from loose to rather stably institutionalised associations are the following:

- Loose and rather idiosyncratic association: Sunday has the meaning of being the laundry/cleaning day for me.
- Shared but rather local association: Within my nuclear family, the sound of a bell on Christmas Eve meant that Santa Claus had just come around.
- Widely shared association: The colour pink conveys the social meaning of girlhood.
- Widely and uniformly shared association: Engaging in practices of male gallantry imparts the social meaning of viewing women as incapable and dependent.
- Rather fixed, stably institutionalised association: The thumbs up emoji has the meaning of approval.

If a social meaning is a widely shared association, members of a community of practice will repeatedly coordinate specific actions in specific situations and contexts with reference to it. In particular, social meanings crucially shape which further moves can be made in a social situation, that is, what participants can expect others to do, what they are committed and entitled to. Hence, practices in which a social meaning is operative reveal a pattern that is organised around the social meaning. By engaging in such structured practices, people perform non-verbal and verbal actions. Judith Butler, who like Austin refers to the latter with the terms "performative utterances" or "performatives", famously introduces the idea of performative citationality:

> If a performative succeeds . . . , then it is not because an intention successfully governs the action of speech, but only because that action echoes prior actions, and *accumulates the force of authority through the repetition or citation of a prior and authoritative set of practices*. It is not simply that the speech act takes place *within* a practice, but that the act is itself a ritualized practice. What this means, then, is that a performative "works" to the extent that *it draws on and covers over* the constitutive conventions by which it is mobilized. (Butler 1997, p. 51)

Butler's remarks with regard to speech acts equally apply to non-verbal performances.

The activity of citing the same (patterns of) actions in similar situations over and over again changes what members of the community of practice hold each other accountable for. The more the same (patterns of) actions with reference to a shared social meaning are performed, the more uniform they are, and the more this social meaning becomes institutionalised. The shared association thereby moves from being one of several possible associations towards becoming the only association available across several situations.[7] Members of a community of practice thus increasingly hold each other accountable to a specific set of

---

[7] Butler's notion of citation provides useful resources to address issues of individual responsibilities in language use. Following her understanding of responsibilities as tied to speech as repetition instead of origination, Medina has developed the idea of echoing responsibility: 'a speaker must assume responsibility for his or her speech acts as a contribution to a chain of linguistic performances that *echo* one another. Accordingly, discursive responsibility concerns how to contribute to performative sequences, how to continue or discontinue them, how to echo voices and their performances in and through these chains . . . The issues we are confronted with as speakers are whether a legacy of use is worth maintaining and in what way' (Medina 2008, p. 103). My discussion here shows how (ir)responsible citations of certain actions with certain social meanings can lead to (ir)responsible linguistic performances.

commitments and entitlements that comes along with certain moves centring around the social meaning in question.

Since we use terms to describe actions and statuses with social meanings, the institutionalisation of these social meanings affects the linguistic meaning of the terms involved. If the terms that we use for these institutionalised meaningful actions and statuses are *exclusively* used for these actions and statuses, the institutionalisation enters their linguistic meaning. There is a continuum between terms which are used to describe many different actions and statuses and those which exclusively refer to something with one fixed social meaning.

Take the expressions "pink", "inner city" and "thumbs up" as illustrations. Even though the colour pink is widely and strongly associated with femininity, the literal meaning of "pink" does not include any claim with regard to femininity. Colours are everywhere and there are too many other uses of pink that do not make any reference to gender. A less clear case is "inner city" in its current use in the USA. The term used to describe the geographical area at the core of a city. In the last several years, however, the invocation of inner cities has increasingly served the purpose of activating the racist stereotype of Black people who are poor because they are allegedly lazy and criminal. This social meaning of being an inner-city citizen has become so pervasive that, by now, many people refrain from using the term "inner city", or if they do so, they do it with great caution. As a tentative explanation of the meaning of "inner city" in the USA, I suggest that we understand it as a term that is currently undergoing a change in meaning from a purely geographical to a rather pernicious socially categorising one. Finally, if we assume that "thumbs up", or even the emoji for thumbs up, originated in the gesture of putting one's thumbs up, they are examples of the final stage of uniformly institutionalised social meanings.[8] We do not merely associate the meaning of "thumbs up" with approval, by now the formerly social meaning of a specific action has become the literal meaning of the expression.

By uniformly citing the same performative verbal and non-verbal actions in similar situations over and over again, we allow for the social meanings to settle and become rather fixed. Thereby, some social meanings turn into the conventionalised, literal meanings of terms. This process of sedimentation is best understood as analogous to the process of dying metaphors. Take Donald Davidson's remarks on dead metaphors such as "bottle mouth":[9]

---

8 This interpretation might be contested. Historical sources remain rather silent about the primacy of the gesture or the linguistic expression. Thanks to Anna Goppel for pointing this out.
9 Davidson's use of "dead metaphor" means that the expression in question has lost its metaphorical function of making people aware of similarities between two different things. His understanding of dying metaphors only prima facie contradicts Lakoff's and Johnson's view on the

Once upon a time, I suppose, rivers and bottles did not, as they do now, literally have mouth. Thinking of present usage, it doesn't matter whether we take the word "mouth" to be ambiguous because it applies to entrances to rivers and openings of bottles as well as to animal apertures, or we think there is a single wide field of application that embraces both. What does matter is that when "mouth" applied only metaphorically to bottles, the application made the hearer *notice* a likeness between animal and bottle openings. . ... Once one has the present use of the word, with literal application to bottles, there is nothing left to notice. (Davidson 1978, p. 37)

Just as the conventionalisation of a former metaphor frees it from its metaphorical function and turns its metaphorical meaning into a literal meaning, the stably institutionalised social meanings of actions and statuses may enter the literal meaning of the terms that describe these actions or statuses.

My discussion so far can be applied to the question of how the linguistic meaning of many derogatory terms come about and change. The use of many slurring terms centres around a stereotypical description of their targets. Through repeated citations of stereotypical ascriptions in the use of certain terms, the core of the stereotype which is applied to a (specific behaviour of a) group enters the literal meaning of the derogatory term for them. Slurs usually do not neutrally describe some characteristic and add a negative value to it, their meanings tend to be much more nuanced. For example, the meaning of the French *"b·che"*[10] for German people at the time of WWI connected being German with likeliness to be cruel – and not, for example, with some undesired physical appearance. "Sl·t" is a term for women who engage in, or are assumed to engage in, despicable non-stereotypical sexual behaviour, rather than for women in general. Recently, *"rosbifs"* (Eng. "roast beef") is being used metaphorically among French people to describe British people who spend their vacation time at the beach in France and, presumably because they deal with the sun irresponsibly, tend to get sunburnt very quickly.

---

metaphorical nature of almost all terms we live by. They call those metaphors living metaphors: 'Expressions like "wasting time", "attacking positions", "going our separate ways", etc., are reflections of systematic metaphorical concepts that structure our actions and thoughts. They are "alive" in the most fundamental sense – they are metaphors we live by' (Lakoff and Johnson 1980, p. 473) In contrast to Davidson, they are not interested in the literal and metaphorical meaning of individual terms, but in the underlying conceptual structure. They situate metaphors on this conceptual level of language use.

10 I introduce '·' whenever I do not consider a term to be part of my vocabulary, primarily in order to avoid the mobilisation of harmful associations, and secondarily because there is no neat mention/use distinction regarding derogatory content (see Tirrell 1999a, p. 292).

The upshot for political action is straightforward: if pernicious social meanings can enter the literal meaning of terms, we have strong reason to cut at the roots of a pernicious linguistic practice – i.e., its social meanings – and not only the use of specific words as the most salient fruits thereof. The n-word is far more pernicious than "l·mey" because of their different underlying symbolic and material structures. We cannot perniciously derogate (groups of) people if there are no pernicious social meanings with reference to which we organise our social practices. Moreover, there are no pernicious social meanings if the underlying material, for example, economic and legal, structures are such that the target is in a dominant social position.

Here, too, constructive contestations are key to preventing pernicious social meanings from turning into the linguistic meanings of our terms. Apart from criticising the pernicious language use of certain communities of practice, political contestations should also prevent such communities of practice from dominating public discourse. To this end, building and engaging in counter practices and thus constructively contesting pernicious social meanings is equally important and perhaps more efficacious than focusing on criticising the use of a few highly pernicious terms in which these social meanings are manifest. As we have seen, such constructive contestations consist both in engaging in positive practices which involve alternative social meanings and in changing the material conditions of our social practices.[11]

---

11 Thanks to Christine Bratu, Anna Goppel, Rebekka Hufendiek, Renée Inaiá, Dominique Kuenzle, Nico Müller, Michael O'Leary, Matthieu Queloz, Melanie Sarzano, Christine Sievers, Samuel Tscharner, Marie van Loon, Ingrid Vendrell Ferran, Markus Wild, and Friederike Zenker for helpful comments on earlier drafts. Special thanks to Sally Haslanger for many inspiring discussions about issues taken up here. I am grateful to the audiences at the *Who's Got the Power?* Conference in Reykjavík, the workshop on *Feminist Issues in Practical Philosophy* in Bern, and the workshop on *Language and Power* in Barcelona for engaging with parts of this paper. Thanks to the anonymous reviewer whose comments significantly improved the content and clarity of this paper.

Bibliography

Anderson, Elisabeth (2010): *The Imperative of Integration*. Princeton: Princeton University Press.

Anderson, Luvell (2018): "Calling, Addressing, and Appropriation". In: David Sosa (Ed.): *Bad Words: Philosophical Perspectives on Slurs*. Oxford: Oxford University Press, pp. 6–28.

Blow, Charles M. (2014): "Paul Ryan, Culture and Poverty". In: *The New York Times* https://www.nytimes.com/2014/03/22/opinion/blow-paul-ryan-culture-and-poverty.html, accessed 28 November 2017.

Bourdieu, Pierre (1982): *Ce que parler veut dire. L'économie des échanges linguistiques*. Paris: Fayard.

Brandom, Robert B. (1994): *Making It Explicit. Reasoning, Representing, and Discursive Commitment*. Cambridge: Harvard University Press.

Brandom, Robert B. (2001): *Articulating Reasons. An Introduction to Inferentialism*. Cambridge: Harvard University Press.

Butler, Judith (1997): *Excitable Speech. A Politics of the Performative*. New York: Routledge.

Celikates, Robin (2015): "Against Manichaeism: The Politics of Forms of Life and the Possibilities of Critique". In: *Raisons politiques* 57. No. 1, pp. 81–96.

Davidson, Donald (1978): "What Metaphors Mean". In: *Critical Inquiry* 5. No. 1, pp. 31–47.

Frye, Marilyn (1983): *The Politics of Reality. Essays in Feminist Theory*. Trumansburg: The Crossing Press.

Gooding-Williams, Robert (Forthcoming): "Ideology, Social Practices, and Anti-Black Concepts". In: Robin Celikates, Sally Haslanger and Jason Stanley (Eds.): *Ideology: New Essays*. Oxford: Oxford University Press.

Haslanger, Sally (2005): "What Are We Talking About? The Semantics and Politics of Social Kinds". In: *Hypatia* 20. No. 4, pp. 10–26.

Haslanger, Sally (2017): "Racism, Ideology, and Social Movements". In: *Res Philosophica* 94. No 1, pp. 1–22.

Haslanger, Sally (2019): "Cognition as a Social Skill". In: *Australasian Philosophical Review* 3. No. 1, pp. 5–25.

Jäggi. Rahel (2014): *Kritik von Lebensformen*. Frankfurt a. M.: Suhrkamp.

Jäggi. Rahel (2015): "Towards an Immanent Critique of Forms of Life". In: *Raisons politiques* 57. No. 1, pp. 13–29.

Kukla, Rebecca (2018): "Slurs, Interpellation, and Ideology". In: *The Southern Journal of Philosophy* 56 (Spindel Supplement), pp. 1–26.

Lakoff, George, and Mark Johnson (1980): "Conceptual Metaphor in Everyday Language". In: *The Journal of Philosophy* 77. No. 8, pp. 453–486.

Langton, Rae (2002): "Beyond Belief: Pragmatics in Hate Speech and Pornography". In: Ishani Maitra and Mary K. McGowan (Eds.): *Speech and Harm. Controversies Over Free Speech*. Oxford: Oxford University Press.

Langton, Rae (2018): "Blocking as Counter-Speech". In: Daniel Fogal, Daniel W. Harris and Matt Moss (Eds.) *New Work on Speech Acts*. New York: Oxford University Press, pp. 72–93.

Lessig, Lawrence (1995): "The Regulation of Social Meaning". In: *The University of Chicago Law Review* 62. No. 3, pp. 944–1045.

McConnell-Ginet, Sally (2011): *Gender, Sexuality, and Meaning. Linguistic Practice and Politics*. Oxford: Oxford University Press.

Medina, José (2006): *Speaking from Elsewhere: A New Contextualist Perspective on Meaning, Identity, and Discursive Agency*. Albany: State University of New York Press.

Medina, José (2008): "Whose Meaning? Resignifying Voices and Their Social Locations". In: *The Journal of Speculative Philosophy, New Series* 22. No. 2, pp. 92–105.

Medina, José (2013): *The Epistemology of Resistance: Gender and Racial Oppression, Epistemic Injustice, and the Social Imagination*. Oxford: Oxford University Press.

Mills, Charles (2007): "White Ignorance". In: Shannon Sullivan and Nancy Tuana (Eds.): *Race and Epistemologies of Ignorance*. New York: State University of New York Press, pp.13–38.

Mühlebach, Deborah (2021): "Semantic Contestations and the Meaning of Politically Significant Terms". In: *Inquiry* 64. No. 8, pp. 788–81.

Mühlebach, Deborah (Manuscript): "Neopragmatist Inferentialism and the Meaning of Derogatory Terms – A Defence", (unpublished manuscript, August 16, 2021), Microsoft Word Format.

Queloz, Matthieu, and Damian Cueni (2020): "Left Wittgensteinianism". In: *European Journal of Philosophy*. Published online. DOI: https://doi.org/10.1111/ejop.12603, accessed 20 September 2021.

Saul, Jennifer (2018): "Dogwhistles, Political Manipulation, and Philosophy of Language". In: Daniel Fogal, Daniel W. Harris and Matt Moss (Eds.): *New Work on Speech Acts*. New York: Oxford University Press, pp. 360–383.

Scheman, Naomi (1997): "Queering the Center by Centering the Queer: Reflections on Transsexuals and Secular Jews". In: Diana Meyers (Ed.): *Feminists Rethink the Self*. Boulder: Westview Press, pp. 124–62.

Stalnaker, Robert (2002): "Common Ground". *Linguistics and Philosophy* 25, pp. 701–21.

Thompson, John B. (1991): "Introduction". In: Pierre Bourdieu: *Language and Symbolic Power*. Cambridge: Polity Press, pp. 37–43.

Tirrell, Lynne (1999a): "Aesthetic Derogation: Hate Speech, Pornography, and Aesthetic Contexts". In: Jerrold Levinson (Ed.): *Aesthetics and Ethics: Essays at the Intersection*. Cambridge: Cambridge University Press, pp. 283–314.

Tirrell, Lynne (1999b): "Derogatory Terms. Racism, Sexism, and the Inferential Role Theory of Meaning". In Christina Hendricks and Kelly Oliver (Eds.): L*anguage and Liberation: Feminism, Philosophy, and Language*. Albany: State University of New York Press, pp. 41–79.

Tirrell, Lynne (2012): "Genocidal Language Games". In: Ishani Maitra, and Mary K. McGowan (Eds.): *Speech and Harm: Controversies Over Free Speech*. Oxford University Press, pp. 174–221.

Tirrell, Lynne (2017): "Toxic Speech: Toward an Epidemiology of Discursive Harm". In: *Philosophical Topics* 45. No 2, pp. 139–161.

Wenger, Etienne (1998): *Communities of Practice: Learning, Meaning, and Identity*. Cambridge: Cambridge University Press.

Bianca Cepollaro
The Power to Shape Contexts: The Transmission of Descriptive and Evaluative Contents

Abstract: Recently, scholars have been investigating the hidden moral and political valence of apparently non-political forms of communication, by looking at how certain *prima facie* harmless uses of language can spread prejudice and contribute to social injustice. In this chapter I argue that while analyses such as Langton's convincingly explain how *descriptive* contents are transmitted and can contribute to belief formation and knowledge transmission, a different model is required to satisfactorily account for value judgments. I submit that while individuals can come to believe and, in the good cases, know certain descriptive facts about the world only based on what others tell them, testimony cannot transmit evaluative contents in the same way. Informative evaluative presupposition contributes to the creation of beliefs and knowledge, but it does not *determine* them in the way in which testimony typically works with descriptive contents: I argue for this claim by examining taste, moral and aesthetic value judgments. My point does not undermine the idea that presupposing certain value contents may affect one's audience, but challenges the idea that testimony is the right notion to account for how value talk can spread and impose contents on the participants to a conversation.

1 The Political Turn: Propagandistic and Manipulative Language

Philosophy of language and epistemology have long examined the ways in which individuals exchange information, accept, or reject propositional contents, share beliefs and the like. Scholars have recently drawn attention to the mechanisms that convey information implicitly rather than explicitly, encompassing presupposition, conventional and conversational implicature, and so on. Bringing implicit contents into the picture enriches—but also tangles—the question of how communication shapes our beliefs. Furthermore, interesting issues arise if we do not limit our study to the domain of descriptive beliefs but also include evaluative judgments, and investigate how the ways in which we communicate can shape shared values, social hierarchies, and social practices. Having both de-

https://doi.org/10.1515/9783110612318-011

scriptive and evaluative contents under scrutiny, communicated explicitly and implicitly, allows us to address toxic discourse more adequately, i.e., speech that is able to create and reinforce dangerous ideologies. In addition to blatant instances of hate speech, the study of implicit communication is apt to detect more subtle forms of toxic speech. It is within this broad project that scholars like Rae Langton aim to explain why certain prima facie harmless uses of language can in fact spread a good deal of prejudice and contribute to discrimination and social injustice. This line of research is ultimately concerned with the hidden moral and political valence of apparently non-political forms of communication, one of the core traits that mark what may go under the label of "political turn" in analytic philosophy of language.

In this chapter, I argue that, while Langton's analysis convincingly explains how *descriptive* contents (concerning states of affairs) are transmitted and can contribute to belief formation and knowledge transmission, a different model is required to satisfactorily account for value judgments (expressing the values of a given subject). I submit that, while individuals can come to believe and, in the good cases, know certain descriptive facts about the world only based on what others tell them, testimony cannot transmit evaluative contents in the same way. Informative evaluative presupposition contributes to the creation of beliefs and knowledge, but it does not *determine* them like it can when descriptive contents are at stake. I argue for this claim by examining taste, moral, and aesthetic value judgments. My point does not undermine the idea that presupposing certain value contents affects one's audience in crucial ways, especially if we understand presupposition not as a shared assumption, but as an assumption that "*ought to* be shared" (Sbisà 1999, p. 501, emphasis added). However, my argument provides strong reasons to doubt that testimony is the right notion to account for how value talk can spread and impose contents on the conversation participants, even for a cognitivist approach that allows for taste, moral and aesthetic beliefs.

The goal of this chapter is to stress the specificities of the evaluative domain (as opposed to the descriptive domain) which the analysis of the political valence of language and value talk needs to take this into consideration. Whether or not the mainstream testimony model can be extended from the descriptive to the evaluative dimension is a crucial question for those who are interested in the ways in which language can shape world views, social hierarchies, and shared values. In section 2, I briefly present the notion of testimony and illustrate how it applies not only to assertion (explicit communication) but also to presupposition and conventional implicature (implicit communication). In section 3, I contrast descriptive and evaluative contents and argue that testimony can give rise to beliefs and knowledge only for the former. In section 4, I reconsider the ways in which implicit communication can shape contexts in light of the caveats I have discussed, by stressing that testimony (involving both descriptive and evaluative contents) can indirectly—yet effectively—contribute to the formation and the diffusion of value judgments and that implicit communication is particularly capable of surreptitiously slipping contents in the speakers' common ground.

2 Front- and Back-Door Testimony

In recent decades, epistemologists have addressed the question of whether or not knowledge can be transmitted via testimony (see, e.g., Fricker 1987 and 2006; Pritchard 2004; and Adler 2017). This is especially relevant because we acquire so many of our beliefs from the word of others, through newspapers, books, on the internet, or in conversation. Usually, when a speaker presents somethings as true, their audience comes to believe them. Under normal conditions, the audience (the reader, the listener, the addressee, etc.) is not in the position to adequately verify what their interlocutors say, or to test their credibility and sincerity: all these aspects make testimony a challenging phenomenon, especially considering how widespread it is. To put it in Elizabeth Fricker's words:

> The epistemological 'problem of justifying belief through testimony' is the problem of showing how it can be the case that a hearer on a particular occasion has the epistemic right to believe what she is told—to believe a particular speaker's assertion . . . The solution can take either of two routes. It may be shown that the required step—from 'S asserted that P' to 'P'—can be made as a piece of inference involving only familiar deductive and inductive principles, applied to empirically established premises. Alternatively, it may be argued that the step is legitimized as the exercise of a special presumptive epistemic right to trust, not dependent on evidence. (Fricker 1994, p. 128)

The central question in the debate is whether testimony can suffice for knowledge. According to many epistemologists (see, e.g., Fricker 1987 and 2006), to acquire knowledge through testimony does not just mean that the word of others merely *contributes* to the formation of our beliefs (and possibly knowledge). It means that testimony *by itself* can suffice for the transmission of knowledge, thus dispensing the hearer from the need to go and check the received information.

In their study of testimony, epistemologists typically look at assertions. However, Langton (2018), convincingly extends the domain of inquiry to non-asserted content by looking at presupposition and conventional implicature. She is concerned with how implicitly conveyed content can constitute testimony, that is, how people can come to believe (and, in the good cases, know) certain propositions based solely on what their interlocutors *implicitly* communicate. Langton (2018) analyzes the case of informative presupposition. Presupposition is what speakers take for granted in a conversation. For instance, since it is common ground that David visited NYU in 2018, it is felicitous for Victor to utter in 2019 that "David is visiting NYU again this spring." The utterance asserts that David is visiting NYU this spring and presupposes (because of the presuppositional trigger "again") that David has visited NYU at some point in the past. In prototypical cases, the presupposed piece of information is mutually shared among the participants to the conversation. Under the adequate circumstances, however, it is possible to felicitously presuppose pieces of information that were *not* shared among the participants, as long as they are willing to accommodate them in the common ground (see, e. g., Lewis 1979; Stalnaker 2002; Simons 2003; and García-Carpintero 2016). In such cases, we talk about *informative* presupposition (see, e. g., von Fintel 2008; Gauker 2008; and Tonhauser 2015). Consider this scenario: David is talking with his new colleagues about Fiona Apple's new album and says, "My wife bought it as soon as it was released." Even though David's new colleagues did not know he has a wife, they can easily accommodate such a piece of information among the other things that everyone in the conversation knows and knows that the others know, and so on. The exchange can go on without interruptions. Note that the mechanism of informative presupposition may be exploited for casually smuggling problematic contents into the common ground. Suppose that David had the impression that one of his new colleagues is flirting with him; instead of openly saying that he is not available, he casually mentions his wife in order to get this piece of information across without asserting it. Once again, even though David's colleagues did not expect him to be married, they can go on without a hitch.

According to Langton, both asserted and presupposed content, in the right conditions, give rise to proper knowledge: when the participants to the conversation did not know that David has a wife and that she bought Fiona Apple's latest album as soon as it was released, they can just accommodate these pieces of information and go on without challenging them. In general, Langton argues, we can testify not only by means of what we explicitly say, but also by means of what we presuppose. In her view, informative presupposition can constitute a special kind of back-door testimony, if it goes unchallenged. In these cases, the participants to the conversation come to believe (or to know) such contents. If David really has a wife, then his colleagues come to know that he does. If he made that up just to discourage his workmate from flirting with him, then they only come to believe something false about his marital status.

Testimony via presupposition is not as explicit as testimony by assertion but operates covertly. It is, in Langton's terms, a back-door kind of testimony, where the process through which the conveyed contents become part of the interlocutor's beliefs and, in the good cases, knowledge tends to go unnoticed. Implicit contents not only go unnoticed, but they are also much harder and less straightforward to challenge than explicit ones. For instance, if one replies to David "No, that's not true," they succeed in denying the explicit asserted content (that his wife has bought Fiona Apple's album as soon as it was released), but not the implicit one (that David has a wife). To address the latter, one would have to interrupt the natural flow of the conversation, and say something like: "Wait, you have a wife?" or "What are you talking about? Aren't you single?"

As mentioned in the introduction, such mechanisms can be employed to deliberately impose certain contents on one's audience. Take this example, involving descriptive contents. In a rape trail, the victim is asked "Did you or did you not take advantage of later opportunities to resist?" This presupposes that the victim had opportunities to resist. Whether she answers yes or no, the presupposition that she did have the chance to oppose still survives. If no one protests, it simply becomes common ground that the victim had such a possibility (Davies 2014: 512–513). It is of the utmost importance to pay attention to the ways in which implicit contents are communicated under the radar.

We are now ready to turn from descriptive to evaluative contents. Consider the utterance, "Even George could win"[1] (Langton 2018: 148 ff). In uttering it, one is not only explicitly asserting that George could in fact win but is also implicitly conveying the evaluative judgment that George is an unpromising candidate. If they are not common ground, both contents can be the object of testimony, on Langton's account, but while it is explicitly communicated that George could win, it is only covertly conveyed that he is an unpromising candidate. Not only can this piece of information slip under the radar and be accepted by default, but it is also harder to reject, as we have seen: if one replies "No, that's not true," one will only deny the explicit descriptive content (that George could win), but not the implicit evaluative one (that George is an unpromising candidate). To address the latter, one would have to interrupt the natural flow of the conversation, and ask something like, "Wait, are you saying that George is bad?" All in all, Langton treats implicit descriptive and evaluative contents along the same lines.

---

1 Langton analyzes the content associated with "even" as a presupposition, while most scholars take it to be a conventional implicature. This issue does not matter for our current purposes but see Tonhauser et al. (2013) for a taxonomy of projective contents that does without the notion of presupposition and conventional implicature altogether.

My goal in this chapter is to show that the models for descriptive testimony do not automatically apply to evaluative testimony as well, which may need its own model. It is important to acknowledge that, in accordance with Langton, as I will stress in section 4, the mechanisms of implicit communication push the audience to accept the implicit contents. This acceptance, however, is not enough to speak about "knowledge via testimony" when it comes to certain kinds of evaluative contents.

3 Descriptive vs. Evaluative Testimony: Taste, Moral, and Aesthetic Judgments

In this section, I observe three kinds of value judgments (taste, moral, and aesthetic) and point out how the ways in which they may transmit knowledge crucially diverge from the ways in which factual judgments do, whether they are explicitly asserted or implicitly presupposed. Underlining such a disanalogy is sounding a note of caution when we try to apply a model for descriptive testimony to the linguistic spreading of evaluative contents. This disanalogy urges us to distinguish between two independent claims: that implicit communication in general pushes the audience to accept certain things as true for the sake of the conversation on the one hand, and that evaluative testimony can ever lead to proper knowledge, on the other hand. While the former is very reasonable, the latter is harder to defend. Interestingly, however, to a certain extent the two claims are independent, and one can endorse the former without buying into the latter.

Before I discuss this disanalogy to a greater detail, let me underline that, on a non-cognitivist account, the problem of whether we can acquire evaluative knowledge via testimony would not raise at all, because on this view there is no such thing as an evaluative proposition in the first place. In this paper, however, I am adopting a cognitivist perspective that is open to there being evaluative propositions that can be the object of proper knowledge. This allows me, among other things, to analyze asserted and presupposed content in the traditional terms.[2]

An assertion or a presupposition can concern states of affairs (e.g., "I have a car" and "My car broke down") or mental states (e.g., "I love baba ganoush" or "You have realized that I love baba ganoush"). When Ada tells any of these

---

2 Marques and García-Carpintero (2020) have attempted to draw the distinction between assertion and presupposition for non-propositional contents.

things to Victor, in ideal conditions (Ada is sincere, Victor believes her and so on), Victor *comes to know* that Ada owns a car and that she loves baba ganoush. In such cases, it is common to grant the subject with a certain dose of reliability (people tend to know whether they have a car, whether they like a certain dish). However, an assertion or a presupposition could also express a value judgment rather than a mental state of the speaker. For instance, take a taste judgment like "Baba ganoush is delicious" (or one where this content is presupposed, such as "He has finally realized that baba ganoush is delicious"). Suppose Ada says this to Victor, who has never tasted baba ganoush, and thus entertains no belief about how it tastes. She is sincere and he trusts her. Would we say that Victor *has learned* (i.e., that he now *knows*) that baba ganoush is exquisite? It seems more reasonable to maintain that the object of Ada's testimony is that *she* loves baba ganoush, rather than that baba ganoush is delicious. To believe that baba ganoush is exquisite, one needs to have *tasted* it, rather than having *heard* of it, because taste requires first-hand experience in order to entertain a judgment (see, e. g., Pearson 2013; MacFarlane 2014; Ninan 2014; and Bylinina 2017). A person can suspect or expect that baba ganoush is delicious because of what they have heard, but they cannot *know* that a given food is exquisite without tasting it. My point is *not* that acquiring knowledge is easy and automatic for descriptive contents and less so for evaluative ones, but that, under the right circumstances, transmitting knowledge through testimony *can* in some cases be automatic for descriptive contents, but can *never* be automatic for the evaluative contents that require first-person perspective. There are, in fact, many cases in which descriptive testimony does not work; for instance, when what is communicated seems doubtful, or the speaker is untrustworthy. Suppose that Victor—who is known to have no medical knowledge—tells Ada that vaccines provoke autism. She ascribes him no reliability in that domain, and thus does not come to believe that vaccines provoke autism. In contrast, if the entire scientific community holds that there is no connection between vaccines and the autistic spectrum disorder, scholarly testimony will *suffice* for Ada to come to *know* that vaccines do not provoke autism. "Suffices" means that, thanks to testimony, she does not need to go and run studies herself.[3] The problem is whether evaluative testimony, just like descriptive testimony, is ultimately able to relieve the hearers "of the usual burdens involved in coming to know in nontestimonial ways" (Greco 2016, p. 484). The examples I present suggest that it is not. Take baba ganoush once again. Even if the most talented food critic told me that

---

3 See Fricker (2006) on the division of epistemic labor, and John (2011) on non-experts' deference to experts.

baba ganoush is delicious, I could not truly *know* that it is, unless I taste it: at best, I would come to know that *she* likes it or that experts like it. Her positive opinion of baba ganoush can surely encourage me to taste it or positively influence my judgment when I taste it, but I cannot ultimately *know* whether baba ganoush is delicious through testimony. My point is that the fact that the domain of taste requires first-person experience makes the acquisition of knowledge through testimony problematic. This does not mean that evaluative testimony does not play any role in the ways in which we form value judgments: it can surely *influence* people's opinions in many ways, but, unlike descriptive testimony, it just cannot *yield* knowledge.

Any model of how testimony shapes world views and values needs to take into account this disanalogy between descriptive and evaluative testimony, by paying attention to whether testimony can only influence the formation of beliefs or yield proper knowledge.

Similar observations hold for the moral domain. If Ada tells Victor that "Torture is wrong" (or an utterance presupposing it, "I have discovered that torture is wrong"), he cannot just come to *know* that torture is wrong, but just that Ada *thinks* that torture is wrong (see Hopkins 2007; Hills 2009; and Fletcher 2016). A further question is what supplementary conditions would, in this case, turn the belief about Ada's convictions into the belief about torture being wrong. Among the most plausible candidates, take exposure and authority.[4] While for descriptive contents, in ideal conditions, it suffices that Ada says "I have a car" for Victor to know that she does, for moral beliefs, one may need a prolonged exposure to certain moral contents and practices in order to come to embrace a moral perspective. In other words, the automaticity with which we can sometimes come to know descriptive facts contrast with the inevitable gradualness with which moral judgments can be learned. An additional ingredient may transform the belief that Ada disapproves of torture into the believe that torture is wrong, namely, Ada's moral authority. Her chances to transmit knowledge through testimony increase when her interlocutors trust her, have a high opinion of her moral values, and possibly see her as a moral reference point.

Just to repeat it one more time: in the descriptive domain, too, it may be necessary to be exposed to certain contents for a long time in order to come to know them, or to grant the speaker with a special authority, but the disanalogy be-

---

4 I do not consider here a further factor labelled moral understanding, after Nickel (2001), that consists in the understanding of the moral facts involved in a given situation. The idea is that in the absence of this understanding, no moral knowledge can be acquired via testimony.

tween descriptive and evaluative testimony is that the former can yield knowledge by itself, while the latter can at best *contribute* to the formation of beliefs.

Let us finally look at the aesthetic domain. Could Victor come to know that the Alhambra is marvelous, if Ada tells him so or presupposes it? Once again, the object of Ada's testimony—that is, what Victor can, in ideal conditions, come to know—is Ada's appreciation of the Alhambra, not that the Alhambra *is* beautiful. Had Ada been a skilled art critic, Victor would guess that the Alhambra is appreciated by experts, or feel that he is expected to admire it, but he would not come to *know* that the Alhambra is beautiful without seeing it himself. With aesthetic judgments, testimony, once again, does not relieve the hearer from the burden associated with non-testimonial knowledge. As Kant puts it:

> Proofs are of no avail whatever for determining the judgement of taste. . . . [A subject] clearly perceives that the approval of others affords no valid proof, available for the judging of beauty. He recognizes that others, perchance, may see and observe for him, and that, what many have seen in one and the same way may, for the purpose of a theoretical, and therefore logical judgement, serve as an adequate ground of proof for him, albeit he believes he saw otherwise, but that what has pleased others can never serve him as the ground of an aesthetic judgement. The judgement of others, where unfavourable to ours, may, no doubt, rightly make us suspicious in respect of our own, but convince us that it is wrong it never can. Hence there is no empirical ground of proof that can coerce anyone's judgement of taste. (Kant 2007, pp. 113–114)

In the case of aesthetic judgments, all the factors we have been considering seem to play a role in turning the belief about Ada's appreciation into a belief about the Alhambra being marvelous: the first-person experience of the object, the prolonged exposure to certain aesthetic judgments and practices, the authority of the speaker.

To conclude, a mere assertion (or presupposing utterance) does not suffice for the transmission of evaluative knowledge through testimony, while it can be enough to yield knowledge of descriptive facts.

4 The Power to Shape the Context Revisited

As we have seen, according to scholars like Langton (2018), it is possible to transmit knowledge via testimony not just by means of what we explicitly say, but also through what we presuppose. Objectionable contents can get across in this back-door way, and be accepted without being questioned. This mechanism is particularly dangerous for certain kinds of language, such as hate speech—speech that targets individuals based on features like their sexual orientation, nationality, ethnic origin, gender, religion, and so on. If presupposition can spread beliefs, then certain kinds of discourse may have the potential to change and direct people's opinions and feelings. This mechanism, however, does not apply indistinctly to descriptive and evaluative contents.

In this chapter, I have tried to show that value judgments are not the object of testimony in the automatic way in which descriptive judgments can sometimes be, at least on certain occasions. This disanalogy sounds a note of caution in relation to the simple application of the model for descriptive testimony to the linguistic spreading of evaluative contents (both explicitly and implicitly conveyed). Acquiring knowledge through testimony implies that hearers form their belief (knowledge, in the good cases) *solely* based on what they are told; in contrast, expressing value judgments can at best *contribute* to the formation of beliefs, by inviting or even pushing interlocutors to conform to a certain evaluative perspective, without yielding the formation of evaluative judgments.

Being exposed to value judgments (taste, moral, and aesthetic) can surely, under the adequate circumstances, encourage hearers to embrace a certain perspective. However, the mechanism at play does not resemble testimony—understood as the means by which a person *comes to know* something—because the formation of testimonial knowledge requires that subjects form their belief solely on the basis of what they are told and are thus relived from the burden typically associated with non-testimonial knowledge. In the case of value judgments, testimony can at best *contribute* to the formation of beliefs, but cannot determine it, as it cannot ultimately relieve the hearer from getting a first-person perspective.

To see what these observations imply for the general project of investigating the normative power of language and its potential to influence beliefs, attitudes, and norms, three remarks are in order. First, pointing out that the transmission of knowledge works differently within the evaluative and the descriptive domains does not mean to deny the dangers of being exposed to certain value attitudes (for instance, bigoted ones)—quite the contrary. Rather, it calls for the development of new adequate accounts. Stressing that evaluative contents spread in a different way than descriptive ones (and thus that descriptive and evaluative domains require different models of beliefs and knowledge transmission) does not undermine the thesis that certain kinds of communication—such as bigoted speech—are particularly hideous and dangerous because of their toxic and propagandistic power.

Second, implicit communication (especially presupposition) is particularly capable of surreptitiously slipping contents in the speakers' common ground. When evaluative contents are presupposed rather than asserted, they come with a pressure to conform to certain assumptions. To put it in Sbisà's words, presuppositions are not "shared assumptions, but . . . assumptions which *ought to be* shared" (Sbisà 1999, p. 501, emphasis added). Similarly, according to Chilton, presupposition is "at least one micro-mechanism in language use which contributes to the building of a consensual reality" (Chilton 2004, p. 64). This suggests that bigoted presupposition can be very dangerous, even if the process through which its content spreads is not adequately captured by testimony.

Finally, even the most prototypical instances of descriptive testimony can indirectly, yet effectively, contribute to the formation and the circulation of value judgments. Take, for instance, the case of fake news (Mukerji 2018): broadcasting (in good or bad faith) fake news that, thanks to online communication, has the potential to go viral, plays a crucial role in determining or controlling the formation of value judgments. Even if a certain piece of fake news is factual and descriptive in nature, the effect that it may produce on individuals is to create value judgments on someone or something. The remarks I have made in this chapter do not invite one to prize apart the descriptive and the evaluative dimensions, erroneously concluding that expressing value judgments cannot affect our interlocutors' world views; if anything, it encourages to acknowledge the differences between these two dimensions in order to develop more adequate and fine-grained theoretical models that can hopefully address the dangers that lurk in certain kinds of communication.

Bibliography

Adler, Jonathan (2017): "Epistemological Problems of Testimony." In: *The Stanford Encyclopedia of Philosophy* (Winter 2017 Edition), Edward N. Zalta (Ed.), URL = <https://plato.stanford.edu/archives/win2017/entries/testimony-episprob/>, accessed 12 February 2021.

Bylinina, Lisa (2017): "Judge-Dependence in Degree Constructions." In: *Journal of Semantics* 34. No 2, pp. 291–331.

Chilton, Paul (2004): *Analysing Political Discourse: Theory and Practice.* London, New York: Routledge.

Davies, Alex (2014), "How to Silence Content with Porn, Context and Loaded Questions." In: *European Journal of Philosophy* 24, pp. 498–522.

von Fintel, Kai (2008): "What is Presupposition Accommodation, Again?" In: *Philosophical Perspectives* 22, pp. 137–170.

Fletcher, Guy (2016): "Moral Testimony: Once More with Feeling." In: Russ Shafer-Landau (Ed.): *Oxford Studies in Metaethics.* Oxford: Oxford University Press, pp. 45–73.

Fricker, Elizabeth (1987): "The Epistemology of Testimony." In: *Proceedings of the Aristotelian Society Supplementary* 61, pp. 57–83.

Fricker, Elizabeth (1994): "Against Gullibility." In: Bimal Krishna Matilal and Arindam Chakrabarti (Eds.): *Knowing from Words.* Dordrecht: Kluwer, pp. 125–161.

Fricker, Elizabeth (2006): "Testimony and Epistemic Autonomy." In: Jennifer Lackey and Ernest Sosa (Eds.): *The Epistemology of Testimony.* Oxford: Oxford University Press, pp. 225–250.

García-Carpintero, Manuel (2016): "Accommodating Presuppositions." In: *Topoi* 35. No 1, pp. 37–44.

Gauker, Christopher (2008): "Against Accommodation: Heim, van der Sandt, and the Presupposition Projection Problem." In: *Philosophical perspectives* 22, pp. 171–205.

Greco, John (2016): "What Is Transmission*?" In: *Episteme* 13. No 4, pp. 481–498.

Hills, Alison (2009): "Moral Testimony and Moral Epistemology." In: *Ethics* 120, pp. 94–127.

Hopkins, Robert (2007): "What is Wrong with Moral Testimony." In: *Philosophy and Phenomenological Research* 74, pp. 611–634.

John, Stephen (2011): "Expert Testimony and Epistemological Free-Riding: The MMR Controversy." In: *The Philosophical Quarterly* 244, pp. 496–517.

Kant, Immanuel (2007): *Critique of Judgement.* Edited by Nicholas Walker and translated by James Creed Meredith. Oxford: Oxford University Press.

Langton, Rae (2018): "Blocking as Counter-Speech." In: Daniel Fogal, Daniel Harris and Matt Moss (Eds.): *New Work on Speech Acts.* Oxford: Oxford University Press, pp. 144–164.

Lewis, David (1979): "Scorekeeping in a Language Game." In: *Journal of Philosophical Logic* 8. No. 1, pp. 339–359.

MacFarlane, John (2014): *Assessment-Sensitivity: Relative Truth and its Applications.* Oxford: Oxford University Press.

Marques, Teresa, and Manuel García-Carpintero (2020): "Really Expressive Presuppositions and How to Block Them." In: *Grazer Philosophische Studien* 97. No 1, pp. 138–158.

Mukerji, Nikil (2018): "What Is Fake News?" In: *Ergo: An Open Access Journal of Philosophy* 5, pp. 923–946.

Nickel, Philip (2001): "Moral Testimony and Its Authority." In: *Ethical Theory and Moral Practice* 4. No. 3, pp. 253–266.

Ninan, Dilip (2014): "Taste Predicates and the Acquaintance Inference." In: *SALT Proceedings* 24, pp. 290–309.

Pearson, Hazel (2013): "A Judge-free Semantics for Predicates of Personal Taste." In: *Journal of Semantics* 30. No. 3, pp. 103–154.

Pritchard, Duncan (2004): "The Epistemology of Testimony." In: *Philosophical Studies* 14, pp. 326–348.

Sbisà, Marina (1999): "Ideology and the Persuasive Use of Presupposition." In: Jef Verschueren (Ed.): *Language and Ideology. Selected Papers from the 6th International Pragmatics Conference.* Antwerp: International Pragmatics Association, pp. 492–509.

Simons, Mandy (2003): "Presupposition and Accommodation: Understanding the Stalnakerian Picture." In: *Philosophical Studies* 112. No. 3, pp. 251–278.

Stalnaker, Robert (2002): "Common Ground." In: *Linguistics and Philosophy* 25. No. 5, pp. 701–721.

Tonhauser, Judith (2015): "Are 'Informative Presuppositions' Presuppositions?" In: *Language and Linguistics Compass* 9. No. 2, pp. 77–101.

Tonhauser, Judith, David Beaver, Craig Roberts, and Mandy Simons (2013): "Toward a Taxonomy of Projective Content." In: *Language* 89. No. 1, pp. 66–109.

Saray Ayala-López

Hermeneutical Injustice and Conceptual Landscaping: The Benefits and Responsibilities of Expanding Conceptual Landscaping beyond Failure Reparation

Abstract: There seems to be a common timeline when identifying cases of hermeneutical injustice: first, we detect a failure in (dominant collective) conceptual resources; second, we work on some sort of conceptual landscaping to repair the failure and improve resources. I articulate two reasons why we need to expand this sort of conceptual work beyond failure detection and reparation. First, in relation to hermeneutical injustice specifically, cases of interest for a justice-motivated conceptual landscaping do not necessarily involve the discovery of failures in existing dominant collective resources. Second, limiting conceptual landscaping to fixing detected failures only allows add-ons and adjustments in existing resources, preventing radical creativity in our conceptual work. After introducing these two reasons, I propose what a more expansive way of doing conceptual landscaping might look like, using as an illustration the case of gender-open children. Finally, I articulate a caution when expanding conceptual landscaping beyond failure detection and reparation.

1 Introduction

Crafting the conceptual resources we use to think and understand the world (commonly referred to as conceptual engineering, amelioration, or conceptual plumbing)[1] might or might not be morally and politically motivated. Some researchers might find the concept of freedom or the concept of truth as needing crafting (e.g. Sharp 2013 and van Inwagen 2008) for epistemic or semantic reasons. However, recent prominent analyses that prescribe conceptual landscaping correspond to political and moral motivations. Cases of hermeneutical injustice (Fricker 2007) are fertile land for conceptual landscaping. When cases of hermeneutical injustice are identified, we are revealing that there is something in the world that resists conceptualization and sense-making with the existing domi-

---

[1] The label 'conceptual plumbing' appears in Carrie Jenkins (2020). The original comparison with plumbing appears in Mary Midgley (1996), who introduced 'philosophical plumbing'.

https://doi.org/10.1515/9783110612318-012

nant collective available resources, so either we craft the dominant pool of conceptual resources by incorporating resources from the margins (which are already fit to capture those parts of reality) or, if that is not available, we craft (existing or new) concepts. Hermeneutical injustice invites, or rather demands conceptual landscaping.

The timeline in cases of fixing hermeneutical injustice instantiates a pattern that is also present in cases of conceptual landscaping more generally, as follows: first, we detect a failure in conceptual resources in concepts such as, for example, freedom, truth, rape, and woman; second, we put forward conceptual landscaping to repair the failure and improve resources. In cases of hermeneutical injustice specifically, there is an additional justice-related motivation at place. Here, I articulate two reasons why we need to expand conceptual landscaping (both in the service of justice and generally) beyond failure-fixing. First, in relation to hermeneutical injustice, cases of interest for a justice-motivated conceptual landscaping do not necessarily involve the discovery of failures in existing dominant collective resources. Second, limiting landscaping to discovery of failure limits conceptual landscaping to add-ons and adjustments in existing resources, preventing radical creativity in our conceptual landscaping. After introducing these two reasons, I propose what a more expansive way of doing conceptual landscaping might look like, illustrating it with the case of gender-open children. Finally, I articulate a caution when expanding conceptual landscaping beyond failure detection.

2 Hermeneutical Injustice and Conceptual Landscaping

Recent developments in social epistemology are providing more and more sophisticated analyses of the ways in which the tools we use to classify and understand the world and the people living in it embody similarly problematic dynamics of privilege and marginalization that characterize more material parts of our world. The phenomenon of hermeneutical injustice (Fricker 2007) is of critical significance for those who care about how the conceptual resources dominant in our societies can fail to be politically, morally, and epistemically satisfactory. Hermeneutical injustice points toward deficiencies in our conceptual tools that are relevant for justice, for (i) they are caused by the structural erasure of groups of people from the process of creation and negotiation of collective meanings, and (ii) they in turn cause the further marginalization of these groups, by either undermining the understanding of experiences and pieces of the world that are important to them (their own or others' understanding), and/or undermining the capacity of nondominant groups to communicate those experiences. Hermaneutical injustice calls for a revision of dominant conceptual resources, moved by the goal of reducing or ideally eliminating the relevant injustice. Considerations about revising our conceptual resources connect the above debate in social epistemology to philosophy of language and metaphysics, for we are also talking about possible changes to the meaning of words, and about the accuracy and also the morality of the relationship between objects, properties and kinds, on the one hand, and our representational machinery (our concepts and terms), on the other.

Conceptual resources like shared concepts, scripts, folk explanations, and social norms constitute the collective tools we use, as members of societies, to navigate and understand reality, and to understand others and ourselves. Let us call these collective resources collective conceptual maps. I will use the term 'conceptual landscaping' to refer to any kind of revision of our conceptual maps, however directly or indirectly related to hermeneutical injustice. A revision might consist of the addition of a new resources (new to the dominant pool), an elimination of an existing one, or the transformation of it (e. g. a change in a concept or meaning of a word). It might consist of substituting dominant resources for non-dominant ones or filling up gaps with already existing nondominant resources. Or it might involve creating new resources from scratch. Revisions might involve changes in concepts, explanations, social norms, and scripts. Since the parts of a conceptual map are connected (e. g. explanations to scripts and concepts to explanations) a revision will usually involve adjustments to several parts.[2] Here, I will focus on concepts. Existing conceptual maps in a society determine that which is intelligible. For each more or less determinate society or group of societies,[3] there are several overlapping maps at work, usually with one that is dominant and potentially several resistant maps that are created in resistant communities. The dominant map is the one associated to dominant practices, which means it is the one institutions—from the medical practice to the law to formal education settings)—rely on.

In the literature on hermeneutical injustice, we find many cases of conceptual landscaping; or example, in relation to the missing concepts sexual harassment or postpartum depression (Fricker 2007).

---

2 This spillover is for some a reason to question both the feasibility and the desirability of conceptual landscaping.
3 Defining the boundaries of societies is beyond the scope of this chapter.

Let us focus on something these and other cases share. They seem to follow what might be called the Failure-Discovery Timeline. First, there is a discovery that something in the world (outside or inside the individual, material or mental) resists conceptualization and sense-making with the existing available (dominant) conceptual map. That is, a discovery that there is something that is not completely intelligible, something that needs to be accounted for (it needs to be named, conceptualized, integrated in our practices and scripts, in our understanding of ourselves and others) that has not yet been accounted for.[4] This mismatch between a piece of reality and the available map is a call for action: we need to change, improve our conceptual map (and therefore also our practices) so that it is comprehensive, for our conceptual maps are supposed to track (at some level, and in some sense to be specified) reality.

A gap in the map means not only a lack of concepts to account for some part of reality, but also leaving that part of reality disconnected from other parts of our conceptual maps (e. g. explanations, scripts, and narratives) and therefore outside of our practices. Take a person who has a series of experiences that we could classify, now that we have the conceptual tools for that, as experiences of being agender. The discovery here would be to realize that these experiences have no representation in the available map. This realization might happen after using existing tools to articulate those experiences and realizing that something is off. Another example is the paradigmatic case of sexual harassment. As we read in Fricker (2007), the discovery process in the case of sexual harassment started with a group of women coming together to fix the problem they had detected (Fricker 2007)[5]: Something was missing from the collective dominant available map, tools and guidance to name and articulate the unwanted sexual advances they were experiencing in the workplace. These advances were more than sexual gestures, comments and invitations that made them feel uncomfort-

---

4 Please note that the discovery is not of a new piece of reality, but of a missing part in our conceptual maps. There might be cases in which the discovery is of a new piece of reality. However, paradigmatic cases discussed in the literature on hermeneutical injustice acknowledge that the reality was already there, but only half-articulated. That is, in the case of sexual harassment, it is not that women came to discover that there was something wrong going on. What was missing was an appropriate set of conceptual tools (specifically a concept and term for it) to articulate it in a smooth and clear way, and to put it in relationship to other concepts and as part of explanations and scripts. Thanks to Dan López de Sa for helping me get clear on this.
5 One can argue that the discovery of the problem did not happen at that time, but much earlier. Working-class women and women of color had been in the working force for a while at the time these white middle-class women got together to brainstorm about what was going on at the workplace that made them give up their jobs. As it happens, the discoveries that gives raise to conceptual landscaping also require certain epistemic and discursive standing.

able. Importantly, they were wrong. Conceptualizing them this way would connect them to other practices—and to the spill over of negative experiences that victims of sexual harassment go through—and would allow the elaboration of explanations of these experiences that appeal to the moral nature of sexual harassment instead of victim's failures of character. Perhaps the most important part of the baptism of sexual harassment was making sure it contains a normative element, which was clearly missing in the available resources in the dominant map, which were resources in the vicinity of flirting.

As mentioned above, our collective conceptual maps are linked to our practices, so that a deficiency in our conceptual map involves a deficiency in our practices. The absence of (dominant) conceptual resources accounting for parts of the material or mental reality means that there is a gap in existing dominant practices, in how people relate to each other and to the material world. For example, a social context with a (dominant) conceptual map that does not include a spot for nonbinary gender identities is a social context in which people relate to each other following exclusively binary-gender scripts, their sexual orientation is defined in binary terms, and their desire is articulated in binary gender terms, too; there are all kinds of social norms in two forms corresponding to each of the genders, from appropriate emotional displays to credibility attribution to assumed skills; the material reality, most of which is usually gendered in some form, is gendered in a binary way, for example, public bathrooms are provided in two forms, and so are clothing sections in stores, and options for gender assignment at birth and in passports.

The discovery of a map failure can happen in different ways and contexts. It might be at the individual level, whether or not the discovery has already been done by, say, a resisting community. For example, a person might be struggling to understand their experiences as an agender person, encountering a deficiency in the conceptual map available to them, while there might already be in circulation in resistant communities' concepts and scripts for agender identity. Access to those resistant communities is not easy for many people living in rural areas or poor environments with no material access to, for example, internet, or minimal social contact, and also to those who are part of certain religious groups. The discovery can also happen as a group discovery by a resistant community after sharing experiences and brainstorming to understand how they relate to existing narratives (as in the case of sexual harassment analyzed in Fricker 2007).

The Failure-Discovery Timeline seems to outline the steps in the process of identifying and attempting to repair cases of hermeneutical injustice: there is a realization that there is a piece of reality that is missing from our conceptual map, and therefore only half-intelligible, which leaves that part of reality disconnected from other parts of our conceptual maps (e.g. explanations, scripts, and narratives) and ultimately disconnected from social practices (including institutional practices). The most relevant feature that drives many cases of conceptual landscaping, specifically those related to hermeneutical injustice, is that they constitute an injustice and not just an obstacle to our epistemological aspirations. That is, the failure discovery reveals both an epistemic deficiency in our conceptual map, and an injustice. First, the deficiency is not just an epistemological error, but a morally relevant wrong, for it is the result of the unjust exclusion of certain groups from map-making (what Fricker calls hermeneutical marginalization, Fricker 2007, 153). Second, the deficiency maintains the unjust marginalization of those same groups, causing epistemic harms (Fricker's primary and secondary harms, pp. 161–166). Under the Failure-Discovery Timeline, however, we have epistemic reasons to proceed with conceptual landscaping, independent of the moral and political reasons that seem to be the motivating factors of the bulk of recent projects in conceptual landscaping (especially those related to denouncing and correcting hermeneutical injustice).

The Failure-Discovery Timeline, however, is not the entire story about conceptual landscaping. It is just the timeline of some, perhaps many cases that call for conceptual landscaping. That is, the actual steps some, perhaps many cases went through. We should resist the temptation of seeing in the Failure-Discovery Timeline a normative guideline, a how to manual for conceptual landscaping. There is more to conceptual landscaping than fixing the gaps we detect in our conceptual maps. In the next section, I offer two reasons to expand conceptual landscaping beyond the detection of failures in our conceptual maps.

3 Expanding Conceptual Landscaping

Mona Simion (2018) proposes to expand concept amelioration, which for our purposes we can identify with conceptual landscaping. She notices that usually conceptual landscaping is limited to conceptual repair. Moreover, she points out that it is common among ameliorative projects that the nature of the justification for the relevant conceptual landscaping is based on the nature of the defect found in the to-be-ameliorated concept (see also Simion and Kelp 2020). That is, the motivation to ameliorate usually reflects the nature of the defect that ameliorators are trying to fix. Instead of a limited focus on repairing defective concepts, Simion and Kelp propose to aim at conceptual innovation. Conceptual landscaping (which they call conceptual engineering) that is focused on conceptual repair is, on the one hand, too demanding, and, on the other hand, too limiting. It is too demanding because we do not need to fix a concept in order to bring benefits; improvement is enough (Simion 2018). It is too limiting because even if, at least in appearance, some of our concepts do not appear to need repair, if we think we can improve it, we should do it.

I add two reasons why we need to expand conceptual landscaping (both in the service of HI and generally) beyond conceptual repair. In section 3.1 I develop one of those reasons: there is more to conceptual landscaping than failure detection.

3.1 Beyond Gap Detection

Not all situations calling for conceptual landscaping involve the discovery of a missing part in our maps. On the one hand, even if there is a triggering psychological discovery-like process, its content might not be that there is a piece of reality unaccounted for in our conceptual map. It could be a distortion in the map instead. On the other hand, it might be that there is zero triggering cause (no discovery, neither of a gap nor a distortion). I address the first case below and the second case in the next subsection.

Our conceptual maps can contain problems other than gaps. They can also contain distorted information; in particular, distorted concepts. I rely here on Rebecca Mason's recent work. While Fricker's definition of hermeneutical injustice only mentions gaps, Mason (2021) identifies distortions in hermeneutical resources as another cause of hermeneutical injustice. To understand what Mason means by distortions, we need to see our maps as containing not only concepts (and explanations, scripts, and norms), but also inferential networks connecting concepts in clusters. In Mason's proposal distortions affect these connections. For example, when a concept like 'homosexual' is inferentially associated in the dominant conceptual map to concepts like 'perversion,' 'sickness,' 'temporary stage,' or 'frivolity,' we can say that this concept is distorted.

Notice that distortion in conceptual resources can happen in other locations, too. They might happen in our explanations. For example, we might use social generics like "Women have trouble getting high level professional jobs" and have the following explanation of the connection between the given category and the property: "because women are risk averse," instead of an explanation that appeals to structural factors (see Vasil and Ayala-López (Manuscript)). When affecting only concepts, distortion can lie not only between concepts but also within concepts. For example, consider the concept of a trans woman. In the dominant (cisnormative) culture this concept is distorted and means "a man who lives as a woman" (Bettcher 2013, p. 235). According to Talia Bettcher, this is not a problem of a few people misunderstanding the meaning of the concept. This is widespread and this is the meaning that we see at work in institutions and policies, like law enforcement agencies, or domestic violence and homeless shelters (Bettcher 2013).

While in Fricker's definition of hermeneutical injustice the central problem is that affected subjects do not possess the relevant concepts, due to its absence in the dominant collective map, in Mason's redefinition affected subjects might possess the relevant concepts (as they are part of the available map) but these are distorted, and therefore subjects might fail to apply them to themselves (and others). Take the cinematic example of the movie *Carol* (Haynes 2015) analyzed by Esa Díaz León (2017) as an example of hermeneutical injustice. Therese has strong feelings for Carol and is trying to understand what those are. There is certainly something wrong in the dominant map available to Therese in that North America of the 50s, but instead of a complete absence of conceptual tools to make sense of her experience, Therese does have access to a concept that in principle seems to be apt for the job. This concept is however distorted.

While from the third-person perspective, and with the benefits of historical perspective, we can easily claim that the concept homosexual was distorted in the maps of the time (and sadly, it still is in many maps in many places), when we take Therese's perspective there are at least two readings we can make. One way to read what possibly happened in Therese's mind is this: in her attempts to understand her feelings, Therese inquires[6] and she *discovers* something about the available conceptual map: she realizes that the dominant map is distorted, that it misrepresents reality; in particular, homosexuality. What she feels has nothing to do with what homosexuality is said to be. Another, perhaps less psychologically ambitious reading is to say that Therese simply realizes that the existing concept associated with the term homosexual, and possibly other tools related to that concept (e.g. explanations of homosexuality), do not apply to her. Even though the dominant map seems to have something to say about her feelings for Carol, she realizes the map is not actually covering the terrain it claims to be covering. In this second reading, Therese finds a gap. We can imagine her saying some like this: "if homosexuality is about *that* (accompany this with an emphatic disapproving gesture), then it definitely does not apply to what I'm feeling." In the first reading the map is off mark. In this second read-

---

6 It is fair to say that what she does is not a very active inquiry: she just asks her boyfriend whether two women or two men could fell in love with each other; she did not go to the library to look for literature on homosexuality, which she would have found, even if on some hidden, semi-forbidden self; she did not ask any of the peculiar-looking women who looked at her in half-complicity and half-judgmentally at the music store, nor went to any of those dark, nocturnal places where homosexuals were known to be found.

ing, the map is gappy. In both readings, Therese makes a discovery, but only one is a discovery of a gap.

The example Mason analyzes is more explicit about the realization of a mismatch between the experience of the affected subject and the (distorted) concept in principle available for them to articulate that experience. In the novel *A Boy's Own Story* by Edmund White, we find the protagonist hesitating about the available concept for homosexuality. He does not want to apply it to his feelings. The subject is realizing that there is something wrong with the conceptual map available to him. Detecting this kind of distortion might involve a sense of disorientation, and perhaps distortions are more insidious than gaps, for the distortions, so to speak, "seem to do the job." A distorted map is still used to understand ourselves and others, and to coordinate our behavior, infecting everything (our understanding of others and ourselves, and our practices) with the distortion. There is a lot to say about this kind of distortion and the psychological processes of detecting them and coping with them, but we leave that for another work.

In cases of distorted maps, conceptual landscaping is necessary and beneficial. Crafting the concept homosexual in dominant maps would improve reality-map accuracy (which means epistemic gains), and has obvious moral and political benefits. In sum, cases of distorted maps, and specifically distorted concepts (in the sense of distorted inferential associations) call for conceptual landscaping without necessarily involving the discovery of a gap in existing maps.

3.2 Beyond Detection

So far, we have covered cases that call for conceptual landscaping that involve a triggering psychological (individual or group) moment, one in which we discover either a gap in the dominant map or a distortion. There are, however, situations that call for conceptual landscaping that do not involve any of these triggering psychological moments. Expecting the discovery of a failure in our maps limits the prescription of conceptual landscaping to only a subset of the cases which would require so. What if there is no detection of a problem? We might fail to detect failures in our map for several reasons. In particular, our map might be such that it makes it very difficult to detect certain kind of failures. And that should not be a reason to think that conceptual landscaping is not needed. Let me unpack this.

Sally Haslanger (2019) identifies a relevant limitation in the way hermeneutical injustice is usually presented. As originally introduced by Fricker (2007), HI is presented as an injustice that is linked to the process of struggling to make sense of an experience for which the map lacks appropriate concepts. That is, to what we have here described as the discovery of failures in the map. However, Haslanger reminds us that our maps shape us, they shape what we pay attention to, what we desire and memorize, how we understand ourselves and others, how we parse reality and understand how its parts relate to each other and ourselves. Maps are not (just) the result of how we think. They feed our thinking, they shape how we understand ourselves, and therefore the identities that we develop. Demanding the discovery of a failure is already putting a high standard for conceptual landscaping, for it demands that we take some distance from the map, that we find a crack between the map and our thinking, between the map and our identity. But this crack might not be apparent. I might not detect any failure, and still be suffering a hermeneutical injustice. More generally, we might not detect any failure in our maps, and still be required to do conceptual landscaping. For example, a woman suffering from sexual harassment might not go through any triggering state of failure discovery. She might just live those experiences as part of what it is to be a woman; no surprise, no failure, nothing wrong (in the sense that this is what the map says and prescribes).[7] Similarly, an entire community might not discover failures in the map they rely on, but we might still want to say that their map needs revision. Those would be cases of ideological oppression, in which people are not being explicitly coerced into thinking or being in a particularly problematic way but have internalized that way of thinking or being.[8] They would, however, not count as cases of hermeneutical injustice according to Fricker's definition, something we might want to revise.

Haslanger is making room here for what she calls symbolic power. Hermeneutical Injustice, as defined by Fricker, is a form of symbolic power at work, but it is too limited.

In Haslanger's words, "The problem is not just that there are gaps in our hermeneutical resources, but the power of the hermeneutical resources we are given" (Haslanger 2019, p. 15). We can understand Haslanger here as saying

---

7 We find a relevant example of this in Amia Srinivasan (2020). In the scenario she calls domestic violence, a woman is regularly beaten up by her husband in a social context in which everyone around her thinks this is her fault, and she herself, after reflection, concludes this must be her fault and something women deserve when they misbehave, something that is part of being a woman. This is a case of bad ideology, in which false beliefs help maintain systems of oppression. Thanks to an anonymous reviewer for directing me to this example. Examples of this kind bring to mind what theorists call adaptive preferences, preferences that people suffering oppression end up with after habituation or acceptance of their conditions (see Nussbaum 2001 and Khader 2011).

8 Ideological oppression or injustice contrasts with repressive oppression or injustice. While the latter is forced upon people with coercive methods, the former is "enacted unthinkingly or even willingly by the subordinated and/or privileged" (Haslanger 2019, p. 5).

that the way we understand ourselves, the identities we acquire, can be traced back to our map, and if the map has problems, we can expect those to be part of how we understand ourselves. Importantly, because of the intricated relationship between the map and how we understand ourselves, we might not detect those problems. They are part of who we are. It is possible to frame all this in terms of distortion, as it is one more (sneaky) way concepts can be distorted: so that we *do* apply them to ourselves and distort our own selves. If so, it should be easy to count them as cases of hermeneutical injustice according to Mason's definition.

We can now revise our examples above: Therese and the protagonist of White's novel might not find it conflicting to apply to themselves the concept of homosexual that was part of the dominant available map. They might see themselves as perverted, as "going through a stage," as having feelings that are not relatable to (default, heterosexual) romantic feelings as dominantly understood. They might not detect a crack between their understanding of their experiences and the distorted guidelines in the map. Their understanding of themselves has been shaped by those distorted guidelines in the first place. In these cases, we might still want to say that conceptual landscaping is necessary, even if no single subject detects a failure in the available map. In sum, I propose not to limit conceptual landscaping to failure-detection.[9]

4 An Alternative: The Creative Critique Framework for Conceptual Landscaping

In this section, I develop the second reason in favor of expanding conceptual landscaping beyond failure detection: limiting landscaping to failure detection limits conceptual landscaping to add-ons and adjustments in existing map, not welcoming radical creativity. Conceptual landscaping that is prescribed in order to fix gaps and distortions is limited to damage control. That is, it uses conceptual landscaping as a tool to solve problems we detect. I think, however, that this damage control approach does not cover all desirable or potentially enriching conceptual landscaping. Let me explain.

In the literature on hermeneutical injustice, the critical move is the identification or discovery of failures (gaps or distortions) in the dominant collective conceptual map. In a sense, we can say that the conceptual tools that correct

---

9 This focus on detection, or rather the lack of it, adds a consideration to the expansion of conceptual landscaping that is missing in Simion (2018) and Simion and Kelp (2020).

the failures were somehow always there, in the sense that they fill in existing problems. They are not invented or created, but somehow discovered, in the same sense we discover a planet and then update the theories and conceptual maps in order to account for the new addition. This forecloses the possibility of conceptual landscaping that is the result of (radical) creative thinking. Margaret Boden's taxonomy of types of creativity comes in handy here. Boden (1994 and 2009) distinguishes between three types of creativity. The first one is taking familiar ideas and making unfamiliar combinations of them. The second and third types are the ones of interest to us. Exploring conceptual space is the second type of creativity. Boden defines a conceptual space as a structured style of thought that we find in a group's culture, from scientific theories to styles of poetry or music (Boden 2009, p. 3). For our purposes, we can take Boden's conceptual spaces to be equivalent to our conceptual maps, or at least parts of our conceptual maps. This second type of creativity means exploring parts of the map that have not yet been discovered, but that have always been there in some sense. Boden offers the following comparison: imagine that, while driving on the highway, you decide to take a secondary road and explore the countryside. Those small roads you are now exploring are marked in the map, but you did not notice them until you went into this "exploratory frame of mind" (Boden 2009, p. 4).[10] In art, exploratory creativity means creating something new within an existing style. Finally, the third type of creativity means transforming the conceptual space. While western music had been, up to nineteenth century, exploring different parts of the tonal harmony space (parts that were, if implicit, still there, even when still unexplored), Schoenberg overtook a transformation of that space that resulted in the new space of atonal music (Boden 1994). Of this transformative creativity, Boden says that "The deepest cases of creativity involve someone's thinking something which, with respect to the conceptual spaces in their minds, they couldn't have thought before" (Boden 2009, p. 5). Boden is here speaking at the individual level, but we can apply the same reasoning at the collective level and say that collective maps that are transformed result in spaces that were not there before. Transformative creativity is highly valued in science, as it means approaching a problem in a new way, for example, by asking a different question instead of providing a different answer to the same question.

In the Failure-Discovery Timeline, we find conceptual landscaping that is the result of either combinatorial or exploratory creativity (either at the individual or

---

10 We could say that those small roads are marked in our collective map in less than explicit ways.

group level). Those conceptual resources that are missing were always there, just waiting to be discovered. For example, sexual harassment was always there, even before white middle-class women joined the work force and discovered the gap and gathered together to give it a name. The damage-control approach of the Failure-Discovery Timeline limits the possibilities for conceptual landscaping. In particular, it prevents conceptual landscaping that is creative in Boden's third sense. I propose we take a different approach to conceptual landscaping, one in which conceptual revision does not require failure detection and welcomes transformative creativity.

Let us call this alternative picture Creative Critique. In Creative Critique, we do conceptual landscaping as part of a commitment to both understanding the complexities of our social world and to do it in a way that acknowledges and addresses the intertwined political and moral aspects of our representational tools.

There are three ideas guiding Creative Critique. First, conceptual maps embody the same injustices that structure our social world. Second, the social world changes, and so our maps need to be constantly updated. Third, we should not limit conceptual landscaping to fixing apparent failures in existent dominant map; rather, we should be open to craft our conceptual resources as part of how we imagine the social reality, given moral and political purposes.

Since, as we pointed out above, conceptual maps are connected to practices and material realities, doing conceptual landscaping based on creative critique opens up new (conceptual and material) spaces and possibilities that were not there before. Being limited to existing possibilities (by exercising combinatorial or exploratory creativity when landscaping) usually leads to augmenting the number of conceptual resources given the map we have. But augmenting the number of conceptual resources might get us only up to a certain point. For example, we have two genders, woman and man, and by exploratory landscaping we might add several non-binary genders. But we could also transform the map in a way that is not limited to adding more concepts for gender. For instance, it could be a transformation such that gender as we understand it disappears or merges with something else. In that alternative transformed map, we do not have to add more conceptual resources. It could be the case that we obtain similar or even better results (in terms of both people's psychological flourishing and social coordination) with a smaller number of concepts.

In a reflection on her past work in conceptual landscaping, Haslanger sees her earlier ameliorative proposal on the concepts of gender and race (Haslanger 2000) as the first step in an emancipatory project that involves at least two steps. Her definition of gender and race (in terms of either a subordinated or a dominant social position) aims at exposing the failures of existing (conceptual and material) practices. This identification of failure is but the first step. We also need to take a second step, "What we might call a visionary moment that gives us resources to create something better" (Haslanger 2020 p. 237). Creative Critique incorporates the vindication of this second, positive, creative element.

Underexplored domains are especially suitable for Creative Critique. One example is gender open children. Gender open children are children who are not assigned a specific gender at birth. When you care for a gender open baby, you do not use binary gender pronouns (in English she and he, in Spanish, *ella y él*) to refer to them (you can use, for example, they singular or zi in English, *elle* in Spanish), and you introduce your baby as a human who does not have a gender yet. The best way to protect the baby from being treated and perceived according to the gender binary (as either a girl or a boy) is not to reveal their genital status, which is commonly taken to be critical in the determination of sex and to be straightforwardly associated with their gender. Among the many questions we might want to address in relation to gender-open children is what concept(s) of gender, and what specific gender identities might serve gender-open children best. Gender open children might grow up to identify within the binary, or they might identify outside of it. In both cases, the possibilities might surpass any conceptual adjustment we might make to the existing map. A person who has remained, in a more or less pure way, outside gendered practices and gendering education, might require conceptual resources radically different from the ones we have. We could try creative critique to brainstorm and craft conceptual spaces (which would facilitate material spaces) that could fit the experiences of those children when they grow up. In doing this, caution is needed. I elaborate what this caution might require in the final section, where I also address a couple of concerns.

5 Cautions and Responsibilities

The discovery of inadequate dominant conceptual resources is usually done by people who suffer those inadequacies in profound ways. They are usually the ones who raise the alarm and trigger conceptual landscaping in what we labeled the Failure-Discovery Timeline. If Creative Critique were to remove the requisite of that discovery, important questions arise: What is going to guide conceptual landscaping? Can we do politically motivated and politically effective conceptual landscaping without the politically situated nature of failure detection? Expanding conceptual landscaping beyond failure-detection risks making it an exercise of mental gymnastics, doable from any armchair, that trivializes the importance of politically situated work. Within a creative conceptual landscaping in the line of creative critique, we need to find a place for politically situated work, and also be sensitive to the perspective of those who might be epistemically better positioned. A second concern is: what is the role left for members of marginalized groups in Creative Critique? Creative Critique, as described above, might actually be a burden on marginalized groups. If we want to put members of marginalized groups at the center of the transformative process, how can we do so if the stage of detecting failures is removed (and therefore also the associated acknowledgement of how these failures affect them)? It might seem as if Creative Critique is a platform for theorical and imaginative work suitable to do from a position that can be oblivious to existing failures and how these affect the lives of people.

Responding to these worries requires at least two moves. First, we need to clarify

the relationship between failure-discovery and Creative Critique. Second, we need to provide more guidelines on how to do this transformative landscaping. In relation to the first move, it is important to clarify that Creative Critique does not require ignorance or obliviousness to failures. There are indeed failures in dominant conceptual resources, and we learn from identifying and naming them, and we improve things when we fix or at least try to fix them (e. g. we improve things when we un-distort conceptual associations in the vicinity of the concept of homosexual). Creative Critique includes attention and sensitivity to failures and goes beyond that.

A transformation of the conceptual space does not need to be independent of failure-detection. It just takes conceptual landscaping a step further. Let us recall the lessons from the negative part of this work: (i) that failure-detection should not be restricted to gaps (but should also include distortions in Mason's sense), (ii) that conceptual landscaping should not be restricted to the detection of failures, for there are failures that might escape our radar, and (iii) that our conceptual landscaping can do much more than repairing (detected) failures. Creative Critique takes these lessons to go beyond failure-detection, without necessarily ignoring what we learn from the failures we do detect, but rather incorporating them.

Moreover, failure-detection can serve as a trigger for the radical transformation that is part of Creative Critique, that is, it can serve as the motivation to transform the existing dominant map in ways that go beyond fixing failures. This, I think, acknowledges the central role of members of marginalized groups in the process of transformation (since they are more sensitive to failures in dominant resources), while advocating for conceptual landscaping that goes beyond the reparation of such failures. The key point here is to distinguish between, on the one hand, failure-detection that is focused on the outcome, that is, repairing the detected failure, and, on the other hand, failure-detection that is rather focused on the process, that is, exposing the need for conceptual work. Exposing this need for conceptual work does not specify the character of the work that is called for, which can be restorative of existing tools or radically transformative.

In relation to the second move, I address it by drafting two principles that regulate Creative Critique. Creative Critique needs a research ethics, an epistemic humility, and sensitivity principle. The need for a research ethics is especially salient when we are talking about other people's experiences and identities. This is how I defined a research ethics, in the form of an ethical avowal, which is a desideratum when philosophizing about gender-open children:

> when philosophizing about other people's identities, I will have ethical principles guiding my inquiry, and I will use these principles as I reflect on my interest on the subject matter, the goals of my inquiry (e. g. why am I interested in this? What do I want to attain with my research?), and the way I proceed in my research. This does not imply that my philosophizing will be spoiled or coerced towards specific positions. It means I will examine my subject matter not from an impossibly abstract and supposedly neutral armchair, but acknowledging and being transparent about the values that are guiding my inquiry in more or less implicit ways. (Ayala-López 2020, p. 13)

When engaged in creative critique in relation to gender-open children, we might be guided by different purposes and motivations, but whatever those are we need to be mindful that we are not crafting conceptual (and eventually material) spaces for new breeds of abstract entities or new artistic expressions. We are talking about people. In the same way scientific research follows ethical principles, philosophical work benefits from that. Following a research ethics does not mean we are imposing a limit on the intellectual work we want to pursue. As I argue elsewhere, the starting point is never a limitless philosophy. Any philosophizing has limits:

> It's only the overly optimistic, or perhaps rather, vain philosophers who think that there is no actual limit to the questions that philosophy explores. There are limits to philosophy: the limit is set by our interests, our concerns, our desires and available resources. Every once in a while, we should wonder why we inquire into the questions we do (again, as individuals and as a society). This meta-inquiry can reveal a lot about the very issues we are investigating. (Ayala-López 2020, p. 13)

A research ethics does not limit our intellectual adventures, it actually enriches the intellectual work we are set to pursue by incorporating meta-questions, which inform the analysis in important ways (e. g. the meta-questions we ask can serve to justify the specific landscaping we propose or the very project of engaging in landscaping). The other principle for Creative Critique refers to epistemic humility and sensitivity. The idea is that when engaging in Creative Critique, it is desirable to always be open to learning things that we do not know we can learn, and to be sensitive to the perspective of those who might be better positioned to inform and pursue a creative crafting of resources.[11]

Bibliography

Ayala-López, Saray. (2020): "(Philosophizing about) Gender Open Children." In: *American Philosophical Association Newsletter in Feminism and Philosophy* 19. No 2, pp. 45–49.

Betcher, Talia (2013): "Trans Women and the Meaning of 'Woman.'" In: Alan Soble, Nichola Power, and Raja Halwani (Eds.): *Philosophy of Sex: Contemporary Readings*, 6th ed. Lanham, Maryland: Rowan & Littlefield. pp. 233–250.

Boden, Margaret A. (1994): "Precis of The Creative Mind: Myths and Mechanisms." In: *Behavioral and Brain Sciences* 17. No. 3, pp. 519–570.

Boden, Margaret A. (2009): "Creativity in a nutshell." In: *Think* 5. No. 15, pp. 83–96.

Díaz-León, Esa (2017): "Feminist Metaphysics and Philosophy of Language." In: Carol Hay (Ed.): *Philosophy: Feminism*. Gale: Macmillan Reference USA, pp. 251–271.

Fricker, Miranda (2007): *Epistemic Injustice: Power and the Ethics of Knowing*. Oxford, New York: Oxford University Press.

Haslanger, Sally (2000) Gender and race: (What) are they? (What) do we want them to be? *Noûs* 34 (1):31–55.

Haslanger, Sally (2019): "Cognition as Social Skill." In: *Australasian Philosophy Review* 3. No. 1, pp. 5–25.

Haslanger, Sally (2020): "Going on, not in the same way." In: Alexis Burgess, Herman Cappelen, and David Plunkett (Eds): *Conceptual Engineering and Conceptual Ethics*. Oxford: Oxford University Press, pp. 230–260.

Jenkins, Carrie (2020): "When Love Stinks, Call a Conceptual Plumber." In: Elly Vintiadis (Ed.): *Philosophy by Women*. New York: Routledge, pp. 44–53.

Khader, Serene J. (2011): *Adaptive Preferences and Women's Empowerment*. New York: Oxford University Press.

Mason, Rebecca (2021) "Hermeneutical Injustice." In: Justin Khoo and Rachel Sterken (Eds): *The Routlegde Handbook of Social and Political Philosophy of Language*. New York: Routledge, pp. 247–258.

Midgley, Mary (1996). *Utopias, Dolphins and Computers: Problem in Philosophical Plumbing*. New York: Routledge.

Nussbaum, Martha (2001): "Adaptive Preferences and Women's Options." In: *Economics and Philosophy* 17, pp 67–88.

Scharp, Kevin (2013): *Replacing Truth*. Oxford: Oxford University Press.

Simion, Mona (2018): "The 'should' in Conceptual Engineering." In: *Inquiry: An Interdisciplinary Journal of Philosophy* 61. No. 8, pp. 914–928.

---

11 I thank Dan López de Sa for his comments on the very first version of this paper, and to Fernando Broncano, Manuel de Pinedo, and Neftalí Villanueva for conversations on the questions discussed here and related ones. I also thank an anonymous referee who helped me craft a better final version. A previous version of this paper was presented at the Madrid Philosophy Network in October 2020. I thank all participants for their questions and suggestions.

Simion, Mona, and Christopher Kelp (2020): "Conceptual Innovation: Function First." In: *Noûs* 54. No. 4, pp. 985–1002.

Srinivasan, Amia (2020): "Radical Externalism." In: *The Philosophical Review* 129. No. 3, pp. 395–431.

van Inwagen, Peter (2008): "How to Think about the Problem of Free Will." In: *Journal of Ethics* 12. No. 3, pp. 327–341.

E. Díaz-León

The Meaning of 'Woman' and the Political Turn in Philosophy of Language

Abstract: In this chapter, I review some arguments for contextualist theories of the meaning of 'woman' and discuss and defuse some recent objections against contextualism. I also show how contextualist views can help to show what is at stake in the debates between trans-inclusive views about the meaning of 'woman' and so-called gender critical views. Moreover, I argue that normative considerations are the contextual factors that contribute to fix the referent of 'woman' in different contexts, and that in many contexts, normative considerations favor a notion of gender identity or gender as a class, rather than a notion of womanhood as biological sex.

1 Introduction

What does the term 'woman' mean? This question has received growing interest in recent decades in analytic feminist philosophy. Here, the political turn in analytic philosophy is manifest in two ways: first, the very choice of the question is motivated by the political significance of the topic; and second, the reasons and motivations for defending one semantic theory over another about the meaning of this term can also include moral and political considerations, as we will see.

Why is this question important for feminism? The answer, in short, is that feminism is the movement (both theoretical and practical) that aims to explain and resist the oppression and discrimination of women, and in order to engage in this project, we need an account of *women*. Recently, Mari Mikkola (2016) has argued against this orthodox view, claiming (i) that feminism does not need to provide a substantive account of the meaning of 'woman' in order to describe and fight the oppression of women, and (ii) that feminism does not need to provide a substantive account of the meaning of 'woman' to normatively justify why the oppression of women is *wrong*. Regarding the second point, I am sympathetic to Mikkola's claim that, in order to show why the oppression of women is normatively unjust, it is not necessary to identify a social kind that is constituted by oppressive social practices and which can explain why the oppression of women is unjust, but rather, it is sufficient to appeal to a concept of dehumanization in order to explain why the oppression of women is wrong, as she does in her book. But regarding (i), it is not clear to me that we can properly describe and explain

https://doi.org/10.1515/9783110612318-013

the oppression of women without a substantive account of the meaning of 'woman'. Mikkola argues that it is sufficient to appeal to our extensional intuitions about the application of the term 'woman', and this is enough to make sense of our claims about the oppression of women. Indeed, according to externalist accounts of meaning, we can be competent users of a term whose reference is determined by external factors, without knowing the criteria of application of the term, that is, without knowing necessary and sufficient conditions for something to fall under the term. This is true, but the problem in the case of 'woman' is that different speakers can have different intuitions about the extension of the term that yield different extensions, precisely with respect to politically significant cases.[1] For instance, some speakers may believe that all trans women fall under the concept, whereas others may believe that some trans women do not fall under the concept. It is not clear to me how appealing to mere extensional intuitions can settle this important debate. This does not mean that we should reject an externalist theory of meaning and search for something akin to a Fregean description associated with the term. Indeed, externalism about meaning can be of help here (see Haslanger 2006), but even if we assume externalism, we need to investigate which external factors determine the referent in order to give an answer to those politically significant debates about the extension of 'woman'. For this (tentative) reason, I will assume, following many feminists, that it is important to give a substantive account of the meaning of 'woman' in order to properly understand claims about the oppression of women.[2]

2 The Sex/Gender Distinction

There seem to be two views about the meaning of 'woman' that are most salient (see Saul 2012). According to the first view, woman works as a sex-term, that is, 'woman' refers to those individuals who are biologically female (and 'man' refers to those individuals who are biologically male). According to the second view, 'woman' works as a gender-term. Gender is a technical concept introduced by feminist scholars in order to capture the idea that many of the behavioral, social, and cultural differences we can observe between men and women are the product of social norms, social expectations, and acculturation, rather than being de-

---

1 See Jenkins (2018b) and Cull (2020).
2 As I said above, I agree with Mikkola that the account of woman does not have to make it constitutive of women that they are oppressed. What we do need is a set of criteria for the application of woman in order to be able to settle disagreements about the extension of woman.

termined by purely biological and anatomical differences (the classical source of this idea is Simone de Beauvoir's influential book, *The Second Sex*).[3] Hence, feminists conceive of gender as the different social positions that men and women occupy, due to those social norms and expectations dictating how men and women should behave. According to this idea, there are at least two genders, namely, men and women.[4] (It is standard to use the terms male and female to refer to the aforementioned biological sexes, and man and woman to refer to these different gender properties.) In this way, we could understand the gender category of woman as referring to those individuals who share the social position assigned to women.[5] A tentative characterization of this social position could be the following: those individuals who are more likely to be subject to norms of feminine appearance, receive more pressure to do most of the housework and most of the childcare, are more likely to be employed in low-pay jobs, are more likely to suffer sexual harassment and sexual assault, and so on.[6]

Many feminist philosophers have argued that these two views face serious problems. I will explain some of the main problems with each view in turn.

---

3 Hence, I will assume that gender is a category introduced by feminist scholars, whereas 'man' and 'woman' are part of ordinary speech. According to externalism, many terms of ordinary speech are natural kind terms, that is, terms that aim to refer to the natural kind (or scientifically substantive property) underlying the paradigmatic instances of the kind that we competent speakers identify in virtue of their appearance properties, and whose nature can be revealed to us only empirically (as famously defended by Putnam 1975 and Kripke 1980.) In response to Kripke and Putnam, Dupré (1981) argues that there are many terms from ordinary speech, such as names for kinds of fruits, vegetables, animals, fish, and so on, that are not co-extensional with the scientific terms used by scientists to refer to the underlying biological properties shared by the paradigmatic instances. So, on this view, 'whale' as used by ordinary speakers and 'whale' as used by biologists would not have the same extension. Here I am assuming, though, a standard externalist framework, according to which an ordinary term such as 'woman' and a theoretical term such as 'woman' (qua gender) are co-extensional, and where feminist scholars can tell us what the real nature of the (social) kind is, to which we were referring all along.

4 Some feminists suggest that there might be more than two gender categories, such as being trans or being genderqueer (see, e.g. Witt 2011, pp. 41–42). As I will explain later, drawing on the insights of transfeminism (e.g. Bettcher 2017), I believe most trans women do belong to the same gender as cis women (in particular those who self-identify as women), and therefore they all belong to the gender category of woman, not a third one. On the other hand, it seems plausible to understand the category of genderqueer as an additional gender category (see Dembroff 2020 for a very interesting discussion of the category of genderqueer).

5 This is often understood as the social position assigned to those who are (perceived to be) biologically female, but more on this later.

6 See Mikkola (2019) and Hay (2020) for useful surveys of feminist work about the sex-gender distinction.

First, regarding the view that 'woman' works as a sex-term, many feminist philosophers have argued that this view cannot capture our use of the term woman. As Jennifer Saul (2012) has argued, it is not clear how the sex-based view would classify intersex people. Intersex individuals are those that have some biological or anatomical features corresponding to the male biological sex, and some features corresponding to the female sex. For instance, an intersex person might have XY chromosomes, but due to insensitivity to male hormones, they might develop female genitalia and female secondary sex characteristics such as breasts and very little facial hair. If the sex-based view assumes that the term 'woman' refers to those who have all or most of those biological and anatomical features associated with the female biological sex, and the term 'man' refers to those who have all or most of those biological and anatomical features associated with the male biological sex, then this intersex person would count as neither a man nor a woman. This result seems problematic, for two reasons. First, from a descriptive point of view, one might argue that this is not how we use the ordinary concept of woman. Indeed, many competent users of the term 'woman' would be disposed to apply the term to an intersex individual who was assigned female at birth and has female genitalia and female secondary sex characteristics, even if she has XY chromosomes and does not have a uterus. So, this analysis does not give the intuitive result in this case. Second, from an ameliorative point of view, one could argue that this is not how we *should* use the concept of woman: arguably, there are normative considerations, including moral and political considerations, in favor of the claim that we should use the term 'woman' in our linguistic community so that it includes intersex people who have been raised as women and who identify as women in its extension.[7]

There is an additional worry: if the term 'woman' works as a sex-term, then trans people would also be incorrectly classified. (Trans people are those individuals who do not identify with the gender assigned at birth.) Again, we can understand this point in terms of either a descriptive or an ameliorative consideration. One could argue that, given how we use the ordinary concept of woman, that is, given what we mean by it, the ordinary usage does apply to many trans women, but this sex-based view would yield the result that many trans women are not women, since they do not have all or even most of the traits associated with the female biological sex. Here we should be careful, since it is not clear exactly how we should understand the phrase "having *most* of the traits associated with

---

7 See Haslanger (2006) for a useful distinction between descriptive and ameliorative projects in philosophical analysis.

the female biological sex." Indeed, there are reasonable interpretations of this phrase according to which many trans women who have had sex-reassignment treatments do count as women, even on this sex-based view, since they have *most* of the traits (being flexible about how many are necessary), but many other trans women would not count, if they have not had sex-reassignment treatment. This gives rise to the question of how we should understand biological sex.

In my view, the most useful understanding of biological sex is in terms of a *cluster* of biological and anatomical features that are associated ether with the female biological sex or with the male biological sex. This cluster of features would include chromosomes, hormonal levels, internal sex organs (i.e. uterus, ovaries, prostate, etc.), external genitalia (i.e. having a penis or a vagina), and secondary sex characteristics (e.g. having breasts, or having a beard). Many of these features are not *discrete*, that is, all or nothing features, but rather a matter of a spectrum, like hormonal level. As we suggested above, these traits tend to be co-instantiated (that is, individuals who have XX chromosomes tend to have all the other traits of the female cluster, whereas individuals who have XY chromosomes tend to have all the other traits of the male cluster), but they are not always co-instantiated, like in the case of intersex individuals (who represent at least 1.7% of the general population). It is important to realize that the existence of this cluster of features, many of which are a matter of a scale (e.g. hormonal levels), implies that there are *different* ways of classifying individuals as male or female, and opting for one system of classification over another is a matter of theoretical choice.[8] Even features that might seem to be discrete, such as having a penis or a clitoris, are rather a matter of a spectrum since there are individuals who have what could be classified both as a micro-penis or a macro-clitoris. How to characterize a 'functional' penis is a theoretical question that admits of several answers, and it is not clear that there is a characterization that carves nature at its joints here. Given all these considerations, whether trans women who have had sex-reassignment treatment fall under the term 'woman' on the sex-based view is a question that depends on what theoretical choices we make, with respect to the notion of biological sex that we choose (that is, how many traits of the cluster are necessary, and to what extent). And one could argue that there are normative considerations to include at least some trans women under the notion of 'biologically female.' In any case, those trans women who have not had sex-reassignment treatment will not count as women on the sex-based view.

---

8 See Stein (1999) and Fausto-Sterling (2000) for very useful elaborations of this idea.

This is a problematic result for a theory about the meaning of 'woman', again for two reasons. From a descriptive point of view, one could also argue that this result is in tension with how the term is used in some linguistic communities. Many competent users would be disposed to apply the term 'woman' to trans women. As we have seen, this is clearly the case with respect to trans women who have had genital surgery. Maybe some competent users would hesitate with respect to the case of trans women who have not had sex-reassignment treatment, but here it is useful to draw on the work of Talia Bettcher (2013). She argues that there exist different linguistic communities with different uses. There is a dominant usage of the term 'woman' who would perhaps not apply the term to trans women who have not had sex-reassignment treatment, but there is also a resistant usage of the term according to which trans women who identify as women do fall under the term regardless of whether they have had sex-reassignment treatment or not. The crucial idea here is that we can understand this point as being relevant to the descriptive project: Bettcher argues that the resistant usage is also part of how people in fact use the term, and that it would be methodologically biased to focus only on patterns of use by speakers that belong to the dominant group. So, to sum up this point, the idea is that there are descriptive considerations to understand the term 'woman' as including all trans women who identify as women, regardless of whether they have had sex-reassignment treatment or not. And if so, then the sex-based view of the meaning of 'woman 'is clearly wrong.

Furthermore, from an ameliorative point of view, even if one is doubtful about these descriptive considerations, one could argue that there are normative considerations in support of the view that we should use the term 'woman' so as to include all trans women in its extension. If so, then we can conclude that the sex-based view is incompatible with our best normative considerations.

What about gender-based views? According to this view, 'woman' refers to those individuals who occupy the social roles assigned to women. A first worry has to do with the problem of intersectionality. This term was introduced by black feminists (Crenshaw 1989) in order to explain the idea that it is not possible to separate the discrimination suffered by women from that suffered, say, by racialized people, or disabled people, or non-heterosexual people, etc. That is, we cannot explain the discrimination of black women just in terms of the addition of the discrimination of white women and the discrimination of black men, since the discrimination of black women is sui generis and cannot be figured out just by examining the social positions of white women and black men. We need to study the discrimination of black women themselves in order to be able to understand their discrimination. This idea implies that it is not possible to identify the social role that all women share, since women instantiate very different social positions depending on other social identities such as race, class, age, nationality, disability, sexual orientation, and so on.[9] Therefore, this challenges the claim that the term woman refers to the social role that all women share. Feminist philosophers have tried to answer this challenge in different ways. One especially prominent strategy is that carried out by Sally Haslanger (2000), who aims to identify some abstract social patterns that all women share in virtue of being oppressed. Haslanger (2000) offers a characterization of gender according to which to belong to a gendered group is a matter of occupying a social position of either privilege or subordination along some axis, such as economic, political, legal, or cultural axes of discrimination or privilege. In this way, women who differ with respect to other social identities, such as race, class, nationality, sexual orientation, etc., will differ with respect to the axes of discrimination that they suffer (for instance, white middle-class women will tend to be wealthier than, say, working class immigrant women of color), but they all have in common the fact that they are discriminated under some axis of oppression, with respect to men. Within this framework, Haslanger suggests that we can define 'woman' as the social position of discrimination along some axis (political, legal, economic, or cultural), in virtue of being perceived to play a female role in biological reproduction.[10] This is one way in which feminist philosophers defend the idea that 'woman' refers to a social position, compatible with the phenomenon of intersectionality.

However, there are problems still lurking. First, there is some debate about whether Haslanger's account of woman could correspond to either the *descriptive* concept of woman that we actually have, or the *ameliorative* concept of woman that we should associate with the term. Regarding the first question, Haslanger (2006) has argued that there is space for that option,[11] although Ron Mallon (2017) and others have presented serious objections.[12] Regarding the second question, Jenkins (2016) and others have raised important objections. For all these reasons, there is no consensus within feminist philosophy about how to understand the claim that 'woman' works as a gender-term.

Moreover, N. G. Laskowski (2020) has suggested that accounts of the meaning of 'woman' should satisfy what he calls the *usage constraint*, that is, the fact that there seem to be different patterns of usage of the term 'woman', including

---

9 See Spelman (1988). See also Mikkola (2006) for a response.
10 See Haslanger (2000, pp. 39–43) for a more elaborated account.
11 See Díaz-León (2012) and (2020) for further discussion of the claim that 'woman' refers to something like the Haslangerian social property characterized above, as a matter of a descriptive inquiry.
12 See Díaz-León (2019) for a response to Mallon's (2017) challenge.

mainstream uses of the term according to which it seems to refer to those who are biologically female, and non-mainstream uses of the term according to which it seems to refer to other features such as social role or self-identification.[13] If this is correct, then this suggests that the term 'woman' could sometimes work as a sex-term and sometimes as a gender-term (perhaps with further different meanings within each option). There is a theory of the meaning of a term that can capture this variation very well, namely, contextualism.

3 Contextualism

Contextualism about an expression claims that the meaning of that expression changes form context to context. Some expressions that are clearly context-sensitive include pronouns such as I, you, we, and they. For instance, the term I refers to me, Esa Díaz-León, when I utter the sentence "I am tall," but it refers to Kamala Harris when she utters the sentence "I am tall." The term I is an indexical expression that refers to whoever is the utterer of that very expression. That is, the referent of the term is determined by a certain factor of the context, namely, who is the utterer of the term. Content-sensitive expressions are said to have two dimensions of meaning, one called *character*, which has to do with the conventional rules that tell us which factors of the context fix the referent, and how; and another called *content*, which corresponds to the referent that the term has in each context.[14] In this way, the term I has a character that corresponds to the rule "I refers in a context C to the utterer of this expression in C" and the content will be the referent in each context (that is, either me, or Kamala Harris, or whoever).

---

13 This claim is similar to Saul's appeal to a "collection of ordinary usage data" (Saul 2012, p. 200), and also to Bettcher's (2013) proposal that there are both dominant and resistant patterns of usage of the term woman, which have different extensions. Bettcher (2013) argued that the resistant usage of the term is the correct one since it involves a worldview that is more accurate. Laskowski's suggestion seems to be that, from a descriptive point of view, a theory of what the term "woman" means should account for the different patterns of use in a community. This is compatible, of course, with the claim that some of these patterns of usage are problematic. In addition, Laskowski argues, drawing again on Saul (2012), that theories of the meaning of "woman" should satisfy not only the "usage constraint" but also the "communicative constraint", that is, the idea that communication among these different communities is possible. He argues that a polysemy account can better satisfy these two criteria than a contextualist view.

14 See Kaplan (1989) and Mount (2012) for elaboration of the distinction between character and content.

Jennifer Saul (2012) develops a contextualist proposal about the term 'woman', although she does not endorse it. Motivated by the problems of both the sex-based view and the gender-based view about the meaning of 'woman' that I summarized above, Saul suggests that it might be worth exploring the prospects of a contextualist view, since this might perhaps do justice to the different intuitions at play. However, as we will see, she ultimately argues that the view should be rejected since it does not do justice to all the aims that we should take into consideration when theorizing about the meaning of 'woman'. In section 4, I will briefly summarize my response to Saul in previous work,[15] where I argue that a modification of the contextualist view can overcome Saul's objections. Moreover, in the remainder of the chapter, I show that there are additional reasons to motivate this view. First, in section 5, I argue that contextualism of the sort I advocate can help to explain what is at issue in the heated debate between trans-inclusive accounts of the meaning of 'woman' and trans-exclusive accounts and can help to move the debate forward. Second, in section 6, I respond to some recent objections to my formulation of contextualism, according to which my version of contextualism would collapse into invariantism,[16] and conclude that these objections are not lethal to my account.

As we have seen, contextualism about 'woman' claims that the term changes the referent from context to context. The crucial question that a contextualist view about 'woman' should answer is what are the factors of each context that determine the different referents. That is, a contextualist view needs to offer an account of the character of the expression. Saul (2012) does precisely this. Her proposal (which she will later reject) is as follows:

(CP): "*X is a woman* is true in a context C iff X is human and relevantly similar (according to the standards at work in C) to most of those possessing all of the biological markers of female sex." (Saul 2012, p. 201)

According to this view, a sentence of the form "*X is a woman*," for any individual *X*, would have the following truth conditions. Such a sentence will be true if and only if *X* is human and *X* is relevantly similar to a certain group of people (always the same), with respect to some criteria of similarity (which vary from context to context). In particular, *X* has to be relevantly similar to most individuals of the following group: the class of individuals who possess all of the biological markers of female sex that we mentioned in the previous section. (In what follows, I will sometimes just say "individuals who are biologically female," to abbreviate.)

---

15 See Díaz-León (2016).
16 See Bettcher (2017), Laskowski (2020), and Zeman (2020).

So, *X* has to be similar to most of the individuals in this group (not similar to all of them, only similar to most of them). But similar with respect to what criteria? For any two entities, they are always similar with respect to some criterion. So, the account has to specify the *criteria* of similarity. And this is what changes from context to context. That is, *X* has to be similar to most individuals in the comparison group, with respect to some criteria of similarity. That is, it must be the case that most individuals in the target group have property *P* and *X* also has property *P*. What this property *P* amounts to is what changes from context to context. To clarify: property *P* is not a property that all members of this comparison group need to share, only most of them. And *X* does not need to belong to the comparison group (although she might): the account only requires that *X* shares property *P* with most of the individuals in the comparison group, for some *P* that is determined by the context.

This will become clearer with an example. We can imagine the context of a feminist conference where the organizers want to keep a register of how many women have registered at the conference, for equality purposes. So, at the registration desk, they ask people to fill in a registration form where people have to indicate their gender. Lena is a trans woman who has not had genital surgery, and she completes her registration form. One of the organizers says to another, putting the form in a folder:

(1) "Lena is a woman."

In this context, we can assume, the relevant criterion of similarity is whether people sincerely self-identify as women. Indeed, this is the criterion that the conference organizers take into consideration. So, according to (CP) above, this utterance of (1) would turn out as true since Lena is similar to most individuals who are biologically female with respect to the relevant criterion of similarity in this context, namely, identifying as a woman. (As we can see, most, but not all, individuals who are biologically female self-identify as a woman, and Lena also does, so they are similar in this respect. Whether Lena herself belongs to the comparison group in this case depends on how we characterize the notion of biologically female, as we saw above. But this does not matter for this view, since someone can fall under the extension of the term 'woman' regardless of this, according to CP.)

Let us compare this case with the following example, drawn from Saul. We can imagine the context of a clinic that has received a notice from health authorities to test all women over 45 years old for vaginal diseases. In this context, let us imagine a nurse who is going through patients' files and utters (1): "Lena is a woman." Saul's claim is that, on contextualist view (CP), this utterance of (1) would turn out to be false in this context, since the criterion of similarity that seems salient in that context is having a vagina, and Lena has not had genital surgery. (We can imagine another nurse quickly correcting the first, and uttering (2): "no, Lena is not a woman," which would turn out to be true in the context of figuring out which patients should receive calls in order to make an appointment for the test, but would be false in other contexts—for instance, if Lena intends to use the women's toilet—since in this context the relevant criterion of similarity would amount to self-identifying as a woman.)[17]

So far, the contextualist view seems to yield intuitively plausible results. But Saul argues that the view also faces serious objections and, because of these, she ultimately rejects the view. The main worry is as follows. Contextualism is indeed very flexible, and this can render true the claims of the advocates of trans women (as seen in the feminist conference example above), but it can also render true the claims of the opponents of trans women. For this reason, the view is unsatisfactory. We can better understand the problem with an example given by Saul. We can imagine the context of a transphobic community where most people believe that what matters in order to use women's toilets is having female genitalia.[18] In this context, if someone utters (1), this will turn out to be false, since Lena does not have female genitalia. Indeed, if someone utters (2) in this context, this will turn out to be true. Saul claims that contextualism cannot avoid this consequence and therefore we should reject it. She also suggests that the only way to do justice to our moral and political considerations regarding the use of the term 'woman' is engage in what Haslanger (2000 and 2006) calls the ameliorative project, and ask not what our ordinary concept of woman is, but what it should be.

In previous work (Díaz-León 2016), I have argued that the contextualist view can be modified so that it can overcome this problem. I also argued that moral and political considerations are relevant with regard to the project of what 'woman' *actually* means, not only with regard to what it *should* mean. In this

---

17 The issue of how to use terms such as man and woman in medical contexts is complex, and I cannot do it justice here. My initial claim is that it seems intuitively plausible to interpret these utterances by the nurses as involving a sex-based reading. Below I will further discuss some moral and political considerations regarding how we should use gendered terms in medical contexts. See Freeman & Ayala-López (2018) for further discussion of ethical considerations regarding this question.

18 Unfortunately, this context is not far from reality. In North Carolina, a bill restricting the use of bathrooms according to the sex assigned at birth was approved in 2016. Fortunately, this bill was rescinded in 2017, but similar legislations are enforced or under discussion in other parts of the world.

way, I believe, the political turn in philosophy of language is more wide-ranging that it might seem at first sight. In the remainder of this chapter, I will briefly rehearse my response to Saul on behalf of contextualism, and I will examine some recent critiques to my modified version of contextualism. I will also explain how my version of contextualism can help to make sense of what is at issue in the debate between trans-inclusive and trans-exclusive accounts of the meaning of 'woman', and how to make progress in this debate.

4 A New Version of Contextualism

My response to Saul's objection is as follows. Saul characterizes the relevant criteria of similarity in each context in terms of the criteria that speakers have in mind. That is to say, the advocates of trans women would have in mind criteria of similarity that are inclusive of trans women (such as self-identifying as a woman), whereas opponents of trans women would have in mind criteria of similarity that are not so inclusive (such as having a vagina, which some but not all trans women do). But in response, and drawing from the literature on contextualism about knowledge, I argue there is another way of characterizing the relevant factors of the context, namely, in terms of the features that are relevant given the situation of the subject of the utterance, that is, the subject that the utterance of the form "X is a woman" (or similar utterances) is about.[19] In particular, I defend a contextualist view, according to which the relevant criteria of similarity in each context are determined by the best moral and political considerations involving the subject of the utterance.

---

19 In making this distinction, I drew from work in contextualism about knowledge, where a distinction between so-called *attributor*-contextualism and *subject*-contextualism has been made. According to this distinction, attributor contextualism has it that the standards that determine the different extensions of the term 'knowledge' depend on features of the speaker, namely, what the speaker has in mind, whereas subject contextualism has it that the standards that determine the different extensions of the term depend on features of the subject that is predicated to know (or not to know), that is to say, the environmental features of the subject. For instance, according to attributor contextualism, my utterance "Luisa knows that there is a zebra in front of her" would be false if someone near me has raised the possibility that those zebras might be mules disguised as zebras, regardless of whether there were mules disguised as zebras in the zoo Luisa was visiting, whereas according to subject contextualism, what determines the standards of justification has to do with whether there were really mules disguised as zebras in Luisa's vicinity, not whether that possibility has been raised to the speaker. See DeRose (2009) for further discussion of the distinction and its significance.

For example, in the case of the transphobic community described above, my version of contextualism would yield different results. When a transphobic speaker utters (2): "No, Lena is not a woman," this utterance would turn out to be false, because Lena is similar (to most of those who are biologically female) with respect to the criterion of similarity that is morally and politically relevant in this context (where transphobic speakers are discussing whether trans women should be allowed into women's toilets, and they conclude that only those who have a vagina are allowed), namely, the criterion of self-identifying as a woman, not the criterion of having a vagina. Why? Because what determines the salient criterion of similarity is a matter of the moral and political considerations that result from our best moral reasoning, not the criteria that the transphobic speakers have in mind.[20]

Therefore, my proposed version of contextualism also follows Saul's (CP) schema. The only but crucial difference is how to understand the phrase "relevantly similar (according to the standards at work in *C*)." On my view, the best way of interpreting these is in terms of the best (the most relevant) moral and political reasoning involving the situation of the subject of the utterance. As we have seen, if we are discussing who should be allowed into women's toilets, our best moral reasoning will establish that all individuals who sincerely self-identify as women should have access to women's toilets. Another way of formulating my proposal (which I did not think of in my (2016) article but has since been suggested to me) is that the concept of woman is indeed a *thick* concept.[21] Or, in other words, the schema (CP) is in part descriptive, in part evaluative. This seems plausible, since the phrases "*X* is human" and "most of those possessing all the markers of female biological sex" are supposed to be *descriptive*, whereas "relevantly similar" is *evaluative*, in my view. Since this is an evaluative term, it is no surprise that we need to appeal to our best moral reasoning concerning the situation of the subject that the utterance is about, and what criteria of similarity are relevant in the context of the subject.

So, how would my version of contextualism deal with the other examples mentioned above? On the one hand, in the case of the feminist conference, it is clear that the best moral and political considerations would yield the result that the relevant criterion of similarity is self-identifying as a woman (so uttering (1) in this context would turn out true). On the other hand, in the case of the clinic that plans to test all individuals with vaginas over 45 years old for vaginal diseases, it seems okay to me to assume that the relevant criterion of similarity is

---

20 See Kapusta (2016) for a very compelling argument about the harms of misgendering.
21 See Roberts (2013) for an excellent discussion of this notion.

having a vagina (at least when we examine a conversation between two nurses going through the patients' files in order to figure out who to call for an appointment), so uttering (1) in this context would turn out false.[22]

5 A Contextualist Response to Gender-Critical Views

Questions about the extension of the term 'woman' in different contexts have raised heated debates in recent years. In several countries, proposals for new legislation concerning legal recognition of gender identities different from those assigned at birth have received strong opposition both by far-right parties and by some feminist organizations.[23] Within feminist philosophy, views according to which the term 'woman' does not in fact include all trans women are becoming more prominent and have attracted a lot of debate. These debates have also translated to the political arena and the social media, causing a lot of polarization. In what follows, I will describe some of the arguments for the trans-exclusive position about the meaning of 'woman', and I argue that my modified version of contextualism can help us see how the trans-inclusive position can respond to the arguments by the trans-exclusive camp.

In a detailed blog post, a group of so-called "gender-critical" academic philosophers attempt to explain their position and their arguments as clearly as possible. I would like to quote a longish passage from this blog post, since I believe it clearly summarizes their view:

---

22 As I said above in n. 17, this question is complex. Given my version of contextualism, the moral and political considerations involving the case are brought to the front. One might argue that, according to the best moral reasoning, it would not be appropriate for the nurses to use the term 'woman' in this context, and they should instead use a term such as 'individuals with vaginas,' since they want to test all people with vaginas (over 45) for vaginal diseases. I am sympathetic to this suggestion, but I also believe that using the term 'woman' to refer to humans with vaginas in this very specific context could have some utility. The crucial question is how to weigh all these different considerations.

23 For instance, in the UK, LGBT associations have lobbied for a reform of the Gender Recognition Act (which was last voted in 2017), but so far the government has not committed to make all the changes solicited by LGBT associations. In Spain, LGBT associations have similarly supported a new "trans bill," and a new proposal for a law reform has been made by the Equality ministry, but this is still pending discussion in parliament, as I finish this chapter in March 2021. This proposal for a new Spanish trans bill has received support from some left-wing political parties and many feminist associations, but it has also received a lot of criticism from other feminist associations, right-wing parties, and even some segments of the socialist party.

> We're a group of gender-critical and radical feminist academic philosophers. In our work, some of us argue that women, by definition, are adult human females. On this view, since no trans woman is an adult human female, no trans woman is correctly categorized as a woman. The rest of us are currently agnostic between i) exclusively taking the former position, and ii) also taking a position that says that there is an additional, meaningful sense of 'woman', understood as applying to those who occupy a certain feminine social role, on the basis of perceived membership of the female sex category. Unlike i), ii) entails that a limited number of trans women count as women, in at least one sense. Still, ii) entails that many trans women aren't correctly categorized as women, since many trans women don't occupy a feminine social role on the basis of perceived membership of the female sex category. Either way, we are all skeptical of the political value of accounts of womanhood that identify it as essentially involving possession of a feminine 'gender identity'. We also all insist that it's politically essential to retain a clear conceptual differentiation between males and females, in order to continue to be able to name and refer to sex based patterns of oppression, and harmful sociocultural stereotypes about the 'right' ways for males and females respectively to be. (Stock et al. 2019, para. 1)

In my view, this excerpt makes vividly clear what is at issue in the debate between gender-critical views and trans-inclusive views about the extension of 'woman'. For example, one of the claims made by gender-critical philosophers is that the notion of a feminine gender identity (or in our terms, self-identifying as a woman) has little political value. In addition, they claim that it is *politically essential* to appeal to the notion of biological sex, and the distinction between those who are biologically female and those who are biologically male, in order to be able to explain sex-based patterns of oppression. That is to say, those philosophers are appealing to claims about the *political* value of some notions over others, in order to defend a claim about the meaning of 'woman'. As we can see, in that passage they say that some of them hold a view according to which the term 'woman' means something like "adult human who is biologically female," and the rest are agnostic between this view, and another view according to which 'woman' means something similar to the Haslangerian social property that we characterized above, that is, the class of individuals occupying the social position of subordination in virtue of being perceived to be biologically female. As they say, the term 'woman' leaves out some trans women, on either of these views. What I am interested in is the kind of considerations that are put forward in defense of the gender-critical position. They do not make explicit whether this view is intended as a descriptive or an ameliorative inquiry, but since they say that "women, by definition, are . . . ," it seems plausible to interpret this as a claim within the *descriptive* project. If so, then these philosophers are using moral and political considerations in order to argue for a claim about the descriptive meaning of the term 'woman'.

Therefore, the methodology they appeal to is compatible with my version of contextualism, since according to my contextualist view, what determines the criteria of similarity that are relevant in different contexts, in order to determine the extension of 'woman', amounts to moral and political considerations involving the situations that the subjects of the utterance are in. This methodology seems to correspond pretty well with the reasoning behind the passage above. In this section, I will argue that, if we focus on the relevant moral and political considerations involving the cases that they discuss, I believe that we have good reasons to deny both of the semantic views that they suggest, namely, that 'woman' always refers to biological sex, or that 'woman' sometimes refers to biological sex and sometimes to a Haslangerian social position (but never to gender identity). One of the main political goals that they discuss is which notion of woman is more useful in order to explain sex-based patterns of oppression. I will argue that, with respect to this aim, the semantic position that they put forward fails, precisely because it has *less* political value than some alternatives.

I understand the aim of explaining sex-based patterns of oppression in terms of the generalizations that are useful in order to explain and predict the oppression suffered by women, including generalizations such as the following:

(3) Women are more likely to be subject to sexual assault and domestic violence than men.
(4) Women are more likely to be subject to demanding norms of beauty than men.
(5) Women are more likely to be employed in low-pay jobs than men.

Many feminists would agree that these are true and useful generalizations. That is, these claims are explanatorily useful in part because they involve useful kinds. So, what are the explanatorily useful kinds that will render these politically useful generalizations true? This is the question that seems to be at issue in the passage above. That is to say, the question is about what kinds are more relevant, from a moral and political point of view, for the situation that the subjects of these utterances are in, that is, the situation of oppression that these very claims try to explain. Well, what are these kinds?

I first want to argue that the notion of biological sex is a non-starter. Biological sex is not the kind that is the most explanatorily useful with respect to those generalizations. As Haslanger (2016) argues, the relevant kind of explanation, in order to explain the oppression of women, is structural explanation. What explains the patterns of oppression that women suffer is a matter of complex social structures, constituted by social practices and social schemas (as Haslanger (2016) helpfully explains). Another way to put this point is this: biological sex in itself cannot explain why women are more likely to suffer domestic violence or be employed in low paid jobs because we can easily imagine possible worlds where men and women are biologically identical to the actual world but there is no domestic violence against women and there is no salary gap between men and women. This is clearly *conceivable*, and it also seems *nomically possible* (at least this is the aim of feminism, the world that we fight for). This shows that an explanation of oppression in terms of biological sex is incomplete. It might be true, statistically speaking, that those individuals who are biologically female are more likely to suffer domestic violence and be employed in low-pay jobs.[24] This might be true *statistically* speaking but this is not a *complete* explanation. If what we want is the best explanation of these patterns of oppression, we need to appeal to social structures of the sort Haslanger talks about, because these social structures are more explanatorily relevant than biological sex.[25] Thus, it is clear that our best moral and political considerations do not support the claim that the term 'women' in utterances (3), (4), and (5) refers to the class of biologically female individuals.

Given this reasoning, so far we get the preliminary result that the term 'women' in utterances (3 – 5) can refer to something like a Haslangerian social property, since this is the kind that seems more explanatorily useful. But, as the passage above suggests, this semantic view would leave some trans women out. Jenkins (2016) also argues that Haslanger's (2000) characterization of woman in terms of a social position of subordination in virtue of being perceived as being biologically female (or more precisely, to have a female role in

---

24 We should be careful here: if we are making generalizations about those who are biologically female and those who are biologically male, some trans women and intersex women might count as biologically male, but their patterns of domestic violence and salary are more similar to cis women than to cis men (indeed, the rate of violence and the rate of unemployment suffered by trans women are much greater than that of cis women). However, it might be the case that these statistical generalizations about biological sex still come out as true, if we assume that intersex people are about 1.7% and trans people are about 2% of the general population. My main point, though, as I explain later, is that these generalizations involving biological sex might be true but are not *explanatorily relevant*.

25 Another way of putting this point is the following. Sentences (3 – 5) are put in terms of probability ("more likely to be subject to . . .") but they could also be put in terms of 'because' sentences, since they express explanations. For example: "Women are more likely to be employed in low-pay jobs *because* they are biologically female" is clearly false, or inaccurate, as an explanation, given the semantics of because statements. The following sentence expresses a much more accurate explanation: "Women are more likely to be employed in low-pay jobs *because* they occupy positions of social subordination within a social structure that marks those who are perceived to be biologically female as being aptly positioned in that social role of subordination." See Haslanger (2016) for further discussion of structural explanations.

biological reproduction) would leave out some women, namely, those who are not perceived to be biologically female (do not "pass" as cis women). A first response to this point is that, given the possibility of contextualism, it might be the case that the term 'woman' in those utterances refers to a Haslangerian social position, but that the term 'woman' refers to other kinds in other contexts. (As I will argue below, I believe there are contexts where the term 'woman' clearly refers to those who self-identify as a woman, and other contexts where it can refer to biological sex, among other options.) So the fact that 'women' in those utterances leaves some trans women out does not imply that it always leaves some trans women out, precisely because the possibility that 'woman' is context-sensitive is an open possibility.

But I think there is another possible line of response against this objection. Indeed, I believe that, when focusing on sentences about explanations of the patterns of oppression of women, the most explanatorily useful notion might be trans-inclusive after all. Let me explain. Let us focus on the example of generalizations about women being employed in low-pay jobs. The salary gap between men and women is an important generalization that is well documented. The question at issue here is: what is the notion of woman that is involved in these generalizations? According to my version of contextualism, this is determined by the moral and political considerations in the vicinity, that is, the notion of woman (or in other words, the criterion of similarity) that is more explanatorily useful in the context of these generalizations. This is also the method that the advocates of gender-critical views quoted above put forward: they invoke considerations about the political value of different notions with respect to the project of explaining the oppression of women. Well, it seems clear that the notion of biological sex is not the most explanatorily useful, for the reasons I gave above. The remaining notion that could do the job here seems to be the notion of social position à la Haslanger. It seems plausible to say that the Haslangerian notion of social position of privilege or subordination in virtue of being perceived to be biologically female is very explanatorily useful. In my view, this notion is explanatorily useful, but there is another notion in the vicinity that is even more explanatorily useful, or so I will argue.

The notion of being perceived to be biologically female depends on the notion of gender expression, since gender expression is the fallible guide that is commonly used to ascertain biological sex (and therefore, we are marked as aptly being positioned in the social role of women in terms of our gender expression, which is a property that can be easily ascertained just in virtue of appearance). Therefore, there will be a strong (but not perfect) correlation between gender expression and the Haslangerian social position. For this reason, gender expression will be a good predictive factor in order to explain salary differences.

Gender expression is clearly different from biological sex. In my view, gender expression is an explanatorily useful kind that can appear in useful generalizations. As Mallon (2003) argues, these robust social kinds are explanatorily independent of biological facts. For this reason, even if the Haslangerian social property that we have been appealing to is characterized in part in terms of the social position of subordination occupied by those who are *perceived* to be biologically female, the social property that is constituted by these social practices (that is, discriminating those who are perceived to be biologically female) turns out to be a stable, robust social kind that is explanatorily useful independently of the explanatory efficacy of the notion of biological sex (and, I want to add, independently of the notion of being perceived to be biologically female). What is doing the explanatory job is the social structure, not the biological sex. And once this stable social kind is created, it acquires a life of its own, and it can be explanatorily useful on its own. That is to say, we can characterize another social kind that is strongly correlated to the Haslangerian social property, but can be conceptually distinguished from it, namely, the property of occupying a social position of subordination in virtue of having a feminine gender expression.[26] One can argue that this social property is more trans-inclusive than the Haslangerian social property, precisely because it does not require that the subjects are perceived to be biologically female, so trans women who do not pass as cis women but have a feminine gender expression will fall under the concept. For this reason, it seems to me that there can be trans women who do not pass as cis women but still share enough of this social position, so as to say that the notion of woman that is the most explanatorily useful in the contexts of utterances like (3–5) is actually trans-inclusive, and does not leave trans women who do not pass as cis women out, contra what the advocates of gender-critical views argue.

Another claim made by gender-critical philosophers is that the notion of gender identity has little political value. I disagree. I believe that there are some contexts where the most relevant notion in the vicinity involves gender identity. That is to say, there are contexts where people make utterances involving the term 'woman', and in those contexts, given the most pressing moral and political considerations involving the situation that the subjects of the utterances are in, the criteria of similarity that are relevant in order to determine the extension of 'woman' amount to self-identifying as a woman. A clear example is posed by statements describing psychological dimensions of the oppression of women.

---

26 This notion is different from the notion of self-identifying as a woman, which I discuss in the following paragraph.

I will focus on the case of descriptions of hermeneutical injustice.[27] This form of injustice consists in the existence of gaps or lacunas in our collective conceptual repertoires regarding concepts or representational devices that would be useful in order to conceptualize the experiences of members of subordinated groups that are in their interest to render intelligible, but are missing. My claim, then, is that, with respect to statements describing cases of hermeneutical injustice concerning experiences of women that are difficult to communicate due to androcentric bias in the introduction and dissemination of concepts, the most explanatorily useful notion of woman is a notion of gender identity in terms of self-identifying as a woman. In my view, it is clear that all trans women can suffer from hermeneutical injustice with respect to their experiences concerning their self-identification as women, including trans women who are not out as trans women yet and may have a masculine gender expression and thus still occupy a social position similar to cis men, or perhaps they are out as trans and have a feminine gender expression but do not pass as cis women and thus are not perceived as being biologically female.[28] But all these trans women can suffer from hermeneutical injustice regarding, for example, the difficulties in conceptualizing and communicating their experiences of identifying with feminine social norms and expectations, such as their experiences of being subject to norms of feminine appearance. That is to say, there might be trans women who identify as women but do not pass as cis women and do not share some of the traits of the social position of women who are perceived to be biologically female. However, all these women do have something in common, namely, the fact of suffering hermeneutical injustice regarding their experiences of identifying as women, for example, the experience of dealing with norms of feminine appearance (be it by means of embracing or rejecting them), regardless of whether these norms of feminine appearance are in fact *enforced* onto them or not.[29] Therefore, there can

---

27 See Fricker (2007) for a characterization of the notion of hermeneutical injustice.
28 Jenkins (2016) convincingly argues that those trans women who do not pass as cis women would not share the Haslangerian social property that Haslanger uses to characterize woman, so this characterization of woman would leave some trans women out, which is problematic. In the paragraph above, I mentioned a related social kind that can include those trans women that have a feminine gender expression but do not pass as cis women. However, there may also be trans women who identify as women but as not out as trans yet due to transphobia in their environments and therefore cannot be said to have a feminine gender expression. In my view, the notion of woman in terms of gender identity (or self-identifying as a woman) that I am advocating for in this paragraph would be fully inclusive for this purpose. See Jenkins (2018a) for a very compelling characterization of gender identity.
29 See Jenkins (2016) for a very useful discussion of the distinction between being subject to norms of feminine appearance and *identifying* with those norms.

be generalizations about the hermeneutical injustice suffered by women that apply both to cis and trans women, so the relevant explanatorily kind in this context amounts to the property of self-identifying as a woman (not the property of sharing a social role *à la* Haslanger, let alone the property of being biologically female). Hence, this is the notion that the term 'woman' refers to in the generalizations about hermeneutical injustice suffered by women, contra what gender-critical philosophers say.

6 Does Contextualism Collapse into Invariantism?

Some philosophers, including Bettcher (2017), Laskowski (2020), and Zeman (2020), have recently argued that my modified version of contextualism might overcome Saul's objection only by paying a big price, namely, collapsing into invariantism. Invariantism is a theory about the meaning of a term that has it that the term has the same meaning in all contexts. For instance, terms such as 'triangle' and 'table' might have invariant meanings. Bettcher, Laskowski, and Zeman independently argue that my version of contextualism might collapse into invariantism precisely because the extension at each context is determined by the most relevant moral and political considerations involving the situation that the subjects of the utterance are in. But if the most pressing moral and political considerations involving the subjects that our utterances about women are about are similar enough, then these similar moral considerations will determine similar referents, so it is not clear that the meaning of 'woman' will change from context to context as I claim. In particular, these philosophers suggest that the most pressing moral considerations will give greater moral weight to self-identifying as a woman. In fact, Laskowski (2020) and Zeman (2020) suggest that my treatment of Saul's example about the transphobic community seems to imply such a view: if the most pressing moral and political considerations involving who should be allowed to access ladies' toilets have the result that self-identifying is the factor with greatest moral weight, then it is not clear why in other contexts we would get the result that some utterances of 'woman' have a different extension, if self-identification as a woman is the factor with greatest moral weight.

I agree that this is a serious objection. In response, my conjecture is that although sincere self-identification always provides a pro tanto reason for determining the referent of 'woman' along these lines, in every context, it is not clear that self-identification always trumps other considerations. That is to say, it is not clear that every utterance of 'woman', in any context for any purpose, is such that the moral and political considerations regarding the subject matter of those utterances always gives greatest moral weight to sincere self-identification. And the reason is that we use the term 'woman' in many contexts and for many different purposes. Of course, the referent of these utterances is not determined by the criteria that the speakers have in mind, but rather by the moral and political considerations involving the subject matter and the situation they are in. But there can be contexts such that what is more pressing for the subjects the utterance is about, given the purpose of the utterance, is not just a matter of how people self-identify.

Let us consider some examples that will make this point clearer. Saul offers some examples where she has no problem in saying that the notion of 'woman' at issue could be the notion of individuals who are biologically female. One example is similar to the case of the clinic aiming to test for vaginal diseases I explained above. Two other cases are the following (see Saul 2012, p. 200):

(6) This bone belonged to a woman (uttered in a forensic context, after analyzing some human remains).
(7) Scientists testing COVID-19 vaccines should conduct studies that include women too.

What is the relevant notion of woman involved in utterance (6)? In my view, it is okay to say that the relevant notion here is the notion of biological sex (and in particular chromosome sex). This case seems unproblematic, because the subject the utterance is about is long dead. What are the moral and political considerations that are relevant in this case, then? I believe the relevant moral and political considerations are those involving the aims and purposes of forensic sciences.

What about utterance (7)? This utterance is motivated by the fact that some scientific studies to test the efficacy of certain drugs leave women out (and other subordinated groups such as people of color).[30] What is the relevant notion of woman here? One issue is which notion of woman would make this claim true. Another question is which notion of woman is the most explanatorily useful here. In my view, all the notions of woman that we have mentioned so far, including biological sex, social position *à la* Haslanger, and self-identification

---

30 See Criado-Pérez (2019) for an excellent discussion of many scientific studies that are biased in this way. Another example of bias in science is Lloyd's (2006) excellent discussion of androcentric bias in evolutionary explanations of female orgasm.

as a woman, would make (7) true, since (7) is a scientific generalization that does not have to be true for *all* instances, only for *most* of them. And these different notions of woman are co-extensional for a large part (even if not perfectly co-extensional, of course). In any case, I do believe that the notion that seems more explanatorily useful here corresponds with biological sex, since there can be biological and anatomical differences that make a difference regarding the efficacy of the drug that is being tested. This is not obvious though, since as many philosophers of gender and race have argued, when it comes to the explanation of differences regarding the efficacy of some medical treatments, or the rates of some diseases or medical conditions, the explanation could also appeal to social factors. For example, white people and black people in the US have important differences regarding their access to healthcare, the effects of pollution, access to clean water, safe conditions at work, and so on. All these social factors can also affect their health conditions, or even the efficacy of certain drugs.[31] Likewise, there can be biological and anatomical differences between men and women that are in turn caused by social factors.[32] So, it could be that a notion of social position *à la* Haslanger is actually the most explanatorily useful in this regard, rather than biological sex. Thus, about this specific case, perhaps it makes sense to be agnostic about biological sex or a social property *à la* Haslanger (but only in the context of explanations like (7) and similar ones). Hence, it seems that the most explanatorily useful notion in this specific context is not self-identification.

Cases (3–5) and (7) all involve examples of generalizations trying to explain some patterns of oppression. I have focused on these cases because this is part of the argument used by gender-critical advocates in order to defend a trans-exclusive account of the meaning of woman. I have argued, in response, that the view that generalizations (3–5) involve a notion of biological sex is not defendable. In my view, in these contexts the notion that is the most explanatorily useful is often a notion of woman in terms of a social position *à la* Haslanger (or in terms of a related social kind correlated with gender expression). But recall that in these cases, we are searching for the most explanatorily useful kind, precisely because these generalizations are in the business of *explaining* patterns of oppression. But of course, the term 'woman' appears in many other contexts that have nothing to do with explanation.

---

31 See Hardimon (2013) for further discussion of these social factors and what this implies for the notion of race.
32 See Haslanger (2003) for further discussion of possible social causes of biological and anatomical differences between men and women.

Some advocates of gender-critical views often use cases such as (7) to argue that 'woman' always refers to individuals who are biologically female. The possibility that the term 'woman' is context-sensitive makes clear that this does not follow: even if 'woman' in (7) refers to biological sex, it does not follow that it refers to biological sex in (3–5). Likewise, the fact that the term 'woman' in utterances (1) and (2) above refers to those who sincerely identify as a woman (for instance in the context of the feminist conference), does not imply that 'woman' always has this reference. Even if we assume my version of contextualism, according to which the referent in each context is determined by the most pressing moral and political considerations, from the fact that in (1) and (2) (uttered in the feminist conference) the most pressing consideration involves sincere self-identification, it does not follow that the same will apply to (3–7). In fact, I have given some reasons to believe that in cases (3–5), the notion that maters the most from a moral point of view amounts to the notion of social position *à la* Haslanger (or perhaps a correlated social position involving gender expression), and in cases (6–7) the most explanatorily relevant kind seems to correspond to biological sex (with the reservations I expressed regarding 7).

But this discussion is a *moral* discussion. It has to do with what are the relevant moral and political considerations in each context. My version of contextualism is a *semantic* theory that claims that the character of the term 'woman' is something along the lines of (CP), appealing to the most relevant moral and political considerations of each context. But what are the most pressing moral and political considerations in each context is a matter of moral deliberation. Someone could agree with my semantic theory but disagree with my moral views regarding each case. I believe that my view can be called contextualism even if, contra what I tend to believe, self-identification was the factor with greatest moral weight in all cases. In some sense, this view still deserves to be called contextualism because there is a difference between the *character* and the *content* of the term woman. An analogy: imagine the meaning of the term I in a possible world where Robinson Crusoe is the only inhabitant. We would still say that I is a context-sensitive term, even if it always yields the same referent.

However, I believe there are good reasons to say that the moral and political considerations that have the most weight can change from context to context. As we have seen, it is plausible to say that in (6–7), 'woman' refers to biological sex. It seems clear that in (3–5), the term does not refer to biological sex, but rather to something like a social position of subordination *à la* Haslanger. And in (1–2) (uttered in the feminist conference, or in a debate about who should be allowed to use women's toilets), it seems that self-identification has the greatest moral weight. Hence, my version of contextualism does not collapse into invariantism.[33]

Bibliography

Bettcher, Talia Mae (2013): "Trans Women and the Meaning of 'Woman.'" In: Alan Soble, Nicholas Power, and Raja Halwani (Eds.): *Philosophy of Sex: Contemporary Readings*, 6th ed. Lanham, MD: Rowman & Littlefield, pp. 233–250.

Bettcher, Talia M. (2017): "Trans Feminism: Recent Philosophical Developments." In: *Philosophy Compass* 12. No. 11, pp. 124–138.

Crenshaw, Kimberlé. (1989): "Demarginalizing the Intersection of Race and Sex: A Black Feminist Critique of Antidiscrimination Doctrine, Feminist Theory and Antiracist Politics." In: *University of Chicago Legal Forum* 1989, pp. 139–167.

Criado-Pérez, Caroline. (2019): *Invisible Women: Data Bias in a World Designed for Men*. New York: Vintage Books.

Cull, Matthew. (2020): *Engineering Genders: Pluralism, Trans Identities, and Feminist Philosophy*. PhD diss., University of Sheffield. Available from White Rose eTheses Online. (uk.bl.ethos.813879).

Dembroff, Robin. (2020): "Beyond Binary: Genderqueer as Critical Gender Kind." In: *Philosophers' Imprint* 20. No. 9, pp. 1–23.

DeRose, Keith. (2009): *The Case for Contextualism*. Oxford: Oxford University Press.

Díaz-León, E. (2012): "Social Kinds, Conceptual Analysis, and the Operative Concept: A reply to Haslanger." In: *Humana.Mente: Journal of Philosophical Studies* 5. No. 22, pp. 57–74.

Díaz-León, E. (2016): "*Woman* as a Politically Significant Term: A Solution to the Puzzle." In: *Hypatia* 31. No. 2, pp. 245–258.

Díaz-León, E. (2019): "On How to Achieve Reference to Covert Social Constructions." In: *Studia Philosophica Stonica* 12, pp. 34–43.

Díaz-León, E. (2020): "Descriptive vs Ameliorative Projects: The Role of Normative Considerations." In: Alexis Burgess, H. Cappelen, and D. Plunkett (Eds.): *Conceptual Engineering and Conceptual Ethics*. Oxford: Oxford University Press, pp. 170–186.

Dupré, John. (1981): "Natural Kinds and Biological Taxa." In: *Philosophical Review* 90, pp. 66–91.

Freeman, Lauren, and Ayala-López, Saray (2018): "Sex Categorization in Medical Contexts: A Cautionary Tale." In: *Kennedy Institute of Ethics Journal* 28. No. 3, pp. 243–280.

Fricker, Miranda (2007): *Epistemic Injustice: Power and the Ethics of Knowing*. Oxford: Oxford University Press.

Fausto-Sterling, Anne. (2000): *Sexing the Body*. New York: Basic Books.

---

33 I have presented this material at workshops and conferences in Vitoria, Berlin, Barcelona, St Andrews, Manchester, Hamburg, and Murcia. I am very grateful to the audiences for their useful feedback. Extra thanks are due to Matt Cull and Katharine Jenkins for their detailed comments, and to the editors of this volume for their patience. I acknowledge financial support from grant number PGC2018–094563-B-I00.

Hardimon, Michael. (2013): "Race Concepts in Medicine." In: *Journal of Medicine and Philosophy* 38. No. 1, pp. 6–31.

Haslanger, Sally. (2000): "Gender and Race: (What) are they? (What) do we want them to be?" In: *Noûs* 34. No. 1, pp. 31–55.

Haslanger, Sally. (2003): "Social Construction: The "Debunking" Project." In: F. F. Schmitt (Ed.) *Socializing Metaphysics: The Nature of Social Reality.* Lanham, MD: Rowman & Littlefield, pp. 301–325.

Haslanger, Sally. (2006): "What Good are Our Intuitions? Philosophical Analysis and Social Kinds." In: *Proceedings of the Aristotelian Society* 80. No. 1, pp. 89–118.

Haslanger, Sally. (2016): "What is a (Social) Structural Explanation?" In: *Philosophical Studies* 173. No. 1, pp. 113–130.

Hay, Carol. (2020): *Think Like a Feminist: The Philosophy Behind the Revolution.* New York: W. W. Norton & Co.

Jenkins, Katharine. (2016): "Amelioration and Inclusion: Gender Identity and the Concept of *Woman*." In: *Ethics* 126. No. 2, pp. 394–421.

Jenkins, Katharine. (2018a): "Toward an Account of Gender Identity." In: *Ergo* 5. No. 27, pp. 713–744.

Jenkins, Katharine. (2018b): "*The Wrong of Injustice*, by Mari Mikkola." In: *Mind* 127. No. 506, pp. 618–627.

Kaplan, David. (1989): "Demonstratives: An Essay on the Semantics, Logic, Metaphysics and Epistemology of Demonstratives and Other Indexicals." In: J. Almog, J. Perry, and H. Wettstein (Eds.): *Themes from Kaplan.* Oxford: Oxford University Press, pp. 481–563.

Kapusta, Stephanie, J. (2016): "Misgendering and its Moral Contestability." In: *Hypatia* 3. No. 3, pp. 502–519.

Kripke, Saul. (1980): *Naming and Necessity.* Cambridge, MA: Harvard University Press.

Laskowski, N. G. (2020): "Moral Constraints on Gender Concepts." In: *Ethical Theory and Moral Practice* 23. No. 1, pp. 39–51.

Lloyd, Elisabeth. (2006): *The Case of the Female Orgasm.* Cambridge, MA: Harvard University Press.

Mallon, R. (2003): "Social Construction, Social Roles, and Stability." In: F. F. Schmitt (Ed.) *Socializing Metaphysics: The Nature of Social Reality.* Lanham, MD: Rowman & Littlefield, pp. 327–353.

Mallon, Ron. (2017): "Social Construction and Achieving Reference". In: *Noûs* 51. No. 1, pp. 113–131.

Mikkola, Mari. (2006): "Elizabeth Spelman, Gender Realism, and Women." In: *Hypatia* 21. No. 4, pp. 77–96.

Mikkola, Mari. (2016): *The Wrong of Injustice: Dehumanization and Its Role in Feminist Philosophy.* New York: Oxford University Press.

Mikkola, Mari. (2019): "Feminist Perspectives on Sex and Gender." In: *The Stanford Encyclopedia of Philosophy* (Fall 2019 Edition), Edward N. Zalta (Ed.) URL = https://plato. stanford.edu/archives/fall2019/entries/feminism-gender/, accessed 30 March 2021.

Mount, Allyson. (2012): "Indexicals and Demonstratives." In: Gillian Russell & Delia Graff Fara (Eds.) *The Routledge Companion to Philosophy of Language.* New York: Routledge, pp. 438–448.

Putnam, Hilary. (1975): "The Meaning of 'Meaning.'" In: *Mind, Language, and Reality: Philosophical Papers*, vol. 2. Cambridge: Cambridge University Press, pp. 215–271.

Roberts, Debbie. (2013): "Thick Concepts." In: *Philosophy Compass* 8. No. 8, pp. 677–688.

Saul, Jennifer. (2012): "Politically Significant Terms and Philosophy of Language: Methodological Issues." In: S. L. Crasnow and A. M. Superson (Eds.) *Out from the shadows: Analytical Feminist Contributions to Traditional Philosophy.* New York: Oxford University Press, pp. 195–216.

Spelman, Elizabeth. (1988): *Inessential Women: Problems of Exclusion in Feminist Thought.* Boston: Beacon Press.

Stein, Edward. (1999): *The Mismeasure of Desire: The Science, Theory and Ethics of Sexual Orientation.* New York: Oxford University Press.

Stock, Kathleen et al. (2019): "Doing better in arguments about sex, gender, and trans rights." https://medium.com/@kathleenstock/doing-better-in-arguments-about-sex-and-gender-3bec3fc4bdb6?fbclid=IwAR25XyfvgA8d-uixC6dKGJ1hmrpVTVg82xXm6uaJJEcA5sDGoUvJCZR2y4Y, accessed 1 October 2021.

Witt, Charlotte. (2011): *The Metaphysics of Gender.* New York: Oxford University Press.

Zeman, Dan. (2020): "Subject-Contextualism and the Meaning of Gender Terms." In: *Journal of Social Ontology* 6. No. 1, pp. 69–83.

Part IV: **Epistemology and Polarization**

Manuel Almagro-Holgado and Alba Moreno-Zurita

Affective Polarization and Testimonial and Discursive Injustice

Abstract: There is strong evidence that the rise of political polarization increases biases, prejudices, and public hostility toward people from certain socially disadvantaged groups, among other things. As a result, cases of testimonial and discursive injustice are more likely to happen. It is more likely that disenfranchised people's claims are given less credibility than deserved and that their words are taken as performing a different speech act from the intended one, or even that they are prevented from performing certain speech acts. The aim of this chapter is to explore which of the two prominent notions of polarization in the literature is better positioned to account for this relation. These notions are ideological polarization and affective polarization. The former is mostly characterized by the distance between certain belief contents in an ideological spectrum, while the latter has more to do with people's willingness to like the ingroup and dislike the outgroup, which is tied to a certain level of credence in the beliefs of the group one identifies with. In particular, we argue that affective polarization is better positioned than ideological polarization to explain the relationship between the increase in polarization and the increase in certain injustices.

1 Introduction

Child sexual abuse is presumably one of the most abhorrent crimes, in part because there is an immense asymmetry between the victim and the perpetrator. The victims of this kind of abuse are extremely vulnerable: they are not in a position to understand what is happening, and the consequences for them are devastating. Child sexual abuse can cause depression, anxiety, and post-traumatic stress disorder, among other mental health problems. This is one of the reasons why this sort of crime presumably triggers enormous anger and aversion. Presumably, the mere suspicion that someone has perpetrated such a crime arouses a visceral reaction. However, and surprisingly for us at first, the very opposite reaction happened recently.

On 24 February 2020, a teacher was arrested in Ceuta, Spain, for allegedly sexually abusing several of his students. On the 27th of the same month, another teacher was arrested in the same city on the same charge. Both cases, with no

https://doi.org/10.1515/9783110612318-014

apparent connection between them, are still under investigation, and therefore there is not much reliable information about them regarding the innocence or guilt of the teachers. All that is known at the moment is that both teachers are free on charges, the first of them after having paid a €5000 bail. However, both cases have generated a reaction that may seem surprising at first glance. What one would expect, given the type of crime they are accused of, and given that in previous years the city has had similar cases where the accused have turned out to be guilty,[1] is that Ceuta's people would express strong condemnation of such an alleged crime, or at least would not support the accused. But that is not what happened.

The day of the second arrest, 27 February, hundreds of teachers and other people congregated under the slogan "Defenseless teachers,"[2] denouncing the existence of a "great mafia" dedicated to falsely denouncing teachers for profit. "If someone wants to denounce in order to make money or whatever, let him or her think about it. We're going to demonstrate as often as possible with everyone. Today for you, tomorrow for me" (León 2020, para 6, our translation), declared one of the protesters. "The "Nescafé salary'—an expression referring to a prize offered by Nestlé consisting in a monthly salary for life—will not be achieved by denouncing a teacher" (El pueblo de Ceuta 2020, para 6, our translation), said another. A few days later, it was revealed that at least the accusation against the second teacher had not been filed by the parents of any of the students; it was the hospital, to which one of the girls went after presenting discomfort, that initiated the process. This fact, together with the lack of evidence that could lead someone to be suspicious about the case, and the kind of crime the teachers are accused of, makes the reaction of these people surprising, regardless of whether the teachers eventually turn out to be innocent or guilty. Given the overwhelming reaction, it is reasonable to think that if the students' families had made any statement, they would have received very low credibility, at least from the population that has demonstrated in defense of the teachers and those who sympathize with these demonstrations, and that their words would not have

---

1 "Procesado un profesor por abusos sexuales, elaborar pornografía infantil e inducir a la prostitución a alumnos en Ceuta" (Europa Press 2020, title, "Teacher prosecuted for sexual abuse, making child pornography and inducing prostitution of students in Ceuta" our translation). https://www.lasexta.com/noticias/sociedad/procesado-profesor-abusos-sexuales-elaborar-por nografia-infantil-inducir-prostitucion-alumnos-menores-ceuta_202003075e639e401ef f8600010400ec.html, accessed 19 May 2020.
2 "Cientos de docentes se manifiestan en apoyo 'a Claudio y Jose'" (León 2020, para 6, "There is a big mafia trying to profit from false accusations against teachers", our translation). https://elfarodeceuta.es/cientos-docentes-manifiestan-apoyo-claudio-jose/, accessed 19 May 2020.

had the intended meaning. It is our contention that this case indeed counts as an instance of testimonial and discursive injustice, insofar as their words would not receive the appropriate credibility and uptake due to the hearer's biases and stereotypes.

One of the keys to understanding people's reaction in this particular case is that several, if not all, of the girls allegedly abused by these teachers are Muslims. Despite the fact that approximately half of the population of Ceuta is Muslim, the racism and marginalization suffered by this part of the population is very high and has grown significantly in recent years. According to the *European Islamophobia Report 2019*, the Muslim population of Ceuta "still suffers segregation, with hundreds of minors without schooling and lacking prosecutors specialized in discrimination and hate crime" (Bayrakli and Hafez 2020, p. 740).

In accordance with the process of polarization toward ethnic minority groups suffered worldwide,[3] polarization in attitudes has increased in Spain, too. For instance, data from the study *Opiniones y Actitudes de la Población Andaluza hacia la Inmigración* (OPIA) indicate a tendency in this line: the data point in the direction that there is polarization of the attitudes of Andalusians toward people forced to leave their countries and a growth in the group of those who strongly oppose the expansion of those people's social rights. Another complementary key to understanding this shocking case is that various political groups have been explicitly and implicitly promoting racism in Spain. One of them, the far-right political party Vox, has received growing support over the last few years in the city, and in fact won the 2019 general election there. The leader of this party, Santiago Abascal, has made statements such as the following, contained in the same report cited above: "Muslims want to destroy Europe and western society by celebrating the fire of NotreDame. Take it into account before it's too late" (Bayrakli and Hafez 2020, p. 47). As another example, Vox used an image of a hijabi candidate of left-wing party Podemos in Ceuta to tweet: "This twenty-year-old is a candidate for Podemos Ceuta. We didn't know that women's liberation consists in wearing a purple hijab." (Bayrakli and Hafez 2020, p. 751). Two of the ideas that can be considered core to this political party, at least during 2019 – 2020, are that Muslims want to invade Spain (as if being Muslim makes you not Spanish!) and take advantage of the country's social aids, and that there are numerous false denunciations that are used to make a profit (see Bayrakli and Hafez 2020).

---

3 See Carothers and O'Donohue 2019; Johnston et al. 2015; Mounk 2018, p. 166; and Wojcieszak and Garrett 2018.

We can see this and other similar potential cases of testimonial and discursive injustices, we argue, as a consequence of the rise of polarization. We believe that the state of polarization in which Ceuta finds itself is part of what explains the early, disproportionate reaction of the protesters, and it promotes the occurrence, as in this case, of testimonial and discursive injustices. Of course, racism is not a new thing, and someone might object that the demonstrators' disproportionate reaction could be simply explained by appealing to their racism. We agree on that score, but we think that it misses an important point: even though racism has been present in the city since long ago, the current high level of polarization not only increases racism but also makes a group of people, racist people, to feel legitimize to publicly display such attitudes. It is not a new thing that there are racists, but perhaps it is relatively a new thing that racists feel they have the right to blatantly express their racist attitudes.

Scholars have intensively studied how inter-group division fuels conflict and promotes negative attitudes towards outgroup people (see, for instance, Mason 2018). Regarding ethnic minorities and people forced to leave their countries, which are one of the greatest challenges and one of the most tense issues in contemporary polarized societies,[4] it has been shown that when national identity is activated, those who oppose people forced to leave their countries publicly exhibit greater negative trait evaluations, greater relative ingroup favorability, and lower common intergroup identity with those people (see Wojcieszak and Garrett 2018). Hence, although it has not yet been explicitly claimed, it is clear that, at least theoretically, polarization promotes testimonial and discursive injustices by increasing the prejudices, stereotypes, and implicit biases that generate these harmful situations.

As noted, there is solid evidence to think that polarization increases biases, prejudices, and public hostility toward people from certain socially disadvantaged groups, and these kinds of attitudes are arguably those causing testimonial and discursive injustices. We assume such a connection here. Thus, instead of arguing about the connection between polarization and testimonial and discursive injustices, the main thesis of this chapter is that, unlike the notion of polarization that has mainly to do with what people say they believe (ideological polarization), the notion of polarization concerning what people say about what they feel toward people from another group because of their high level of confidence in some core belief of their ideological identity (affective polarization) can successfully explain the connection we have assumed. What we do in this work is

---

4 See Carothers and O'Donohue 2019; Johnston, Newman, and Velez 2015; Mastro 2015; and Mounk 2018.

thus to explore which of these two notions of polarization (ideological and affective) allows us to adequately explain the relationship between the increase in polarization and the increase in certain injustices, and to argue that affective polarization is in a better position than the other to do that. In this sense, this chapter follows the spirit of the political turn in analytic philosophy understood as this book does: we evaluate our theories in virtue of their power to account for, and to intervene in, particular harmful situations.

The plan is as follows. In section 2, we introduce ideological and affective notions of polarization, the two notions we focus on here to explore which of them is better positioned to account for the relation between increased polarization and increased testimonial and discursive injustices, exemplified in the case presented above. Section 3 introduces and discusses the phenomena of testimonial and discursive injustices and explains why the case presented above counts as such. Section 4 aims to argue in favor that affective polarization enables us to explain the connection between political polarization and testimonial and discursive injustices through the analysis of the case above.

2 Ideological and Affective Polarization: Two Different Stories

Traditionally, political polarization has been conceived in terms of what has been called ideological polarization, that is, as the tendency towards the extremes of an ideological spectrum in which people's content beliefs about different political ideologies, and mostly about specific political issues, are represented. The idea is that the closer people's content beliefs are to the extremes, the greater the division or rupture, and therefore the polarization.[5] For instance, if the two opinions on abortion "Abortion should be legal and not restricted in any way" and "Abortion should be illegal in any situation" are the most extreme opinions in a given spectrum, then the closer a society's views are to these opinions, the greater the polarization.

It is important to note that there is more than one way to conceive and measure ideological polarization: there are different ways of analyzing the representation of different groups' content beliefs. In this respect, Aaron Bramson et al. (2017) have recently done some excellent conceptual clarification work and have distinguished nine different senses in which ideological polarization can be understood and measured. Instead of distinguishing between different senses of

---

5 See DiMaggio, Evans, and Bryson 1996; Hetherington 2009; and Fiorina 2017.

ideological polarization, they explicitly say that they distinguish nine senses of polarization in general (Bramson et al. 2017, p. 117), but inasmuch as what these different senses measure has to do with the distribution of a society's opinions and content beliefs on an ideological spectrum, these nine senses can in fact be seen as different senses of ideological polarization.

This general way of understanding the rupture that characterizes polarization has been called extremism (Sunstein 2017), and, as we have said, it mainly deals with the way in which belief contents are represented in a given distribution (Talisse 2019). The main idea behind the notion of extremism could be understood in terms of the way the YouTube's algorithm works (see Tufekci 2018), or what has been called "the YouTube model" (Almagro and Villanueva 2021, p. 60): if for instance you start by watching videos about Trump's speeches in YouTube, the platform ends up recommending videos about Holocaust negationism and white supremacism. The point of this way of conceiving polarization is that polarization is closely related to the idea of ending up believing contents that are one step further than the initial ones. Thus, if you believed that abortion should be legal but with some restrictions and now you believe that abortion should be legal and not restricted in any way, and the resulting opinion is located more near of the extremes in a spectrum, then you have become more extreme in your beliefs in the sense of extremism.

Thus, although there are different particular ways of conceiving it, ideological polarization has to do with how the contents of the beliefs of at least two groups are distributed in a given ideological spectrum. Specifically, ideological polarization has to do with the idea of extremism, which can be defined as follows:

> Extremism: If, at $t1$, agents $X1 \ldots Xn$ and $Y1 \ldots Yn$ respectively hold attitudes $A1$ and $A2$, where $A1$ and $A2$ have conflicting contents, then their attitudes polarize if, at $t2$, $X1 \ldots Xn$ and $Y1 \ldots Yn$ respectively hold attitudes $A3$ and $A4$, where the contents of $A3$ and $A4$ are situated nearer the extremes in a given ideological spectrum.[6]

However, the notion of ideological polarization is just one way of defining and measuring polarization. In a foundational work of 2012, Shanto Iyengar et al. point out that the analysis of political polarization was extremely focused on

---

6 We are aware that some of the notions distinguished by Bramson and other scholars, as well as the asymmetrical conception of polarization according to which the increase of polarization is due to the shift of the content beliefs just of one group, do not exactly fit with this definition of extremism. However, this definition captures the general idea associated with the concept of ideological polarization and, in this sense, it is sufficient for the goal of this work.

policy preferences, and that there are other important aspects related to polarization that should be considered. From their study, it has been widely recognized that measuring polarization also requires taking into account other issues, such as how people dislike their opponents. Thus, Iyengar and his colleagues introduce an affect-based notion of polarization, called affective polarization, on the basis of the concept of social distance (Bogardus 1947). According to this notion, the higher the hate and other negative feelings toward people from opposing groups, the greater the affective polarization in a society.

To understand the rupture that characterizes polarization from this approach, it is no longer relevant to look at the contents of the beliefs of two groups, but to the way people in a group evaluate those who do not belong to their own group and therefore do not think like them. The touchstone of polarization then no longer has to do with what people say about their political views or how these views are distributed across an ideological spectrum, but rather with the negative attitudes people in one group have toward people in other groups.

Here it is also important to note that there is more than one way to conceive and measure affective polarization (see Druckman and Levendusky 2019; Harteveld and Wagner Manuscript; and Iyengar et al. 2019), but we are more interested here in the similarities between some different conceptions of it. Affective polarization is often conceived as based on group affiliation, that is, on the pertaining to a social or ideological group. As Iyengar, Good, and Lelkes hypothesize, "the mere act of identifying with a political party is sufficient to trigger negative evaluations of the opposition" (Iyengar, Good, and Lelkes 2012, p. 3). These evaluations include considering members of groups deemed as opposite as hypocritical, selfish, and closed-minded (Iyengar et al. 2019). Mason, who calls affective polarization "social polarization," argues that it comprises three phenomena: implicit biases, emotional reactivity, and activism (Mason 2018, pp. 4, 17). In particular, Mason argues in favor of the idea that it is the internal alignment of a group, especially in social terms, that fosters the rise of polarization by creating what she calls "mega-identities" (Mason 2018, p. 14). So, identity and social sorting seem to be key aspects of the notion of affective polarization.

When the group to which you belong is increasingly homogeneous, deliberation enclaves (Bordonaba 2020 and Sunstein 2017) and echo chambers (Jamieson and Capella 2010) are more easily generated, that is, situations wherein you are exposed to a limited and biased set of arguments that mainly support your previous beliefs are more likely to happen. These scenarios contribute to the increase of your identification with your group and with the ideas that define it (Almagro and Villanueva 2021 and Sunstein 2017), which is closely related to the increase of the expression of negative attitudes towards others: if I am absolutely certain that what characterizes those of us who group together is true, then it makes no sense for me to pay attention to the alleged evidence of those who think otherwise, and it becomes rational not to consider the reasons of those who think otherwise as worthy of respect. In this sense, there seems to be a very close connection between the increase in adhesion to a political identity, the increase in the credence attributed to the central ideas—which are salient at a particular time—of that identity, and the expression of negative attitudes towards others.

This general way of understanding the rupture that characterizes polarization has been called radicalism (Sunstein 2017), and it mainly deals with the *degree of belief* rather than with belief contents (Talisse 2019). The main idea behind some sorts of affective polarization could be understood in terms of the way the Spotify's algorithm works, or what has been called "the Spotify model" (Almagro and Villanueva 2021, p. 60): if, for instance, you start by listening to classic 90s American rap on Spotify, the platform constantly recommends music to you that fits that genre and period. After 10 hours of listening to this type of music on Spotify, you will not end up being a fan of trap or other genres; you will simply love the music you already liked more. The point of this way of conceiving polarization is that polarization is closely related to the idea of increasing the confidence in your own previous beliefs, rather than ending up believing contents that are one step further than the initial ones. Thus, according to this conception of polarization, if you believed that abortion should be legal but with some restrictions and now you believe the same content but with much more confidence, then you are more polarized in the sense of radicalism. When the belief in which you have significantly increased your level of confidence is a strongly salient belief of your identity group at a particular time, and your group is increasingly homogeneous, then negative evaluation and hostility toward people who identify with a group with a different opinion on that issue becomes almost inevitable. Bordonaba and Villanueva call "impervious reasoning" a type of affective polarization that captures this idea well: "it is only when the credence that we attribute to the core beliefs of the group that we identify with goes considerably higher, that we start thinking about our political adversaries as unworthy of our respect" (Bordonaba and Villanueva 2018a, p. 1; see also Bordonaba 2019). This seems to be part of the story behind the notion of affective polarization.

It is possible to be a radical without being an extremist, if the beliefs to which one holds fast are at the center of the political spectrum, and also to be a non-radical extremist, if the kind of social changes you think are necessary place you in an extreme of the political spectrum but you are willing to listen to others' reasons and to question your beliefs (see Sunstein 2017, pp. 74–75). Thus, affective polarization has to do with the idea of radicalism, which can be defined as follows:

> Radicalism: If, at $t1$, agents $X1 \ldots Xn$ and $Y1 \ldots Yn$ respectively hold conflicting attitudes $A1$ and $A2$, then their attitudes polarize if, at $t2$, $X1 \ldots Xn$ and $Y1 \ldots Yn$ give more credibility to their respective attitudes at $t2$.

One of our assumptions is that a suitable notion of polarization should be able to explain the connection between political polarization and testimonial and discursive injustices. The reason is that our theoretical tools must be evaluated in virtue of their capacity to explain and identify certain injustices. Since the increase of negative attitudes toward certain identity groups seems to be one of the outcomes of political polarization, a notion of polarization capable of accommodating such a connection is required in order to explain why political polarization is dangerous. In the following sections, we will argue that, unlike ideological polarization, affective polarization can explain how the increase in political polarization can generate more situations of testimonial and discursive injustice and, therefore, can explain one important point concerning the dangerousness of political polarization.

3 Testimonial and Discursive Injustice

In this section, we analyze the main example of this paper with the objective of presenting it as a case of testimonial and discursive injustice and see to what extent affective polarization can explain this kind of case. The general idea that can be derived from the notions of testimonial and discursive injustices is that belonging to a particular social group makes one more vulnerable to suffering at least two kinds of injustice—one related with the credibility of what we say (testimonial injustice), and the other with how what we say is interpreted (discursive injustice).

As noted in the introduction, according to the *European Islamopohia Report 2019* (Bayrakli and Hafez 2020), there are alarming data in Ceuta concerning the level of polarization and the difficulties that the Muslim population faces in accessing the social and economic life of the city. Despite the fact that more than 40% of the population of Ceuta is Muslim, there is a high percentage of segregation and hate crimes against this population, mostly promoted by the far right. What the data show is that this population suffers discrimination that affects practically all areas of their lives, which is intensified with the raise of polarization. Such social discrimination provides a perfect scenario for a deficit of credibility or misinterpretation of what the Muslim population in Ceuta might mean through their words. The social environment in which these events take place is relevant in order to analyze this case from the perspective of testimonial and discursive injustice.

First, let us begin with the notion of testimonial injustice. According to Miranda Fricker, testimonial injustice refers to the phenomenon whereby the credibility of a speaker is undermined because she is not considered as a subject of knowledge by the hearer (see Almagro, Navarro, and Pinedo 2021 for a recent discussion about the epistemic nature of this phenomenon). In this context, "social identity" usually refers to those characteristics related to geographic origin, race, accent, gender, sexual preferences of speakers, and so on (Fricker 2007, p.17). The classic example that Fricker proposes as a clear case of testimonial injustice is based on the novel *To Kill a Mocking Bird*. A good part of the plot of this novel revolves around Tom Robinson, an African American who, in a social context that is strongly racist toward the African American population, is accused of having committed a rape. In the course of the Robinson trial, the victim's lawyer asks him about his reasons for going to the victim's house on the night the alleged rape occurred, and Robinson's response is that he believed that the victim needed help to perform a certain task in the house. However, the fact that Robinson publicly expresses the belief that a white person needed help from an African American is one of the triggers for the tribunal to find Robinson guilty, despite all the physical evidence indicating that Robinson could not have committed the crime.

Fricker's analysis of Robinson's case shows how the implicit prejudices and biases held by white people toward the African American population are decisive for the decision of the tribunal in the development of the judicial process against Robinson; the tribunal's deliberation is not based on facts but on racial prejudices. These biases and prejudices explain that under no circumstances could the African American population perform certain acts without a negative interpretation. In a similar way, we argue, in the case presented at the beginning of this chapter, the prejudices and biases prevent giving any credibility to the people who have suffered the abuses since the whole story has been built on the testimony of the people who have committed the crime and not of the victims.

According to Fricker, for a situation to be considered a case of testimonial injustice in the relevant sense, it must meet the following criteria. On the one hand, situations of credibility deficit have to be systematic, that is to say, the harm perpetrated, based on hearers' prejudices related to the social identity of the speakers, must affect several areas of the victim's life. When prejudices related to the social identity of speakers operate in a considerable number of social experiences and practices, we are faced with a case of systematic injustice (Fricker 2007, pp. 27–28). On the other hand, the lack of credibility, in addition to being systematic, must be persistent. This means that the credibility deficit that speakers face because of their social identity must be repeated continuously over time. Moreover, the harm caused to the victim by the perpetrators must constitute a violation of some moral principle (Fricker 2007, pp. 21–22). For example, a speaker who, due to some feature of her social identity, has been blocked from an epistemic practice is ethically damaged as an agent, and will be more vulnerable to failing victim of other kinds of social injustice related to the economic or institutional sphere. So, the cases that consist in a credibility deficit that is not caused by identity prejudice are cases of epistemic bad luck, but not cases of testimonial injustice in the relevant sense (Fricker 2007, p. 29).

In our main example, however, the deficit of credibility does not occur at the moment the families of the victims decide to report the crime to the police. There does not seem to be any problem at this step of the process, since according to what we know about the case, the corresponding police protocols were activated. At first glance, hence, it does not seem to meet the criteria to count as a testimonial injustice case, and therefore it is not at all obvious why it could count as an instance of such. However, we think that it can be indeed considered as a case of testimonial injustice if we take into account the demonstrators' reaction and the way in which the events were treated by the different media and focus on the resultant context in relation to what the families will be able to say.

The demonstrators' interpretation of the events was as follows. The families who had denounced the alleged sexual abuse were motivated by financial compensation that would be due to them for having been victims of sexual abuse. They had made false reports, and therefore the possibility that the accused teachers might have committed the crime of which they were accused was automatically ruled out. In the media's treatment of the case, we only found information about how the work of teachers had been impaired. Furthermore, media coverage suggests that it is the teachers who are in a vulnerable situation because they are exposed to an unfair judicial process, preventing them from carrying out their professional work. All this, together with the data obtained from the *European Islamophobia Report 2019* (Bayrakli and Hafez 2020) and the data regarding the rise of polarization, suggests that the situation in Ceuta is prone to cases of testimonial injustice suffered by Muslim people. The point is that, despite the fact that there has not been any particular statement to which less credibility than deserved has been assigned, in this case one can still speak of testimonial injustice to the extent that the conditions for it have been promoted.

Besides the injustices related to the testimony of the agents, that is, the credibility we give to the speakers according to their social position, we can distinguish another type of injustice that is restricted to the way we publicly express ourselves. It could be the case that, when people from certain groups that have been systematically discriminated perform a particular speech act in certain public contexts, the automatic interpretation by their audience is that what they are expressing is something different from what they intended to. This is the idea that Quill Kukla (Writing as Rebecca) (2014) develops through the notion of discursive injustice. This notion accounts for situations in which the ability of speakers belonging to a non-privileged group to successfully perform certain speech acts is systematically undermined, even though the speakers are qualified to perform such speech acts. For example, when a woman who has a high rank in a company gives an order to her employees, the order is systematically understood as a request (Kukla 2014, pp. 442–443).

In these cases, according to Kukla, despite the fact that the speaker meets the conditions required to be taken as performing a certain speech act, their interlocutor places them at a certain point in the social space that prevents their speech act from being interpreted as an order. So, speakers seem to be unable to successfully perform certain speech acts, not because their audience resist to give the appropriate uptake, but because the audience situates them in a node of the social structure (Ayala 2016) that undermines their capacity to perform that speech act, that is, a speech act with a certain effect. The question that immediately comes to mind is whether or not the case counts as one of discursive injustice.

The most obvious sense in which this case could be considered as a case of discursive injustice would be to consider that at the moment when the families of the minors decided to file a report at the police station this speech act could not have been carried out satisfactorily. However, families of the alleged victims can perform the action of denouncing by filling a report, so here we could not speak of a case of discursive injustice. What is the problem, then, if the families were able to denounce? As we have pointed out above, the socially constructed account of the case suggests that the families who have filed the reports are motivated by the economic compensation associated with this type of crime. Furthermore, it is assumed that the families are aware of the legal channels that will be activated at the time when the reports become effective, which also implies a legal advantage on their part.

All of these reasons, along with the idea that the accused teachers are respectable people who only intend to do their job, support the majority interpretation that the reports are false. When we say that a report is false, what we mean is that the facts that are being reported did not occur, and, furthermore, that the people who are reporting are not entitled to use this legal procedure. All of this means that the affected families are unable to report or publicly express the crimes they denounce. Recall that it is not possible to find any public testimony about the families of the victims. We consider this to be a case where speakers find it difficult to perform certain speech acts and this may fit within the notion of discursive injustice. The majority reception of the reports in the media aims not to cover an alleged crime, but the claim of certain economic compensation, and this has as a direct consequence the restriction of the range of actions, including linguistic ones (Ayala 2016 and 2018), that the minors and their families can perform.

Before looking at another case of discursive injustice, similar to the one we have just discussed, let us pay attention to this quote from the writer Caitlin Moran:

> It all depends on the tone: if a man sees a woman who is angry, he is going to encourage her speech in which she is angry and not listen to what she says. If she says the same thing, but in a slow and unpassionate way, there is a chance that someone will listen to what you have to say. It is very easy to look at an angry woman and say she's hysterical. (Castillo 2020, para 10, our translation)

Moran advocates that, if you are from an oppressed group, it is better not to say things in an angry tone. What she describes is a situation of discursive injustice where the gender of the person who has something to say plays a key role in getting her audience to correctly react to what she is saying. Moran's statement allows us to pay attention to a fact that occurs before speakers perform a speech act, that is, in order to perform a certain successful speech act, speakers are restricted in the ways (being angry, sad, or overwhelmed) in which they can express themselves. Another way of claiming this is by saying that people have a particular set of speech affordances in virtue of their social position, and some people have unjustly restricted their set of speech affordances (Ayala 2016). For example, a couple of years ago, when Professor Christine Blasey Ford[7] accused Conservative Judge Brett Kavanaugh of sexual abuse, Ford's expression was so restrained at all times that it is natural to think that she could not perform her speech act if her discomfort was visible. This did not happen because Ford was not a competent speaker, but because she had no right to be angry, even though she had many reasons to be so if she wanted her speech to be successful.

So, there is no speech act or claim that has not been interpreted as it should. What we have are several cases in which the harmed persons do not have the right to publicly present themselves as being victims of certain crimes, and

---

7 We would like to thank Neftalí Villanueva for this example.

this is a form of discursive injustice. In our main case, the reason why demonstrators and some public coverage maintain, without hesitation, that the denunciations are false reports is due to increased affective polarization. In particular, increased confidence in certain core beliefs of a particular ideological identity causes a part of the Ceuta population to maintain without any doubt that no crime was perpetrated and that the families were trying to economically take advantage with the denunciations.

4 Affective Polarization Facilitates Testimonial and Discursive Injustices

In this section, we will argue that only by adopting the notion of affective polarization can we account for the connection between the rise of polarization and the increase of the space for testimonial and discursive injustice, because it is affective polarization that has to do with our attitudes toward others. To see this, let us first consider the concept of ideological polarization and analyze its suitability to explain such a connection.

Remember that, according to this notion, polarization consists in the increase of the distance of the content beliefs of at least two groups in an ideological spectrum, that is, extremism. That is, if two groups of people think, for instance, that (a) economic decisions should be made by individuals and that (b) economic decisions should be made by collective institutions, respectively, then these groups are more polarized than if they think that (c) economic decisions should be made mostly by individuals but also by public institutions and that (d) economic decisions should be made mostly by public institutions but also by individuals, to the extent that (a) and (b) are nearer of the poles of an ideological spectrum than (c) and (d). Despite the intuitive nature of this idea of polarization, its allegedly harmful condition is not quite obvious. Why is having more extreme beliefs, in the sense of belief content or extremism, necessarily harmful to democracy? Arguably, the less two sets of views have in common, the more difficult it will be to reach agreement between them. But the matter is more complicated than it might appear. To see this, consider the following idea. If the members of a group are willing to listen to the arguments and reasons that support the beliefs of those belonging to the opposite side and are also willing to reach agreement with them when it is required, then it is not clear how having beliefs located at the extremes of an ideological spectrum diminish the quality of a democracy. In fact, the very opposite can be argued: the farther away the extreme views are in a society, the more diversity of opinions will be available in that society (Aikin and Talisse 2020). Insofar as none of these views attacks the basic assumptions of any democracy, the distance or whatever parameter deemed relevant in order to conceive polarization in terms of how beliefs and opinions are represented in an ideological spectrum does not sound per se very compelling to explain the harmful nature of political polarization. A different point is that the further away one group's beliefs are from the beliefs of another group, the more likely it is that both groups will develop hostile and exclusionary attitudes toward people in the opposite group as a result of the increase of bigotry and group identity (see, e.g. Mason 2018 and Klein 2020). But even in that case, having beliefs whose contents are further away from the beliefs' contents of the other part is still not harmful per se; it is harmful only to the extent that in fact there is a correlation between the distance between two beliefs and the increase in hostile attitudes toward the other party, which does not seem to us a necessary consequence.

Regarding the purpose of this chapter, instead of just focusing on how extreme beliefs could be harmful to democracy in general, we have to deal with the particular question of how beliefs at the extremes of an ideological spectrum facilitate the increase of testimonial and discursive injustice, and here the answer does not seem obvious at all. The problem is that, if people have the right spirit, then it does not seem obvious that those who belong to one of those groups whose beliefs are at one end of the ideological spectrum will give less credibility than they deserve to people in the other group simply because their beliefs are at the other end in the spectrum. Think about a friend of yours whose beliefs on different topics are located far from yours. It is not hard to think that, despite that the contents of her beliefs are different from yours, you love to talk with her. In fact, you might learn much from people with different views, and you might often reach agreement with them. That is exactly what is advocated by those who maintain that disagreement is a source of learning and coordination (see Bordonaba 2017), or by those who defend that acknowledging that the truth or falsity of our evaluations is relative to our standards is what enables us to explain moral progress (Pérez-Navarro 2019). Thus, it is not clear how polarization, understood as the increase of extreme belief contents, might facilitate the increase of testimonial and discursive injustices. Disagreement is an essential feature of any valuable democracy (Aikin and Talisse 2020).

Affective polarization seems to avoid this objection and therefore to be better positioned to account for the pernicious consequences of the rise of polarization. Polarization understood as radicalization, which, as we have seen, is the conception behind affective polarization, does not explain the harmful nature of the phenomenon in terms of the distance between the contents believed by two groups. According to this notion, the harmful nature of polarization would rather be linked to the adoption of a very high level of confidence in one or more previous beliefs that are central to the identity of one's ideological group. Once one becomes radicalized, it is when one comes to adopt such a level of trust in the salient beliefs that define their group, at a certain moment, that it is rational to ignore the others' reasons. A high level of identification with one's own group, and in particular with some of its core beliefs, increases the possibility of being impervious to the others' reasons and to negatively evaluate them. Becoming an ardent supporter of Vox, whose leader claims things like that Muslims want to destroy Europe, is becoming in someone who can easily commit testimonial and discursive injustices; someone who contributes to the maintenance of unjust social norms that make it impossible for certain people to perform certain speech acts. This first observation points in the direction that the notion of affective polarization is in a better position than the ideological one to account for what we intend in this work.

As we have said, having a very high level of credence in the beliefs that essentially characterize the group a person identifies with leads to perceive people who belong to and identify with opposing groups, and specially their reasons and arguments, as not worthy of respect. It is important to note that this outcome is not necessarily a matter of irrationality. In fact, when one believes a proposition with much confidence, it becomes rational not to pay attention to opposing views, and, in that case, it is perfectly understandable that one gives less credibility to people who advocate opposite views. For instance, if you believe with much confidence that the Earth is not flat, or that climate change is the result of human activity, then it becomes rational to you not to pay much attention to the flat-Earthers or to those who deny global warming. Presumably, you will not give these people much credibility. However, when it becomes rational not to pay attention to what is said by people who have and propose a different way of living, or whose social and ideological organization is different from yours, simply because your credence in your identity core beliefs has been viciously reinforced,[8] democracy suffers, because it makes coordination and agreement harder to achieve. That is, certain mechanisms that have an evolutionary purpose (Tajfel et al. 1971) have undesirable effects when the situation changes rapidly through cultural evolution without allowing time for biological evolution to take its course.

---

8 Some mechanisms through which a person can viciously reinforce her beliefs are being exposed to crossed disagreements (Bordonaba and Villanueva 2018b and Osorio and Villanueva 2019), increasing the capacity to filter the information received (see Sunstein 2017), and going through processes of group polarization (see Talisse 2019).

Thus, polarization promotes hostile and disrespectful attitudes towards people from other ideological and identity-based social groups. These attitudes, which reflect the view one has about the world—one's convictions and the form of life to which one belongs—are the very same in nature as those attitudes that cause testimonial and discursive injustice. When a person gives less credibility to a woman than a man others things being equal, says that an African American is guilty and that a white person who has done the same is not, or disbelieves a Muslim person just because of her being Muslim, that person expresses her sexist, racist and xenophobic attitudes, which shape his mind; in particular, he is expressing his high level of confidence in some of his beliefs—for example, the belief that African Americans commit crimes much more often than white people do.

Let us consider the case presented at the beginning again. We can now see more clearly why the reaction of a part of Ceuta's population to the cases of alleged sexual abuse has been from the outset one of unconditional support for the accused. The city has undergone a process of strong polarization in recent years. The population is roughly divided between people who identify with extreme right-wing groups and fascism, and those who are the object of criticism, racism, and hostility from the former. This division, in terms of extremism, is not new. The high material inequality that exists in the city has always reinforced a clear division between the contents believed by two different groups, which mostly coincide with the division between the population that is Muslim and that which is not. One of the consequences of the recent process of polarization was an increase in the confidence and level of credence that each person has in the core beliefs of his or her group. For example, one of the explicit statements that has guided the election campaign of the ultra-right-wing Vox party that won the last election in Ceuta was that Muslims and African people are invading Spain for destructive purposes. According to Vox, people who belong to a different culture from the one that predominates in Spain want to destroy the country and take advantage of its resources. But not only this. Another central idea of extreme right-wing parties that has also been increasingly common in other countries is the belief in the existence of gender laws that leave men unprotected against certain allegedly false claims. The high level of credence in these beliefs is what allows us to understand why the reaction of part of the Ceuta population has been as it has. It is shocking that a group of people shows their support, without any evidence to do so, for a person accused of sexually abusing a girl. The simple fact that the girl belongs to a Muslim family was enough to make it clear for the demonstrators that this was a false report aimed at financial gain, and their high credibility in these beliefs led them not to hesitate to express such attitudes. Hence, certain actions, verbal and non-verbal, performed by any person of this family have received less credibility than it would have if they were not Muslims, and certain actions, verbal and non-verbal, performed by them have been taken as a different action. The increased affective polarization has facilitated the appearance of testimonial and discursive injustice, as it has occurred in this case. This is one way in which the kind of polarization that is dangerous for a democracy and testimonial and discursive injustices are connected: polarization is the result of increased confidence in identity beliefs, which promotes the emergence of the attitudes that cause these injustices.

Bibliography

Aikin, Scott, and Robert Talisse (2020): *Political Argument in a Polarized Age: Reason and Democratic Life.* Cambridge: Polity Press.

Almagro, Manuel, Llanos Navarro-Laespada, and Manuel de Pinedo (2021): "Is Testominial Injustice Epistemic? Let me Count the Ways." In: *Hypatia* 36. No. 4, pp. 657-675.

Almagro, Manuel, and Neftalí Villanueva (2021): "Polarización y Tecnologías de la Información: Radicales Vs. Extremistas." In: *Dilemata: International Journal of Applied Ethics* 34. No. 13, pp. 52 – 69.

Ayala, Saray (2016): "Speech Affordances: A Structural Take on How Much We Can Do with Our Words." In: *European Journal of Philosophy* 24. No. 4, pp. 879 – 891.

Ayala, Saray (2018): "A Structural Explanation of Injustice in Conversations: It's about Norms." In: *Pacific Philosophical Quarterly* 99. No. 4, pp. 726 – 748.

Bayrakli, Enes, and Farid Hafez (2020): "European Islamophobia Report 2019". [SETA] https://www.islamophobiaeurope.com/wp-content/uploads/2020/06/EIR_2019.pdf, accessed 28 October 2021.

Bogardus, Emory (1947): "Measurement of Personal-Group Relations." In: *Sociometry* 10. No. 4, pp. 306 – 311.

Bordonaba, David (2017): *Operadores de Orden Superior y Predicados de Gusto: Una Aproximación Expresivista.* Ph.D. Dissertation. Universidad de Granada. https://digibug.ugr.es/handle/10481/48131

Bordonaba, David (2019): "Polarización como Impermeabilidad: Cuando las Razones Ajenas no Importan." In: *Cinta de Moebio. Revista de Epistemología de Ciencias Sociales* 66, pp. 295 – 309.

Bordonaba, David (2020): "Los Peligros de Las Cámaras Eco. Nota Crítica de #Republic: Divided Democracy in the Age of Social Media." In: *Éndoxa. Series Filosóficas* 45, pp. 249 – 260.

Bordonaba, David, and Neftalí Villanueva (2018a): "Affective Polarization as Impervious Reasoning." In: *13th Conference of the Italian Society for Analytic Philosophy (SIFA).* https://sifanovara2018.files.wordpress.com/2018/09/book-of-abstracts-online.pdf, accessed 28 October 2021.

Bordonaba, David, and Neftalí Villanueva (2018b): "Crossed Disagreements: A Quantitative and Qualitative Study on the Minutes of the Sessions of the Spanish Parliament." In:

Proceedings of the IX Conference of the Spanish Society of Logic, Methodology and Philosophy of Science, pp. 101–108.

Bramson, Aaron, Patrick Grim, Daniel J. Singer, William J. Berger, Graham Sack, Steven Fisher, Carissa Flocken, and Bennett Holman. (2017): "Understanding Polarization: Meanings, Measures, and Model Evaluation." In: *Philosophy of Science* 84. No. 1, pp. 115–159.

Carothers, Thomas, and Andrew O'Donohue (2013): *Democracies Divided. The Global Challenge of Political Polarization.* Washington: Brookings.

Castillo, Queralt (2020): "Caitlin Moran: "Los hombres tienen que saber que el feminismo es lo único que los salvará". [El Salto Diario] https://www.elsaltodiario.com/literatura/entrevista-caitlin-moran-nuevo-libro-como-ser-fa-mosa, accessed 28 October 2021.

DiMaggio, Paul, John Evans, and Bethany Bryson (1996): "Have Americans Social Attitudes Become More Polarized?" In: *American Journal of Sociology* 102. No. 3, pp. 690–755.

Druckman, James N., and Matthew S. Levendusky (2019): "What Do We Measure When We Measure Affective Polarization?" In: *Public Opinion Quarterly* 83. No. 1, pp. 114–122.

El Pueblo de Ceuta (2020): "Hay una gran mafia intentando lucrarse con denuncias falsas contra maestros". [El Pueblo de Ceuta] https://elpueblodeceuta.es/art/45504/hay-una-gran-mafia-intentando-lucrarse-con-de-nuncias-falsas-contra-maestros, accessed 28 October 2021.

Fiorina, Morris (2017): *Unstable Majorities: Polarization, Party Sorting, and Political Stalemale.* Stanford: Hoover Institution Press.

Fricker, Miranda (2007): *Epistemic Injustice: Power and the Ethics of Knowing.* Oxford: Oxford University Press.

Gibbard, Allan (1990): *Wise Choices, Apt Feelings: A Theory of Normative Judgment.* Oxford: Oxford University Press.

Gibbard, Allan (2003): *Thinking How to Live.* Cambridge: Harvard University Press.

Gibbard, Allan (2012): *Meaning and Normativity.* Oxford: Oxford University Press.

Hansen, Nate (2015): "Experimental Philosophy of Language." In: *Oxford Handbooks Online.* Oxford: Oxford University Press. DOI: 10.1093/oxfordhb/9780199935314.013.53

Harteveld, Eelco, and Markus Wagner (Manuscript): "Affective Polarization Across Parties: Why Do People Dislike Some Parties More than Others?" (unpublished manuscript, October 16, 202), https://ic3jm.es/wp-content/uploads/2020/11/Harteveld_Wagner_Affpol.pdf, accessed 28 October 2021.

Hetherington, Marc (2009): "Putting Polarization in Perspective." In: *British Journal of Political Science* 39. No. 2, pp. 413–448.

Hetherington, Marc, and Thomas J. Rudolph (2017): *Why Washington Won't Work. Polarization, Political Trust, and the Governing Crisis.* London: The University of Chicago Press.

Iyengar, Shanto, Gaurav Sood, and Yphtach Lelkes (2012): "Affect, Not Ideology. A Social Identity Perspective on Polarization." In: *Public Opinion Quarterly* 76. No. 3, pp. 405–431.

Iyengar, Shanto, and Sean Westwood (2015): "Fear and Loathing Across Party Lines: New Evidence on Group Polarization." In: *American Journal of Political Science* 59. No. 3, pp. 690–707.

Iyengar, Shanto, Yphtach Lelkes, Matthew Levendusky, Neil Malhotra, and Sean J. Westwood. (2019): "The Origins and Consequences of Affective Polarization in the United States." In: *Annual Review of Political Science* 22, pp. 129–146.

Jamieson, Kathleen, and Joseph Cappella (2010): *Echo Chamber: Rush Limbaugh and the Conservative Media Establishment.* New York: Oxford University Press.

Johnston, Christopher, Benjamin Newman, and Yamil Velez (2015): "Ethnic Change, Personality, and Polarization over Immigration in the American Public." In: *Public Opinion Quarterly* 79, pp. 662–686.

Kelly, Thomas (2008): "Disagreement, Dogmatism, and Belief Polarization." In: *Journal of Philosophy* 105. No. 10, pp. 611–633.

Kukla, Rebecca (2014): "Performative Force, Convention, and Discursive Injustice." In: *Hypatia* 29. No. 2, pp. 440–457.

León, Juan (2020): "Cientos de docentes se manifiestan en apoyo "a Claudio y Jose"". [El Faro de Ceuta] https://elfarodeceuta.es/cientos-docentes-manifiestan-apoyo-claudio-jose/, accessed 28 September 2021.

Mastro, Dana (2015): "Why the Media's Role in Issues of Race and Ethnicity Should Be in the Spotlight." In: *Journal of Social Issues* 71, pp. 1–16.

Mounk, Yascha (2018): *The People vs. Democracy. Why Our Freedom Is in Danger and How to Save It.* London: Harvard University Press.

Osorio, Javier, and Neftalí Villanueva (2019): "Expressivism and Crossed Disagreements." In: María J. Frápolli (Ed.): *Expressivisms, Knowledge and Truth.* Cambridge: Cambridge University Press. pp. 111–132.

Pariser, Eli (2011). *The Filter Bubble: What the Internet Is Hiding from You.* New York: Penguin Group.

Pérez-Navarro, Eduardo (2019): *Ways of Living. The Semantics of the Relativist Stance.* Ph.D. Diss., Universidad de Granada. https://digibug.ugr.es/handle/10481/58120.

Sunstein, Cass (2017): *#Republic: Divided Democracy in the Age of Social Media.* New Jersey: Princeton University Press.

Tufekci, Zeynep (2018). "YouTube, the Great Radicalizer" (2018). The New York Times. https://www.nytimes.com/2018/03/10/opinion/sunday/youtube-politics-radical.html, accessed 1 December 2020.

Talisse, Robert B. (2019): *Overdoing Democracy. Why We Must Put Politics in Its Place.* New York: Oxford University Press.

Villanueva, Neftalí (2019): "Expresivismo y Semántica." In: D. Pérez Chico (Ed.): *Cuestiones de la Filosofía del Lenguaje.* Zaragoza: Prensas de la Universidad de Zaragoza, pp. 437–470.

Wojcieszak, Magdalena, and Kelly Garrett (2018): "Social Identity, Selective Exposure, and Affective Polarization: How Priming National Identity Shapes Attitudes Toward Immigrants Via News Selection." In: *Human Communication Research* 44. No. 3, pp. 247–273.

William J. Berger, Daniel J. Singer, Aaron Bramson, Patrick
Grim, Jiin Jung, and Bennett Holman

Philosophical Considerations of Political Polarization

Abstract: This chapter illustrates how philosophy and political science can inform one another by providing an overview of philosophical contributions the
authors have made on the topic of political polarization. The authors outline
three contributions they have made to understanding political polarization, particularly of the epistemic kind, discussing work that gives clearer terminology for
and ways of measuring polarization, precise mechanistic accounts of polarization, and a novel normative view about a possible source of polarization that
casts polarization as a possible outcome of rational, but limited, agents interacting. This last contribution illustrates, contra recent work, how dynamics akin to
epistemic bubbles and echo chambers can develop without associated epistemic
vices. Taken together, these projects can serve as a guide for producing philosophical work which both contributes to a mainstream disciplinary literature
as well as informs cross-disciplinary, empirical literatures in the social sciences.

1 Introduction

The study of community fragmentation has a long history in political science
(see Duverger 1954; Lipset 1960; and Fearon 2003), though the topic has become
more pressing for the study of American politics in recent decades, as the country has experienced growing polarization, and at the same time, rising discord
and acrimony (Iyengar 2019). The attention political scientists have given it at
both elite and mass levels has led to broad and sustained disciplinary scrutiny,
indicated at least in some part by five major review articles on the topic (Hetherington 2009; Layman et. al. 2006; Fiorina and Abrams 2008; Prior 2013; and
Iyengar 2019). As this literature has grown, however, so has terminological confusion, as the expanse of papers has come to discuss polarization in different
terms and offer different dynamical processes to explain the trends. This chapter
illustrates how philosophy and political science can inform one another. We do
this by providing an overview of philosophical contributions we have made on
the topic of political polarization. We will show how our contributions to social
epistemology can help fill gaps and correct inconsistencies in the empirical political science literature. The body of work presented here provides clear exam-

https://doi.org/10.1515/9783110612318-015

ples of how analytic philosophy can borrow from and inform empirical social science, and political science more specifically. We outline three contributions to understanding political polarization, particularly of the epistemic kind. We will discuss work that gives clearer terminology for and ways of measuring polarization, precise mechanistic accounts of polarization, and a novel normative view about a possible source of polarization that casts polarization as a possible outcome of rational, but limited, agents interacting. This last contribution is particularly interesting given recent developments in a literature around epistemic bubbles and echo chambers (Flaxman et. al. 2016; Jasny et. al. 2015; Levy and Razin 2019; Nguyen 2020; and Santos 2020), which understand echo chambers as due to epistemic vice. As we will discuss further, the work of Daniel J. Singer et. al. (2018) shows polarization can arise without such epistemic vices. By gaining a better understanding of how to discuss and measure polarization, what dynamics might give rise to it, and the normative status of those mechanisms, this piece draws connections between the philosophical and social scientific literatures on social justice and oppression.

The next section illustrates how analytic philosophy can contribute more precisely defined concepts and measures to political science. We offer added resolution to research on polarization by presenting our work that provides new senses and measures of polarization. Section 3 uses the conceptual apparatus from section 2 and looks at how approaches in computational modeling shed light on process-based accounts of polarization. Here, we survey plausible mechanisms that give rise to the kinds of polarization we observe in society. Section 4 then offers an additional model with distinctive normative upshots, arguing that under certain conditions polarization can occur even when everyone is listening to one another in good faith. Section 5 concludes by reflecting on the ways that the philosophical work here can contribute to empirical political science. While by no means exhaustive, these accounts demonstrate how philosophy might be informed by political science and contribute to its empirical process in turn. And since our research primarily contributes to social epistemology, we focus here on polarization that looks at political beliefs rather than the fascinating and influential literature on affective polarization, though the latter is by no means less important.

2 Senses of Polarization

Though Maurice Duverger (1954) argues that American politics cannot not support a multiplicity of parties, the two dominant parties that it did produce were famously thought not to have "a dime's worth of difference" between them (the phrase coming from 1968 third-party presidential hopeful George Wallace). That began to change by the 1980s, with Keith Poole and Howard Rosenthal showing diminishing ideological overlap of Members of Congress from Democratic and Republican states (Poole and Rosenthal 1984). While the 1950s and 1960s were marked by considerable legislative common ground, that overlap was slipping away by the late 1970s. Poole and Rosenthal's NOMINATE scores (Figure 1) are one way of measuring partisan drift, providing a standardized location for each congressional member across time, and allowing for a clear articulation of the increasing partisan divide (McCarty et. al. 2006).

Poole and Rosenthal (1984) uses the one-dimensional alignment of congressional votes to estimate elite political drift, but polarization might also be understood differently, say, by looking at the consistency with which disparate political issues get bundled together. This is the approach of Delia Baldassarri and Andrew Gelman (2008), which measures the degree of mass political polarization, operationalizing the concept by assessing the level of correlation between issue pairs. It argues that, in a completely unpolarized population, one's position on one policy issue would not be indicative of one's position on any other—voters' political attitudes would be totally uncorrelated with one another. In a polarized society, however, taking a position on any issue—for or against—would perfectly predict every other political position one holds. Baldassarri and Gelman find that over time there is an increasing correlation between issue pairs in the U.S., which is itself strongly associated with partisanship and ideology. The diminishing heterogeneity of political attitudes among Democrats and Republicans further demonstrates the exacerbating political rift in the U.S.

The work of Poole and Rosenthal along with that of Baldassarri and Gelman are exemplary treatments of an epistemic kind of polarization in political science, among notable others (see Hetherington 2009; Layman et. al. 2006; Fiorina and Abrams 2008; and Prior 2013 for more expansive literature reviews). From even this brief discussion, however, it is evident that identification of polarization can be operationalized in different ways using different measures and conceptual apparatuses. In the absence of careful taxonomy, it is difficult to assess the extent of polarization or establish that accounts are not at odds with one another. The distance between median members of groups is one way of measuring polarization, but intra-group belief uniformity is another. Importantly, these ways of conceptualizing polarization are independent of one another, making it possible for polarization to increase in one sense and decrease in another. The signature elite-level polarization illustrated by NOMINATE scores is itself the confluence of different trends. The ideological middle-ground between the parties evaporates, while partisan extremes increase as well. So, while the fact of polarization is not in dispute, we draw attention to the confusion that can re-

DW-NOMINATE by party of U.S. House: 1963-2013

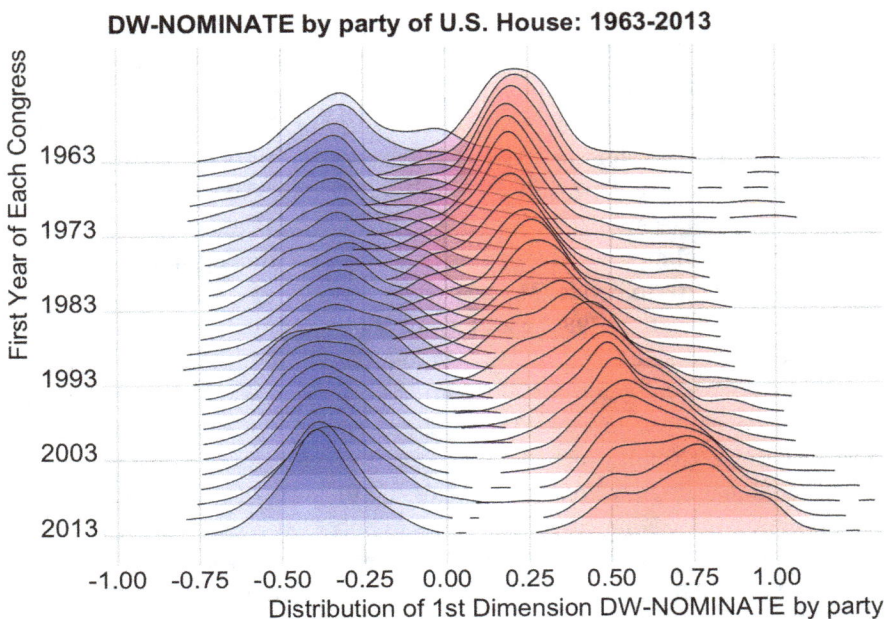

Figure 1: Congressional NOMINATE scores, 1963–2013.

sult from failing to distinguish among a variety of conceptual modes and measures.

Jamie Luguri and Jamie Napier (2013) serves as an example of how muddied discussions of polarization impede research on the topic. In addressing whether abstract or concrete ideation generates polarization, they point to competing evidence on the subject, arguing that though Daniel Yang et. al. (2013) finds abstract thinking decreased polarization, Alison Ledgerwood et. al. (2010) demonstrates that an abstract framing of an issue increases polarization. However, Ledgerwood et. al. (2010) shows that abstract thinking creates greater polarization in one sense (group size parity), while Yang et. al. (2013) shows greater polarization in a completely different sense (distinctness). Such confusion confounds the study of polarization, inhibiting the production of coherent research and illustrating the importance of terminological precision even when just assessing whether or not a given system elicits polarization.

In Aaron Bramson et. al. (2016), we offer a number of distinct senses of polarization in order to enhance conceptual clarity and keep our accounting straight regarding whether polarization, in its various senses, is growing or shrinking. 'Polarization,' we argue, acts like an umbrella term, containing different elements or kinds, rather than as one monolithic concept. In that paper, we offer a taxonomy of and measurement tools for distinct senses of polarization, tracking each type through empirical cases. The thinking here is that in order to understand polarization we must have conceptual clarity fostered by formal measures of the properties we hope to track. Only then can polarization (of this or that type) be evaluated as either increasing or decreasing.

The first and simplest sense of polarization offered in Bramson et. al. (2016) is 'spread,' how far apart the extremes of the distribution are from one another. In the one-dimensional case, where the relevant sense of polarization can be understood by placing people on a line, spread measures how broadly elements (e. g. votes or beliefs) range. As Figure 1 shows, one of the distinctive ways Congress has polarized is through the increased spread of political elites. While Democrats have maintained a left flank around -0.75 for most of the timeseries, the right-most Republicans have moved from a value of 0.75 to greater than 1 over the same period.

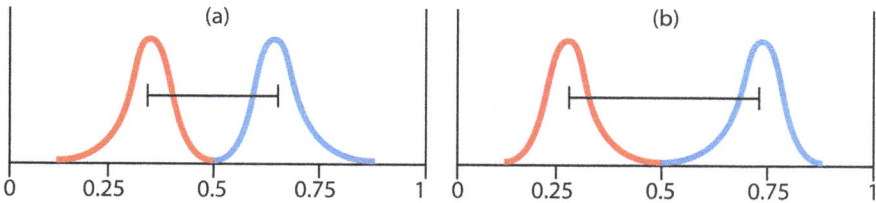

Figure 2: Group Divergence measure, with (b) showing more consensus than (a).

'Group divergence' and 'group consensus' are two further senses of polarization that cleanly pertain to polarization seen in the NOMINATE project. Whereas group divergence measures how far the median members of each group are from each other (Figure 2), group consensus identifies the within-group homogeneity (Figure 3). While spread measures the extremeness of group members, divergence measures the distance of median members of groups, increasing as median members separate, and consensus measures the internal group coherence. An example of consensus is the way party caucuses can deviate from the interests of party leadership. As the ideal points of legislators stray from the parties' centers (and frequently from party leadership), like members of the Freedom Caucus on the right and Progressive Caucus on the left, the consensus of the groups decreases, thereby decreasing the level of polarization.

'Distinctness' is another important conceptualization of polarization (Figure 4). As the overlap between groups diminishes, the level of polarization rises. As indicated above, because in the U.S. Members of Congress have the ability to

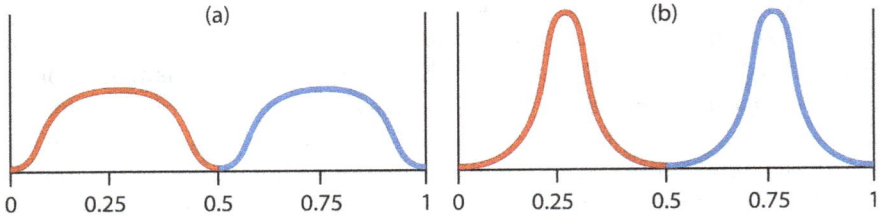

Figure 3: Group Consensus measure, with (b) showing more polarization than (a).

vote against party leadership, it was common for some Democrats to vote with Republican leadership and vice versa. As this tendency diminishes, distinctness increases, leaving less overlap between the parties and thus the prospect for compromise. Political parties, particularly in a two-party system like the U.S., are a coalition of divergent interests. In U.S. representative bodies (unlike in parliamentary systems), members are allowed to vote against their party leadership and often do. Party discipline evaluates how regularly party members vote with leadership (Krehbiel 2000). Increasing party discipline is thereby one realization of increased distinctness polarization. As can be seen in Figure 1, the considerable partisan overlap that existed in the 1960s and 1970s has all but evaporated, leaving two distinct and differentiated parties.

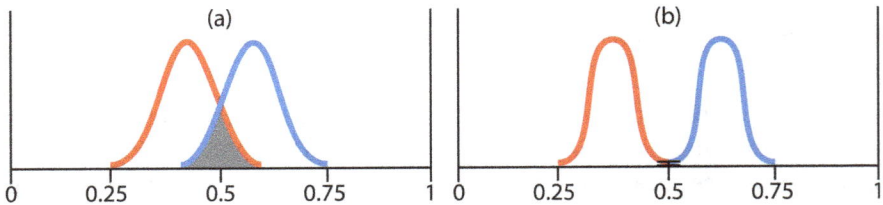

Figure 4: Distinctness measure, with (b) showing more polarization than (a).

A further marker of polarization we discuss in Bramson et. al. (2016) is 'size parity,' or the membership ratio between two groups (Figure 5). Intuitively, lower parity indicates less polarization. Were all but one individual to be a member of the dominant group, we would take that community to be nearly homogenous. Interestingly, Frances Lee (2016) argues, somewhat counterintuitively, that partisan compromise was easier during the middle of the twentieth century when Republicans tended to be a persistent minority in Congress, since vote trading was needed for Republican members to advance their agenda. Her argument gives reason to think that diminished size parity has fundamentally affected political gridlock and polarization.

Not all of the measures we outline in Bramson et. al. (2016) require exogenous identification of groups with predetermined constituencies. 'Community fragmentation,' for instance, looks to endogenously identify the groups by looking at natural splits in the distribution of the members of a population (Figure 6). Often it is perfectly straightforward to mark groups from the start, like Democrats and Republicans in Congress. However, where one seeks to identify congressional factions, like the Tea Party or the Freedom Caucus, community fragmentation is a useful tool to evaluate the number and boundaries of groups in a population, rather than assume the result.

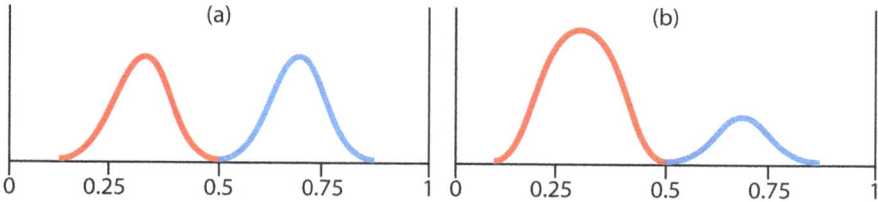

Figure 5: Size parity measure, with (a) showing more polarization than (b).

We see polarization as an umbrella term, containing distinct and independent senses, the measures of which do not always move in the same direction. Increased coherence can often entail members moving inward, which then diminishes spread (as people pull back from extremes) and induces more overlap. Using respondents' beliefs from the General Social Survey (GSS), Bramson et. al. (2016) proceeds to show how the distinct measures of polarization vary over time and relative to one another. Religion turns out to be a good domain to illustrate how these concepts come apart. The paper specifically looks at respondents' belief in the divinity of the Bible as a vivid illustration of this. While there is some correlation between the measures, there also exist clear instances when the measures are anti-correlated.

Analytic philosophy is well suited for this sort of conceptual work. As social scientific literature becomes expansive, it can lose the consistency that initially prevailed. These sorts of definitional inquiries and operationalizations are conducive for philosophers to take on, filling a crucial and treacherous gap in an empirical literature. Though the work in Bramson et. al. (2016) omits certain conceptualizations, such as multi-dimensional approaches like those of Baldassarri and Gelman (2008), it serves as a guide for further clarificatory work, not only on polarization, but in other domains within political science as well.

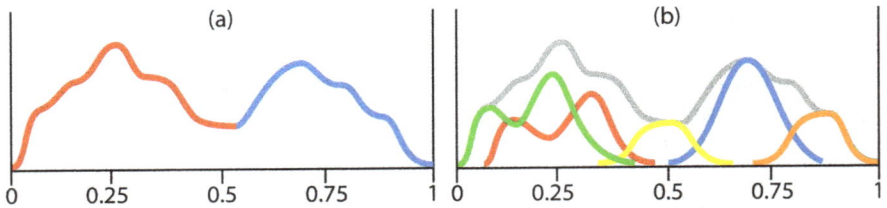

Figure 6: Community fragmentation measure, with (b) illustrating the latent constituencies of (a).

3 Polarization Dynamics

While Poole and Rosenthal (1984) and Baldassarri and Gelman (2008) are primarily concerned with measuring polarization, political science has also offered causal accounts to explain why it has happened. Larry Bartels (2000) demonstrates how partisanship has played an increasing role in U.S. presidential voting since the 1950s, becoming particularly contentious of late. The middle ground among partisans has slowly slipped away, leading to the disappearance of moderate voters and the rise of party extremism. Partisanship spiked similarly around moments like the passage of the Voter Rights Act and the Vietnam War, but the steady trend observed since the 1980s is distinctive. Motivated partisanship is no longer an exception, but a rule. Geoffrey Layman and Thomas Carsey (2002) find similar trends, showing the widening of American political attitudes over the course of three decades, as well as demonstrating that mass partisan polarization is primarily due to increasing issue salience rather than a party realignment. In support of their hypothesis, they find that on issues of race, social welfare, and morality the divide between partisans has intensified over the last half of the twentieth century.

This is not the only causal story on offer for the rise of polarization, though. McCarty, Poole and Rosenthal (2006) argue that the trend is associated with increased congressional turnover, where younger members take the seats of older members. Similar to Geoffrey Layman and Thomas Carsey (2002), Richard Fleisher and John Bond (2004) and Christian Grose and Antoine Yoshinaka (2003) argue that polarization is linked to the increasing ideological purity of Members of Congress. Others point to historical shocks stemming from the Civil Rights Act and the partisan realignment of Southern Democrats to explain the increased homogeneity within the parties (Hetherington 2009; Rae1994; and Han and Brady 2007).

Some discussions of political polarization build on social psychological research that offers finer mechanistic accounts of polarization that might well motivate the political dynamics we observe. Notably, Lord et al (1979) shows how biased assimilation of evidence can strengthen belief polarization. Sustained interest in the processes of group and political polarization, however, is more recent. Hetherington and Weiler (2018) have newly made the case that cultural polarization has cleaved Americans right down to the cars they drive and the beers they drink. Relatedly, Hopkins (2018) has shown that polarization is a national phenomenon—Democrats in Georgia and Oregon are more similar than cross-partisans in the same state. Studies such as these provide evidence of mass polarization, indicating that political polarization is occurring through generic pathways that do not directly relate to the structure or incentives of formal political institutions.

Beyond the historical contingencies that induced specific instances of polarization, we might also want to know the dynamical processes at work "under the hood" that brings about these cleavages. Polarization might, for instance, be brought about by "stretching" existing groups or by "pulling" coalitions apart. Philosophy, and computational philosophy in particular, is well suited to explore these mechanisms. What dynamical processes give rise to the kinds of political polarization above? In Patrick Grim et. al. (2012), we provide an illustration of how such dynamical processes might be designed, while, in Bramson et. al. (2017), we offer a more comprehensive look at computational models of polarization, including Robert Axelrod's dissemination of culture model (1997), Rainer Hegselmann and Ulrich Krause's bounded confidence model (2002), and Frank Harary's structural balance project (1959). Bramson et. al. (2017) looks at these three broad categories of models to interrogate distinctive approaches to polarization that have been taken, as well as offering corresponding empirical examples.

Axelrod (1997) offers a multidimensional approach to the dynamics of polarization. In the model, agents are arranged on a grid, each with an identity vector consisting of elements which take values from $0-9$. Agents are thought to be close to one another insofar as they share element values in the same position of their vectors. If agent 1 has vector $\{1, 2, 5\}$ and agent 2 $\{1, 1, 5\}$, they share two points of similarity. The model operates by randomly pairing neighbors and allowing for harmonization of agents with disparate features. When similar agents are paired, one alters their vector in order to become more similar to the other, thereby moving towards cultural consensus. So, in this example, agent 1 would change its second element from '2' to '1' to be more similar to agent 2. What generally results, however, is a kind of polarization where distinct and stable communities emerge that have no overlapping traits. Those communities, however, are not well-balanced, which is to say that they do not have size parity. There is only a narrow range of initial parameter settings which produce moderately sized groups, and even then, the results are not robust.

Because the project is multidimensional, its outcomes do not neatly conform to our measures of polarization. We can clearly see where the senses of group size, distinctness, and coherence are all relevant for interpreting the model's result with the same measures. For other senses, the features of the model may require a new measure in order to capture these concepts of polarization aside from those described here. The Axelrod model did not set out to study polarization specifically, but when we see polarization emerge as it does here, we are led to ask how to describe it. To that end, the Axelrod model is an illustrative example of how the kind of polarization that social scientists observe might emerge.

It is important that a plausible model of polarization not result in an unrealistic picture in which groups move to the absolute extremes: converging to 0 and 1, or the highest and lowest values that the model offers. The reason for this is primarily empirical—the kind of social and political polarization we see in the world does not paint society into opposite corners, but depicts constituencies pulling away from one another. Figure 1's NOMINATE scores illustrate how U.S. political parties have polarized in a number of senses (spread, distinctness, and group convergences, among others), though groups never pool at the poles, but somewhere in between. Such realistic desiderata, which do not force equilibria to extremes, are particularly relevant for political contexts in which the metrics we have—Likert scales of political ideology, NOMINATE scores, or indices like those in Baldassarri and Gelman (2008)—do not document such stark outcomes.

Bounded confidence models aim to produce polarization with this property. Hegselmann and Krause's model (2002, 2005, and 2006) is an example of exactly this sort of effort, observing how polarization might arise without agents eventually coalescing at the poles (Figure 7). The Hegselmann and Krause model produces polarization through local updating, where agents update their beliefs according to those in their peer neighborhood: those who have beliefs that are sufficiently close. Belief updating occurs here by taking a weighted average of the opinions of those that are "close enough" to one's own—where close enough is determined by a social distance threshold ε.

Hegselmann and Krause position beliefs on a one-dimensional scale from 0 to 1 and vary the social distance threshold to determine how polarization might take hold. As the model runs, agents weigh their beliefs against those of others in their neighborhood, softening their views to take one more in line with those around them. This produces the kind of pattern seen in Figure 7 where low-level clustering occurs at first, followed by higher level clustering. At threshold values

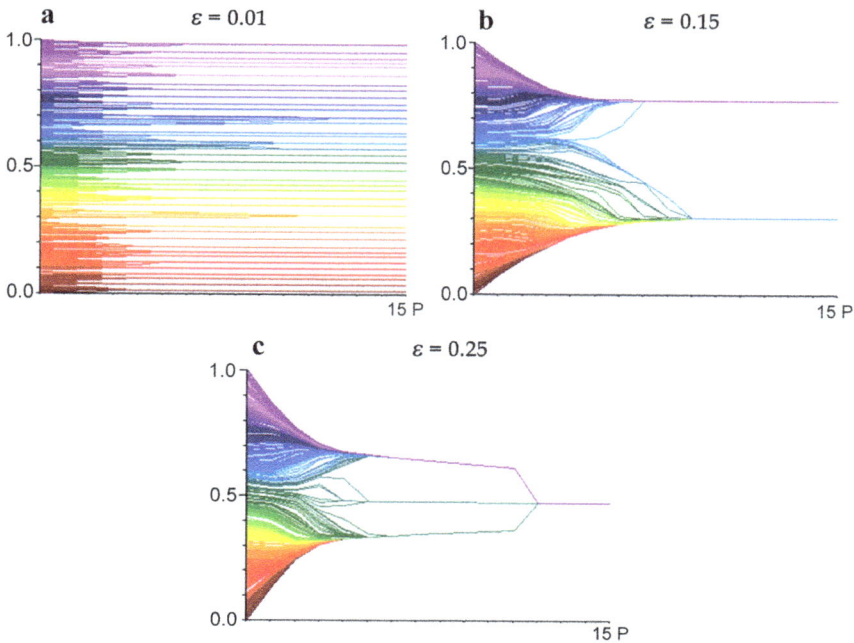

Figure 7: Example changes in opinion across time from single runs with different threshold values $\varepsilon \in \{0.01, 0.15, 0.25\}$.
Hegselmann and Krause (2002).

of 0.25 a single consensus opinion emerges, while at a value of 0.01 individuals cluster into a small number of local groups, exhibiting the kind of community fragmentation seen in Figure 6. At intermediate values, however, of say 0.15, two distinct groups emerge from the whole, clustering at beliefs around 0.25 and 0.75. This result is contingent on the number of agents present and their initial position along the 0–1 measure. If there were too few agents or they were non-uniformly distributed along the range, this would alter the likelihood of polarization as well as the final values at which the system converges.

The Hegselmann-Krause model is distinctive in that polarization *increases* over the runs only in the senses of group distinctness and group consensus, whereas the population-based measures (such as spread) as well as community fragmentation and group divergence measures *decrease*. This contrasts with the Axelrod model in which polarization increases across most of the senses. Though the Hegselmann-Krause model produces outcomes that intuitively strike one as highly polarized, they are polarized in only a few of senses. In some ways, it is right to think of the Hegselmann-Krause model as a model of polarization. For example, the model's distance parameter ε might help us consider how narrow political information bubbles produce greater degrees of political polarization. However, in other senses the model creates greater consensus by decreasing spread, for instance, while leaving distinctness unchanged. The model feature that leads to this outcome—"peeling back from the edges," where groups are formed by the most extreme members of the society being pulled inward—does not conform well to observation where we can see that groups often form and repel one another (e. g. Sunstein 2002) or congressional polarization as observed through NOMINATE scores. So though bounded confidence models offer insight into potential dynamics of polarization, they do not robustly correspond to observed instances of political polarization.

The last category of models addressed by Bramson et. al. (2017) is structural balance models. These systems capture a common sensibility that "the enemy of my enemy is my friend" and have recently been employed to assess the dynamics of congressional polarization (Neal 2020). Structural balance models consist of agents on networks. The links between the agents are either positive (if they are friends) or negative (if they are enemies). A network structure is considered balanced whenever all paths—all unique sequences of links connecting each pair of nodes—have an odd number of negative links connecting enemies. In dynamical versions of these models, links in connected groups will rearrange themselves when the structure is unbalanced, and they continue to do so until the structure becomes balanced.

A clear way of assessing polarization in structural balance is by using a measure of community fragmentation, since we can count up the number of stable clusters in the network after the system has become balanced. As we discuss in Bramson et al (2017), other measures can also be employed to assess the system's polarization prior to it becoming balanced. While structural balance models give us an insightful look at the possible dynamics of polarization, they do not tend to produce the kind of polarization that social systems commonly elicit, where groups are divided, but not maximally so.

Taken together the cultural consensus, bounded confidence, and structural balance models offer promising insights into plausible dynamical processes that bring about polarization, potentially shedding light on empirical accounts like those of Layman and Carsey's findings regarding conflict extension and conflict displacement (2002). It is, of course, possible to achieve polarization simply by pulling groups apart from one another, but the models above each provide more sophisticated dynamics for the process. Simpler models would merely force an outcome of two groups in complete opposition, but such a configuration does not conform to our observational evidence of politics.

4 Rational Polarization

The approaches considered so far are predominantly positive or descriptive, looking at how polarization can be understood, measured, and operationalized. A further consideration, which has received less attention in political science, is whether polarization can be the outcome of a rational process. Though political science is primarily empirical, normative work also plays a role in the discipline (e. g. Zaller 1992; Carsey and Layman 2006; and Druckman et. al. 2012). In Singer et. al. (2018), we provide a new model of a potential mechanism of polarization as well as a reason to think that the polarization in this model is produced by fully epistemically rational agents trying to share their information.

Singer et. al. (2018) offers a minimal agent-based model (ABM) of group deliberation to track key elements of groups sharing information. Like Hegselmann and Krause (2002) and Axelrod (1997), Singer et. al. (2018) provides an agent-based computer simulation that considers how simple, lower-order dynamics interact to produce higher-order phenomena. In Singer et. al.'s model, agents are seeded with a set of reasons for or against some proposition. Reasons are modelled as pairs of numerical weights and valences—positive or negative—for or against some matter p. Each time the model is run, a different set of reasons are randomly given to agents. An agent's belief (at any given moment of the model) is whatever their evidence for and against p supports.

The agents deliberate by randomly sharing the reasons they have. When agents have unlimited memory, eventually all the agents converge on the same belief since they all end up having the same reasons. The paper also considers more realistic cases in which agents' memories are constrained so that they can only hold a limited number of reasons. For agents with limited memory, three different memory management strategies are considered where evidence is dropped when agents exceed their memory capacity: one where they randomly forget a reason ('simple-minded'), one where they forget the reason with the lowest weight ('weight-minded), and one where they forget the lowest weighted reason contrary to the view they support at the time 'coherence-minded'.

The paper measures four senses of polarization in the model output: the time it takes for the group to uniformly agree on a position (for or against p), the divergence of the groups for and against p measured by the distance between group means, the internal subgroup consensuses, and the size parity of the groups. Whereas simple-minded agents, weight-minded agents, and agents with unlimited memory all eventually converge to a single view, coherence-minded agents almost never converge. Moreover, groups of coherence-minded agents typically fracture into evenly sized groups for and against p, and group divergence and subgroup consensus both increase through time (Figure 8). So, groups of coherence-minded agents polarize when they are sharing their reasons for their beliefs.

One might think that being coherence-minded is biased in the same way as Lord et. al.'s (1979) agents who exhibit biased assimilation of new evidence. That is not how Singer et. al.'s coherence-minded agents work, though. Lord et. al.'s subjects are biased because they pre-judge incoming evidence as a function of their previously held beliefs. But coherence-minded agents objectively consider all of their evidence, including the new incoming evidence, before discarding any of it. So, it is not that the new evidence is treated differently or less weighty than it ought to be treated. Coherence-minded agents are doing their best to manage their memory in light of what all of the evidence supports.

Another impetus for thinking that coherence-minded memory management is rational comes from its balancing the epistemic value of an agent having coherent doxastic states against a respect for the weight of their evidence. For agents with limited memories, coherence considerations support favoring evidence that supports the apparently true view and respecting the weight of one's evidence supports keeping the strongest evidence. This leads agents to first forget the weakest reasons in support of apparently (from where they stand) mistaken views.

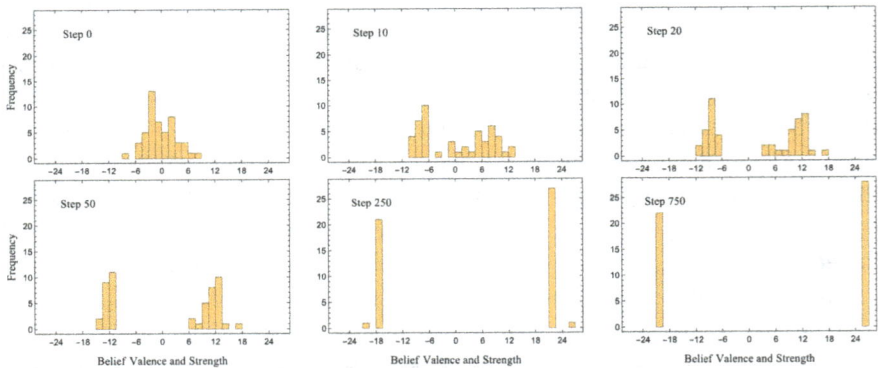

Figure 8: Histogram of beliefs and strengths for a typical run with coherence-minded agents. Singer et. al. (2018).

Treating evidence in this way leads groups of coherence-minded agents polarize in ways that look more like the kinds of polarization we typically see. Groups form endogenously, without dependence on a tuning parameter (like the value ε in the Hegselmann and Krause model) and without polarization that crashes into the extreme values of the model.

One upshot of thinking about polarization in the way that Singer et. al. (2018) does is that it focuses us on the distinctive dynamical effects of echo chambers to polarization, similar to extant work in political science (Prior 2007; Stroud 2010; and Tucker et. al. 2018). If Singer et. al.'s mechanism of polarization is real, then we should expect epistemic bubbles to be particularly debilitating to society, since agents will have their memory filled by reasons that support their view, thereby making it very difficult for them to be swayed by contrary evidence. This means that, insofar as we think that polarization harms political discourse, we would have reason to be particularly wary of echo chambers. Moreover, our work highlights the fact that polarization can happen even when rational agents are sharing their information and listening to others, adding to a growing literature on the rationality of polarization (Lodge and Taber 2000; Redlawsk 2002; Sunstein 2002; Taber and Lodge 2006; and Druckman et. al. 2012). Recent work characterizes echo chambers as groups that selectively exclude people and evidence from epistemic communities (Nguyen 2020 and Santos 2020). There are obvious epistemic vices associated with this, and some cases of polarized groups might include echo chambers. As we demonstrate in Singer et. al. (2018) however, polarized groups can also emerge without those epistemic vices. This shows how polarization and echo chambers can come apart and also illustrates how computational models can be used to generate "how possible" explanations with distinctive normative upshots. This finding gives good reason for political science and public policy to reckon more fully with polarization which is not the product of any malfeasance. Rather than take aim only at irrationality or malice in public discourse, we ought also to address how to avoid polarization caused by how we listen to (and remember) what others say.

5 What Philosophy and Political Science and Learn from Each Other

While political science is increasingly an empirical field, philosophy maintains a commitment to theoretical inquiry, suggesting a basic obstacle for interdisciplinary research. Our work presented here, however, provides a sample of philosophical projects that turn to political science for inspiration and motivation, while producing results that are equally of interest for empirical social science as they are for social epistemology and computational philosophy.

The projects presented here offer conceptual, dynamical, and normative traction on the problem of political polarization, specifically germane to instances relating to people's beliefs. Bramson et. al. (2016) provides a series of definitions and measures of different kinds of polarization. Rather than working with a single concept, we argue for conceptualizing polarization as an umbrella term containing distinct kinds, measured differently, and importantly, which can vary independently of one another. As that paper demonstrates, it is both conceptually possible and empirically commonplace for some measures to increase while others decrease. This would be impossible were we to treat polarization as a singular concept.

Using those distinct senses, Bramson et. al. (2017) pursues a more comprehensive account of the dynamical models that have been offered for polarization. The paper considers three broad classes of models—cultural dissemination, bounded confidence, and structural balance—to probe possible underlying dynamics of polarization and the wider applicability of the polarization measures. These classes of models are not meant to be exhaustive, but broadly indicative of the mechanistic approaches to polarization in which the outcome does not merely result from two groups being pulled away from one another. Like empirical instances of polarization, the measurements of polarization in these computational models do not move in lock step, both indicating that no one model can be considered comprehensive and illustrating a more general lacuna in the literature owing to lack of strong correspondence between any of the three models and observed instances of polarization in politics.

Singer et. al. (2018) goes some distance towards offering a better model of polarization, while also interrogating its descriptive and normative rational features. When rational, forgetful agents in this model exchange information they polarize in ways analyzed in Bramson et. al. (2016). Coherence-minded agents —those that manage their memory by forgetting the least consequential reason that does not conform with their prevailing belief—consistently polarize in natural ways and do so for rational reasons. This sort of modeling exercise buttresses the work of Bramson et. al. (2017) to show further cases of model classes that plausibly produce the kinds of polarization observed by political science. Though the potentially rational aspects of polarization have been largely overlooked in the empirical literature, those aspects have clear and demonstrable implications for both how we understand the phenomenon and the policies offered to countermand the trends. Idealized computational models offer a means of providing explanatory accounts of empirical phenomena which would otherwise be difficult to obtain. Singer et. al. (2018) illustrates how polarization can still occur when everyone is behaving rationally and draws attention to the shortcomings of accounts which rule out rational processes.

The work we have done exemplifies a trend in philosophy to look to nearby fields, in this case the social sciences, to motivate philosophical questions that simultaneously offer clarity on how other fields might theorize and operationalize concepts. Analytic philosophy is well suited for taking on definitional, process-oriented, and normative puzzles. A lack of coherent concepts and measurement undermines good science by muddying conclusions and propagating incompatible results. Bramson et. al. (2016) shows not only that there are numerous kinds of polarization, but that a particular case can yield measurements that move in opposite directions. Incorporating political science in the process of developing philosophical research can make philosophy richer and more naturalistic, while offering stronger interdisciplinary pull on empirically oriented disciplines.

Bibliography

Axelrod, Robert (1997): "The Dissemination of Culture: A MODEL WITH LOCAL CONVERGENCE AND GLOBAL POLARIZATION." In: *JOURNAL OF CONFLICT RESOLUTION* 41. No.2, pp. 203–226.

Baldassarri, Delia, and Andrew Gelman (2008): "Partisans without Constraint: Political Polarization and Trends in American Public Opinion." In: *American Journal of Sociology* 114. No. 2, pp. 408–446.

Bartels, Larry M. (2000): "Partisanship And Voting Behavior, 1952–1996." In: *American Journal of Political Science* 44. No. 1, pp. 35–50.

Bramson, Aaron, Patrick Grim, Daniel J. Singer, Steven Fisher, William Berger, Graham Sack, and Carissa Flocken (2016): "Disambiguation of social polarization concepts and measures." In: *The Journal of Mathematical Sociology* 40. No. 2, pp. 80–111.

Bramson, Aaron, Patrick Grim, Daniel J. Singer, William J. Berger, Graham Sack, Steven Fisher, Carissa Flocken, and Bennett Holman (2017): "Understanding Polarization: Meanings, Measures, and Model Evaluation." In: *Philosophy of Science* 84. No. 1, pp. 115–159.

Druckman, James N., Jordan Fein, and Thomas J. Leeper (2012): "A Source of Bias in Public Opinion Stability." In: *American Political Science Review* 106. No. 2, pp. 430–454.

Duverger, Maurice (1959): *Political Parties: Their Organization and Activity in the Modern State*. New Jersey: John Wiley & Sons.

Fiorina, Morris P., and Samuel J. Abrams (2008): "Political Polarization in the American Public." In: *Annual Review of Political Science*. 11, pp. 563–588.

Flaxman, Seth, Sharad Goel, and Justin M. Rao, Justin (2016): "Filter Bubbles, Echo Chambers, and Online News Consumption." In: *Public Opinion Quarterly* 80. No. S1, pp. 298–320.

Fleisher, Richard, and John R. Bond, (2004): "The Shrinking Middle in the US Congress." In: *British Journal of Political Science* 34. No. 3, pp. 429–451.

Grim et. al. (2012): "Philosophical Analysis in Modeling Polarization: Notes from a Work in Progress." In: *APA Newsletters* 12. No. 1, pp. 7–15.

Grose, Christian R., and Antoine Yoshinaka (2003): "The Electoral Consequences of Party Switching by Incumbent Members of Congress, 1947–2000." In: *Legislative Studies Quarterly* 28. No. 1, pp. 55–75.

Han, Hahrie, and David W. Brady (2007): "A Delayed Return to Historical Norms: Congressional Party Polarization after the Second World War." In: *British Journal of Political Science* 37. No. 3, pp. 505–531.

Harary, Frank (1959): "On the Measurement of Structural Balance." In: *Behavioral Science* 4. No. 4, pp. 316–23.

Hegselmann, Rainer, and Ulrich Krause (2002): "OPINION DYNAMICS AND BOUNDED CONFIDENCE MODELS, ANALYSIS, AND SIMULATION." In: *Journal of Artificial Societies and Social Simulation* 5. No. 3.

Hegselmann, Rainer, and Ulrich Krause (2005): "Opinion dynamics driven by various ways of averaging." In: *Computational Economics* 25. No. 4, pp. 381–405.

Hegselmann, Rainer, and Ulrich Krause (2006): "Truth and Cognitive Division of Labor: First Steps towards a Computer Aided Social Epistemology." In: *Journal of Artificial Societies and Social Simulation* 9. No. 3, p. 10.

Hetherington, Marc J. (2009): "Putting Polarization in Perspective." In: *British Journal of Political Science* 39. No. 2, pp. 413–448.

Hetherington, Marc J., and Jonathan Weiler (2018): *Prius Or Pickup?: How the Answers to Four Simple Questions Explain America's Great Divide.* New York: Houghton Mifflin.

Hopkins, Daniel J. (2018): *The increasingly United States: How and Why American Political Behavior Nationalized.* Chicago, London: University of Chicago Press.

Iyengar, Shanto, Yphtach Lelkes, Matthew Levendusky, Neil Malhotra, and Sean J. Westwood (2019): "The Origins and Consequences of Affective Polarization in the United States." In: *Annual Review of Political Science* 22, pp. 129–146.

Jasny, Lorien, Joseph Waggle, and Dana R. Fisher (2015): "An empirical examination of echo chambers in US climate policy networks." In: *Nature Climate Change* 5. No. 8, pp. 782–786.

Krehbiel, Keith (2000): "Party Discipline and Measures of Partisanship." In: *American Journal of Political Science* 44. No. 2, pp. 212–227.

Layman, Geoffrey C., and Thomas M. Carsey (2002): "Party Polarization and 'Conflict Extension' in the American Electorate." In: *American Journal of Political Science* 46. No. 4, pp. 786–802.

Layman, Geoffrey C., Thomas M. Carsey, and Juliana M. Horowitz (2006): "Party Polarization in American Politics: Characteristics, Causes, and Consequences." In: *Annual Review of Political Science* 9, pp. 83–110.

Ledgerwood, Alison, Yaacov Trope, and Shelly Chaiken (2010): "Flexibility Now, Consistency Later: Psychological Distance and Construal Shape Evaluative Responding." In: *Journal of Personality and Social Psychology* 99. No. 1, pp. 32–51.

Lee, Frances E. (2016): *Insecure Majorities: Congress and the Perpetual Campaign.* London: University of Chicago Press.

Levy, Gilat, and Ronny Razin (2019): "Echo Chambers and their Effects on Economic and Political Outcomes." In: *Annual Review of Economics* 11, pp. 303–328.

Liberman, Akiva, and Shelly Chaiken (1992): "Defensive Processing of Personally Relevant Health Messages." In: *Personality and Social Psychology Bulletin* 18. No. 6, pp. 669–679.

Lipset, Seymour M. (1960): *Political Man: The Social Bases of Politics*. New York: Garden City.
Lodge, Milton, and Charles S. Taber (2005): "The Automaticity of Affect for Political Leaders, Groups, and Issues: An Experimental Test of the Hot Cognition Hypothesis." In: *Political Psychology* 26. No. 3, pp. 455 – 482.
Lord, Charles G., Lee Ross, and Mark R. Lepper (1979): "Biased Assimilation and Attitude Polarization: The Effects of Prior Theories on Subsequently Considered Evidence." In: *Journal of Personality and Social Psychology* 37. No. 11, pp. 2098 – 2109.
Luguri, Jamie B., and Jamie L. Napier (2013): "Of two minds: The interactive effect of construal level and identity on political polarization." In: *Journal of Experimental Social Psychology* 49. No. 6, pp. 972 – 977.
McCarty, Nolan, Keith T. Poole, and Howard Rosenthal (2006): *Polarized America: The Dance of Ideology and Unequal Riches*. Cambridge: MIT Press.
McHoskey, John W. (1995): "Case Closed? On the John F. Kennedy Assassination: Biased Assimilation of Evidence and Attitude Polarization." In: *Basic and Applied Social Psychology* 17. No. 3, pp. 395 – 409.
Munro, Geoffrey D., Peter H. and Ditto (1997): "Biased Assimilation, Attitude Polarization, and Affect in Reactions to Stereotype-Relevant Scientific Information." In: *Personality and Social Psychology Bulletin* 23. No. 6, pp. 636 – 653.
Neal, Zachary P. (2020): "A sign of the times? Weak and strong polarization in the US Congress, 1973 – 2016." In: *Social Networks* 60, pp. 103 – 112.
Nguyen, C. Thi (2020): "ECHO CHAMBERS AND EPISTEMIC BUBBLES." In: *Episteme* 17. No. 2, pp. 141 – 161.
Rae, Nicol C. (1994): *Southern Democrats*. New York: Oxford University Press on Demand.
Redlawsk, David P. (2002): "Hot Cognition or Cool Consideration? Testing the Effects of Motivated Reasoning on Political Decision Making." In: *The Journal of Politics* 64. No. 4, pp. 1021 – 1044.
Plous, Scott (1991): "Biases in the Assimilation of Technological Breakdowns: Do Accidents Make Us Safer?" *Journal of Applied Social Psychology* 21. No. 13, pp. 1058 – 1082.
Poole, K. T., and H. Rosenthal (1984): "The Polarization of American Politics." *The Journal of Politics* 46. No. 4, pp. 1061 – 1079.
Prior, Markus (2007): *Post-Broadcast Democracy: How Media Choice Increases Inequality in Political Involvement and Polarizes Elections*. New York: Cambridge University Press.
Prior, Markus (2013): "Media and Political Polarization." In: *Annual Review of Political Science* 16, pp. 101 – 127.
Santos, Breno R. (2021): "Echo Chambers, Ignorance and Domination." In: *Social Epistemology*, 35, No. 2, pp. 109 – 119.
Singer, Daniel J., Aaron Bramson, Patrick Grim, Bennett Holman, Jiin Jung, Karen Kovaka, Anika Ranginani, and William J. Berger. (2019): "Rational social and political polarization." In: *Philosophical Studies* 176. No. 9, pp. 2243 – 2267.
Stroud, Natalie J. (2010): "Polarization and Partisan Selective Exposure." *Journal of Communication* 60. No. 3, pp. 556 – 576.
Sunstein, Cass R. (2002): "The Law of group polarization." In: *Journal of Political Philosophy* 10. No. 2, pp. 175 – 195.
Taber, Charles S., and Milton Lodge (2006): "Motivated skepticism in the evaluation of political beliefs." In: *American Journal of Political Science* 50. No. 3, pp. 755 – 769.

Tucker, Joshua A., Andrew Guess, Pablo Barberá, Cristian Vaccari, Alexandra Siegel, Sergey
Sanovich, Denis Stukal, and Brendan Nyhan (2018): "Social media, political polarization,
and political disinformation: A review of the scientific literature." In: *SSRN Electronic
Journal*. https://dx.doi.org/10.2139/ssrn.3144139, accessed 28 October 2021.

Yang, Daniel Y. J., Jesse L. Preston, and Ivan Hernandez (2013): "Polarized Attitudes Toward
the Ground Zero Mosque Are Reduced by High-Level Construal." In: *Social Psychological
and Personality Science* 4. No. 2, pp. 244–250.

Zaller, John R. (1992). *The Nature and Origins of Mass Opinion*. New York: Cambridge
University Press.

Notes on Contributors

Manuel Almagro Holgado holds a Ph.D. in philosophy (2021) and is a postdoctoral researcher at the University of Granada, Spain (Department of Philosophy I). His current research includes affective polarization, the evaluative use of language, mental state ascriptions, disagreement and Wittgenstein's philosophy. He is especially interested in the relationship between philosophical topics and practical issues such as social injustice and political polarization. He collaborates with David Bordonaba, Víctor Fernández Castro, Ivar R. Hannikainen, Amalia Haro Marchal, Manuel Heras Escribano, Llanos Navarro Laespada, Miguel Núñez de Prado-Gordillo, Javier Osorio Mancilla, Manuel de Pinedo, Neftalí Villanueva and Jimena Zapata, amongst others.

Saray Ayala-López is an Assistant Professor of Philosophy at California State University Sacramento. They previously worked at San Francisco State University, Universidad Carlos III de Madrid and Universitat Autònoma de Barcelona. After doing research on apolitical cognitive science, they decided to address the social justice hiccup that was deliberately banned from their philosophical armchair, and started including moral questions in their research. They are especially interested in explanations (structural explanations are my favorite) and the absence of them, conversational dynamics and the many things we can do with words, the intricacies of crafting collective conceptual resources, and the challenges of doing (trans)feminism from the margins.

William J. Berger is an Assistant Professor in the department of Political Economy & Moral Science at the University of Arizona. His research contributes both to political psychology and political epistemology and focuses on issues of trust, polarization, and inequality.

David Bordonaba-Plou is a FONDECYT postdoctoral researcher at the Universidad de Valparaíso, Chile. He is an associate researcher at the Centro de Estudios en Filosofía, Lógica y Epistemología (CeFiLoE) of the Universidad de Valparaíso, Chile, and member of the excellence unit FiloLab-UGR and the group Filosofía y Análisis (HUM-975) from the Universidad de Granada, Spain. His main area of research is predicates of personal taste and disagreement, but he also works on experimental philosophy, polarization and echo chambers, the role of intuitions in philosophy, and the relation between predicates of personal taste and aesthetic predicates.

Cristina Borgoni is Professor of Epistemology at the University of Bayreuth. Previously, she held academic positions at the University of Graz, UCLA, UCL, and University of Granada. Her philosophical interests concern aspects of individuals' minds (beliefs, rationality, self-knowledge, fragmentation, and implicit biases) as well as various aspects of interpersonal interaction (testimony, deference to first-person authority, epistemic injustice, and communication). Borgoni has published extensively in both areas. She is currently a member of the executive committee of the ESPP.

https://doi.org/10.1515/9783110612318-016

Aaron Bramson's research specialties are complexity science, methodology for modeling complex systems (agent-based modeling, networks, and mathematics), measuring dynamics in large data sets, and artificial intelligence for geospatial data analysis. He received his Ph.D. from the University of Michigan in 2012 in a joint program with the departments of political science and philosophy. Before graduating, Aaron worked as post-doctoral research fellow in the Rotman School of Management at the University of Toronto teaching MBA courses and developing Bayesian network and agent-based modeling techniques for business applications. Before attending UM, Aaron earned an M.S. in mathematics from Northeastern University. In addition to his academic research, he engages in theoretical, methodological, and practical research and complexity education consulting through his company Complexity Research.

Bianca Cepollaro is a research fellow in Philosophy of Language at the Faculty of Philosophy, Vita-Salute San Raffaele University (2017, PhD in Linguistics, Scuola Normale, Pisa; PhD in Philosophy, École Normale Supérieure, Institut Jean Nicod, Paris). She works on the semantics and pragmatics of loaded and expressive language both on theoretical and experimental grounds. She is the author of Slurs and Thick Terms – How Language Encodes Values (Rowman and Littlefield, 2020), as well as of numerous articles that appeared in journals such as Pacific Philosophical Quarterly, Synthese, Linguistics and Philosophy, Journal of Pragmatics.

E. Díaz-León is an associate professor of philosophy at the University of Barcelona. Before this, she taught at the University of Manitoba, and she received her doctorate from the University of Sheffield. She specializes in philosophy of mind and language, and philosophy of gender, race and sexuality, and her current research focuses on social construction, conceptual ethics, and the nature of gender, race, and sexual orientation. She has published articles and chapters on these topics in journals such as Ergo, European Journal of Philosophy, Hypatia, Journal of Social Philosophy, and Journal of Social Ontology, and in several edited volumes.

Víctor Fernández Castro is a current Juan de la Cierva post-doctoral research fellow at Department of Philosophy I and Filo-Lab at Universidad de Granada. He was previously appointed at LAAS-CNRS (Université de Toulouse, CNRS) and the Institut Jean Nicod, (DEC, ENS, PSL Research University). He has also been research visitant at George Washington University and the University of Edinburgh. His main areas of interest are the theoretical philosophy of mind and psychology and their applications in areas as like social robotics, mental disorders or social philosophy.

Patrick Grim is Distinguished Teaching Professor Emeritus at Stony Brook University and philosopher in residence with the Center for Study of Complex Systems at the University of Michigan. He has published widely in both philosophy and beyond, with work in computational modeling that overlaps a range of disciplines.

Bennett Holman is an Associate Professor of the History and Philosophy of Science at Yonsei University in Seoul, South Korea. His research focuses on areas of scientific inquiry where knowledge claims are actively contested by community members with non-truth-seeking motives (e. g., political values and/or economic values in regulatory science).

Jiin Jung is a postdoctoral researcher in the Department of Psychology at New York University. Her research focuses on belief dynamics under uncertainty and disagreement. She particularly investigates the role of minority dissent in the social psychological processes of social change and diversity.

Emily C. McWilliams is an Assistant Professor of Philosophy, and member of the founding faculty of the undergraduate liberal arts degree program at Duke Kunshan University in Kunshan, China. She works primarily on issues at the intersections of epistemology, ethics and feminist philosophy.

José Medina is Walter Dill Scott Professor of Philosophy at Northwestern University. His work focuses on ignorance, insensitivity, epistemic and communicative injustice, oppression and resistance, and public protest. His primary fields of expertise are critical race theory, communication theory, applied philosophy of language, social epistemology, and political philosophy. His books include The Epistemology of Resistance: Gender and Racial Oppression, Epistemic Injustice, and Resistant Imaginations (Oxford University Press; recipient of the 2013 North-American Society for Social Philosophy Book Award), and Speaking from Elsewhere (SUNY Press, 2006). He is finishing a new monograph in social epistemology and the philosophy of social movements entitled The Epistemology of Protest, which examines the communicative structure and dynamics of public protest and the obstacles that protesting publics face to be heard and to receive proper uptake, arguing for what he terms "epistemic activism."

Alba Moreno Zurita PhD candidate at the Department of Philosophy I, University of Granada. Her interest areas are philosophy of language and social epistemology. Her ongoing dissertation analyzes different presuppositional theories of slurs in order to account for the resistance to cancellation, rejection and retraction exhibited by utterances of sentences featuring these terms. She has recently published two papers in collaboration with E. Perez Navarro, "The resistant effect of slurs: A non-propositional, presuppositional account" in Daimon, and "Beyond the conversation: the pervasive danger of slurs" in Organon F.

Deborah Mühlebach, PhD, is a postdoc at Freie University Berlin. Her research focuses on questions of how power relations and social categories shape phenomena in rather theoretical fields of philosophy such as philosophy of language and epistemology. Her dissertation The Politics of Meaning includes several articles, among them "Semantic Contestations and the Meaning of Politically Significant Terms" (Inquiry) and "Reflective Equilibrium as a Method for Feminist Epistemology" (Hypatia). In her current research project entitled Understanding Critique, she examines the social circumstances under which critical agency may flourish and highlights the importance of different forms of understanding for critical agency.

Manuel de Pinedo (DPhil, Sussex 2000) is professor of epistemology at the University of Granada, Spain. His research has centred on the role of normativity in the philosophy of mind and the cognitive sciences, self-knowledge, political epistemology, philosophy of perception and ecological psychology. He has published, as sole author or in collaboration with Jason Noble, Hilan Bensusan, Manuel Heras, Neftalí Villanueva, Manuel Almagro or Llanos Navarro, in journals such as Synthese, Biology & Philosophy, Frontiers in Psychology or Hypatia. Together with Neftalí Villanueva and María José Frápolli, he leads Filosofía y Análisis, a broad group of researchers at the University of Granada, which assembles five funded research projects ranging from disagreement and polarization to the interplay between expressivism and inferentialism.

Daniel J. Singer is an Associate Professor of Philosophy at the University of Pennsylvania. Along with Patrick Grim, he directs the Computational Social Philosophy Lab. He researches epistemology, the nature of normativity in ethics and epistemology, and social philosophy with an emphasis on diversity and polarization.

Alessandra Tanesini is Professor of Philosophy at Cardiff University. Her latest book is The Mismeasure of the Self: A Study in Vice Epistemology (Oxford University Press, 2021). Her current work lies at the intersection of ethics, politics, the philosophy of language, and epistemology with a focus on epistemic vice, silencing, prejudice and ignorance.

José Ramón Torices is a postdoctoral researcher belonging to the research group Filosofía y Análisis (HUM-975) and the excellence unit FiloLab (Universidad de Granada). His research interests primarily concern debates on the evaluative use of language (expressivism, relativism, contextualism and varieties of disagreement) and the use of language as a mechanism for political manipulation, among other topics. His most recent publications are "Understanding dogwhistles politics" in THEORIA and "Paving the road to Hell: The Spanish word menas as a case study", co-authored with David Bordonaba, in Daimon.

Neftalí Villanueva is assistant professor at the Philosophy I Department, University of Granada. His work ranges from particular issues within the philosophy of language, such as the pragmatics and semantics of attitude ascriptions, to more applied topics, such as evaluative language, disagreement and polarization. His current research is mainly focused on the study of linguistic indirect measures of affective polarization.

Audrey Yap is an Associate Professor in the Philosophy Department at the University of Victoria, working mainly in feminist philosophy, broadly construed, though she also has research interests in the history of analytic philosophy. She is also an immigrant settler who lives on unceded Lekwungen territory.

Index

https://doi.org/10.1515/9783110612318-017

www.ingramcontent.com/pod-product-compliance
Lightning Source LLC
Chambersburg PA
CBHW050644270326
41927CB00012B/2862